BIODIVERSITY AND ENVIRONMENT

BIODIVERSITY AND ENVIRONMENT

Edited by
Prof. A.K. Aditya & Dr. P. Haldar

Department of Zoology
Visva Bharati University, Santiniketan
West Bengal, INDIA

2024
Daya Publishing House®
A Division of
Astral International Pvt. Ltd.
New Delhi – 110 002

First Published, 2000
Reprinted 2024

ISBN: 978-81-7035-226-6 (Hardbound)

Published by : **Daya Publishing House®**
A Division of
Astral International Pvt. Ltd.
– ISO 9001:2008 Certified Company –
4760-61/23, Ansari Road, Darya Ganj
New Delhi-110 002
Ph. 011-43549197, 23278134
E-mail: info@astralint.com
Website: www.astralint.com

Digitally Printed at : **Replika Press Pvt. Ltd.**

Memoir

This book is dedicated to the memory of our respected teacher Late Prof. Sivatosh Mookerjee former Head, Department of Zoology, Presidency College, Calcutta and later Dean & Professor School of Life Sciences, Jawaharlal Nehru University, New Delhi for his untiring encouragement, help and constant inspiration during our studentship and research activities and for his guidance for the betterment and shaping the Department of Zoology, Visva-Bharati, Santiniketan for the present day in various magnitude.

Foreward

THE environmental crisis, confronting the world, is likely to alter profoundly the future of our planet. No one amongst us whatever be our status or strength can remain unaffected. This makes us think over issues of environment seriously and also, to seek solutions for preserving the nature and natural resources without impairing the developmental programmes for human welfare. Environmental conservation must be the common concern of all communities, at all levels from local to global. The creation of an interdisciplinary awareness of what environmental conservation means is a matter of obvious importance.

Conservation of biodiversity and physical limits of environment form now the serious agenda for discussion at national and international levels. The rapid advancement of technologies in agriculture and industry are also associated with adverse and undesirable impacts. One has to realise how that our interference with environment and over exploitation of natural resources for the sake of economic growth and prosperity are associated with ecological calamities in almost all spheres of life. Most of the present day environmental problems and associated ill effects at the level of organisms stem from ecological misappropriation and neglect to recognise the limits of ecosystems and its restoration capacities.

The present text is an attempt to make the people aquainted thoroughly with the knowledge of ecosystem, and the factors concerned with the deterioration of the environment and its valued resources with the view to plan and manage the developmental programmes in such a way that exploitation of natural resources may not upset the balance of nature.

It is high time to reorient our policy of technology development so as to support both sustained economic growth and ecological equilibrium. The end result of such eco-friendly strategy would improve socio-economic and environmental stability, provided human management system and functioning of natural biota are in tune with the local resource base and self regulation capacity of the environment.

The book "Biodiversity and Environment" is provided by what one would like to think of as basic knowledge of ecosystems, productivity and Biodiversity not withstanding this there is the first chapter dealing with the organismal response and bioindicators, chapter II very well illustrates the Human Impact on Ecosystem, the IIIrd one depicts the picture of Invertebrate Biodiversity in Aquatic environment while the last chapter narrates the Invertebrate Biodiversity in Terrestrial Environment. It is hoped that the book will definitely serve developing a proper concern to the readers for our deteriorating environment and also encourage them to become actively involved in matters related to the environment for its proper management. The common goal is to gain a universally better quality of life.

This book is very much a culmination of concepts and research works of scientists from various fields throughout the country and abroad and who have been associated with the field of ecology with special emphasis on Environmental Biology, Toxicology, Developmental Biology, Neurology etc.

This book "Biodiversity and Environment" hopefully will be found stimulating the interests of young scientists and kindle in them a desire to work on environment with keen interest. I feel proud and happy at the outcome of my colleagues efforts in putting together a coherent total picture of Environment through this publication.

Visva Bharati
Santiniketan

D.K. Sinha
Vice Chancellor

Preface

THE papers contained in this book were presented at the National Seminar on Environmental Biology held April 03 to 05, 1998, at the Department of Zoology, Visva-Bharati University, Santiniketan, India. The purpose of this National Seminar on Environmental Biology was to serve as a form for the communication of recent advances on different aspects of Environmental Biology. The meeting had four oral sessions covering the Organismal Response & Bioindicators; Human Impact on Ecosystem; Invertebrate Biodiversity of Aquatic Environment and Inverterbrate Biodiversity in Terrestrial Environment. Each session began with a keynote address followed by oral presentations. The papers are all reviewed by expert member of respective disciplines. We would like to thank the numerous reviewers for sparing their time and effort. Their comments helped the authors in revising the manuscripts.

We are especially grateful to the following people of the local organizing and scientific committees for their invaluable help in organising the meeting :

Local Organising Committee :

D.K. Sinha, Vice Chancellor, Visva-Bharati

Samir Bhattacharya

A.B. Das

N.C. Sukul

Shelley Bhattacharya

A.K. Ray

P. Nath

V.C. Joy

S.P. Sinha Babu

D.K. Mondal

We gratefully acknowledge the financial support of the following agencies and University which enabled us to make the symposium a grand success and also bear the cost of publishing this volume.

Sponsorers :

- University Grants Commission.
- Visva-Bharati University.
- Council of Scientific and Industrial Research, New Delhi.
- National Bank for Agriculture and Rural Development, Bombay.
- Department of Science and Technology, Govt. of India, New Delhi.
- State Bank of India, Santiniketan.

For publication

- Department of Science and Technology, Govt. of India, New Delhi

Besides the local organising and scientific programme committee, many other individuals contributed their time and effort in organising and running the meeting. We would like to thank these individuals for their generous helps.

We are indebted to our respected Upacharya, Prof. D.K. Sinha who has been a source of continuous encouragement and inspiration to us and also for writing a foreword to it.

We wish to express our sincere thanks to –

Prof. P.K. Sensharma, Former Director, Biological Research, Forest Research Institute, Dehradun; Prof. Sujit Kumar Dasgupta, Presidency College, Calcutta; Prof. Sudhendu Mandal, Deptt. of Botany; Prof. N.C. Sukul, Prof. Shelley Bhattacharya, Prof. A.K. Ray, Prof. V.C. Joy, Dr. S.P. Sinha Babu of our department who have greatly enriched the text by providing insightful revision of the manuscripts.

We also wish to thank the following members for their contribution to our work in various ways. Ashoke Kumar Bhattacharya, Deptt. of Botany; Sarasi Das, Amlan Das, Sukla Mondal, Prafulla Kumar Banerjee and Ramesh Chandra Kundu of Zoology Department

Session Chairs :

1. Anjali Mookerjee
2. R. Ramamurthy
3. J.R.B. Alfred
4. S.K. Dasgupta
5. S. Chakraborty
6. A.K. Das
7. R.C. Basu
8. A.K. Ray
9. G.T. Gujar

Finally, we would especially like to thank the participants. The attendance at this meeting by over 110 individuals of different Universities and institutions from different parts of the country including Italy clearly indicates that this symposium series is a pivotal gathering for Environmental Biologists.

A.K. Adîtya
P. Haldar
Conveners

Contents

Section I

Organismal Response / Bioindicators

Proceedings of the National Seminar on Environmental Biology, April 03-05, 1998
***Edited By :* A.K. Aditya & P. Haldar**
DAYA PUBLISHING HOUSE, DELHI

Radioecology by 21st Century

Anjali Mookerjee
Sivatosh Mookerjee Science Centre
Asutosh Mookerjee Memorial Institute
77, Asutosh Mookerjee Road, Calcutta-700025

Radioecological research assumed importance when man learnt to release the nuclear energy, the solar and background radiations being already present. It is to be considered as an applied research field dealing with natural and man produced radioactivity, the movement of radionuclides in the environment and the effects of radiations, ionising and non-ionising, on plants and animal populations.

The main areas of radioecological reasearch are (1) Nuclear weapons production and tests and (2) Peaceful uses of nuclear energy.

Related to these there are three main tasks to be worked out by radioecologists : (i) the improvement of knowledge, (ii) the education of scientists in the field of radioecology, (iii) information for the public. As such, therefore, radioecological investigations draw upon many fields of scientific enquiry.

Also, it has been estimated that the reserves of fossil fuels like coal, oil and natural gas are very limited and will not be lasting more than a few hundred years (if consumed at the present rate). There are other various environmental problems associated with burning fossil fuels (pollution, mining etc.). For these reasons alternative energy sources are necessary.

Nuclear Power

We all know now that energy can be released from nuclear reactions - either by fissioning (splitting) an appropriate heavy nucleus or by combining appropriate light nuclei (fusion).

According to Einstein, equation relating energy E to mass m is : $E = mc^2$, where c is the speed of light. If 'E' is expressed in ergs, and 'm' in g and 'c' in cm/sec, then for 1 g of matter the energy is

$$E = 1 \times (3\times10^{10})^2 \text{ ergs} = 900 \text{ billion ergs}$$

or 25 million Kilowatt hours.

Similarly, one pound of matter will give energy equivalent of 10 million tons of TNT.

Both fission and fusion involve the rearrangement of nucleons (the collective term for neutrons and protons) into different nuclei so that energy is released due to mass defect. If U^{235} is bombarded with slow neutrons then the following reaction takes place :

$$U^{235} + {}_{o}n^{1} \rightarrow Ba^{141} + Kr^{92} + 3\,{}_{o}n^{1}$$

This chain reaction was very important and the names associated with this are : Enrico Fermi, Lise Meitner, Otto Hahn, Hans Bethe etc.

In nature U^{238} is more abundant than U^{235}. It does not fission with slow neutrons, but captures them, making it difficult to maintain a chain reaction.

U^{235} is only 0.7%.

It was found that some mass disappeared in the break up as part of the Uranium atom was converted into energy. As the chain reaction proceeds theoretically we get more and more number of neutrons. In reality as many neutrons fail to split another atom, with time this chain reaction weakens. This happens because most of.these neutrons are fast neutrons. Therefore, to overcome this difficulty moderators like graphite, or heavy water were used. It was known to scientists like Fermi (1942, Uranium-graphite pile) that graphite (pure carbon) did not absorb neutrons but simply bounced off these nuclei like a rubber ball against a hard surface (thereby decreasing the velocity). Also there were substances like cadmium which absorbed (like sponge) the passing neutrons. A combination of the impervious and the absorbing nuclei gave Fermi the tool he needed to control nuclear breakup.

Power of Nuclear Reaction

If 'A' is the number of nuclei undergoing fission,

then $A = nv\,N\,\sigma V$ ....(1)

where v = average speed, so that nv is the average neutron flux

n = average neutron density, i.e., neutrons/cc

N = no. of fissinable (U^{235}) nuclei/cc

σ = cross section for fission

V = volume of the reacter. It is seen that 3.1×10^{10} fissions/sec produce 1 watt of power (P).

$$\therefore P = \frac{n\upsilon N\sigma V}{3.1 \times 10^{10}} \text{ watts.}$$

The reactors require cooling helium gas or air cooling is used.

Solar energy is the only proven very long term major energy source. The vast output of energy by the stars (including our sun) was thought to originate from some kind of nuclear transformations. Prof. Hans Bethe theoretically explained this phenomenon as follows: the extremely high temperatures found in the sun and the enormous pressure exerted by its great bulk cause protons to join together in the process called fusion. In the fusion reaction four protons are combined to form the nucleus of a helium atom. The mass of this nucleus is very slightly less than that of the four protons. The missing mass has been converted into the energy of the sun.

Controlled fusion is the key to another energy source of long duration.

Radiation Hazards

For whatsoever purpose nuclear energy is used - be it for peaceful uses, or nuclear warfare- the physical and bio-environment cannot get away with its hazards. There are the delayed effects of fall out and the increasingly important hazards of the peacetime uses of radiation. Power reactor accidents may occur (already have); the radioactive by-products and spent fuels of reactors will need to be contained and safely disposed off; radiation and radio-isotopes will be finding wider use in technology, as well as, in scientific and industrial research. The medical profession must continue its evaluation of the risks attending diagnostic and therapeutic irradiation of patients and its efforts to reduce these risks to the essential minimum.

Early recognition of the long-term ill effects of low dose radiation and consequent widespread concern led to the formulation of various types of measures during the past few decades which have been responsible in keeping radiation exposure to workers in nuclear power plants and industry, as well as, general population to, as low a level as, can be reasonably achieveable. The present day radiation health protection measures have been the outcome of the systematic documentation and careful analysis of radiobiological research information including epidemiological surveillance that has been accumulated for several years. Probably due to the absence of definitive information on the long-term ill-effects of environmental chemical pollutants, similar efforts were not made to assess the potential health hazards that may be associated with conventional fossil fuel-based energy and other industrial operations. In recent years however there is growing concern that most of the hazardous biological effects induced by chemical pollutants are similar to those by atomic radiation. The community of scientists concerned with public health is now feeling the desirability of an integrated approach for tackling the problem of risk assessment on the same philosophy and pattern by which radiation risk assessment is done and radiation protection measures organised.

Hazards - from Different Radiations and Sources

Types (1) α – particles (or helium nucleus, $2p^+$ and 2n)

(2) β – particles (e^- and e^+)

(3) x and γ – rays (high energy photons).

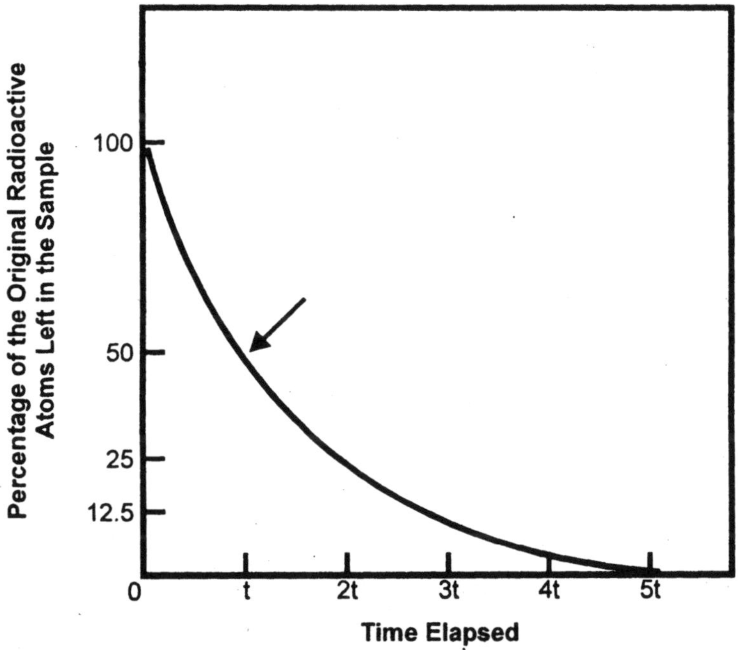

Fig. 1 : The half-life of a radioactive isotope. The arrow indicates a single half-life.

Principal Sources : Radioactive nuclei, X-ray machines and cosmic rays. Background radiation always present due to long half life nuclei and cosmic rays, amount to few mrads.

Effects of Living Matter : These are due to the ionisations produced. Effects can be broadly divided into (a) somatic-damage to the organism exposed i.e. to the body cells; (b) genetic-damage to future generation.

If a sufficient number of ionisations (of the order of 10^6 –10^7) are produced in a single cell, the cell will be killed, α particles are more efficient, because they cause more ionisations, whereas electrons have greater range.

Table - 1 : LD_{50} Doses (ionising radiation)

For smaller organisms		Whole body, 30 days for higher animals	
Organism	**LD_{50} (rads)**	**Species**	**LD_{50} (rads)**
Haploid yeast	1800	Guineapig	250
Diploid yeast	9000	Dog	340
E.Coli B/r^H	2500-2800	Goat	500
Virus T_2 H_r	50,000	Monkey	500
Amoeba	200,000	Man	500
	(Killing dose)	Mouse	500-700
Hydra	75,000	Swine	550
	(Killing dose)	Rat	600-700
		Hamster	700
		Rabbit	800

1. Scientists like Stern Glass suggested that the release of 20,000 megatons anywhere in the world could result in eliminating any surviving infants. He drew this conclusion from the effect of the first series of atomic tests held at Alamogordo (in U.S.). More than 1000 to 1500 miles away from the test site the mortality rates of between 3-4.5 per hundred life births increased by 20-30 per cent.
2. During early 1950s, it was discovered that radioactive strontium became concentrated in cow's milk and was transmitted along with calcium to the bones of the developing human foetus and body. But such effects were slow acting (e.g. in dial painting) and would take long before bone cancer or leukaemia could result.
3. Observations on pregnant women who had been exposed to X-rays indicated the possibility that ova and embryos could be from 20-50 times more sensitive to the possible development of leukaemia than the mature adult. Hence present medical diagnosis does not allow pregnant mothers to be exposed to X-ray analysis, until and unless absolutely necessary.
4. The genetic hazards may be summarised as follows :

Radiation induced mutations, like other mutations however induced, are mostly deleterious. Under a given set of conditions mutation appears to be induced in linear proportion to exposure, thus there is no threshold below which radiation will produce no mutation.

The area of scientific confusion regarding radiation hazards can of course be greatly reduced by the direct procedure of adding to present knowledge. The application of the techniques of microbiology to human cell populations *in vivo* and *in vitro* promises a new approach to questions about human genetics.

Effect of UV Radiation

On absorbing a quantum of UV or visible light the whole of its energy is stored in the molecule which can then undergo one of the number of different reactions, some of which lead to chemical changes (molecular dissociations) and others to physical effects, e.g., fluorescence, heating etc... Action of light rays are much more selective than X (or γ) rays. If U V light of 2600Å passes through an equal mixture of nucleic acid and serum protein, more than 90 per cent of the energy is taken up by the nucleic acid, and less than 10 per cent by the protein. Using X-rays the same amount of energy is absorbed by the protein as by the nucleic acid.

UV Photoproducts in DNA

The dimers formed between adjacent pyrimidines by UV light are chemically the most stable and well-defined lesions readily produced in DNA. Dimers can be formed between T–T and also T–C or between C–C alone (evidence from synthetic polymers). The products formed in the DNA of UV irradiated bacteria appear to comprise of 50 per cent T – T, 40 per cent T – C, and 20 per cent C – C. Their formation involves linking the unsaturated bonds to form a cyclo-butane ring, and must distort the phosphodiester back bone of the twin helix in the vicinity of each dimer.

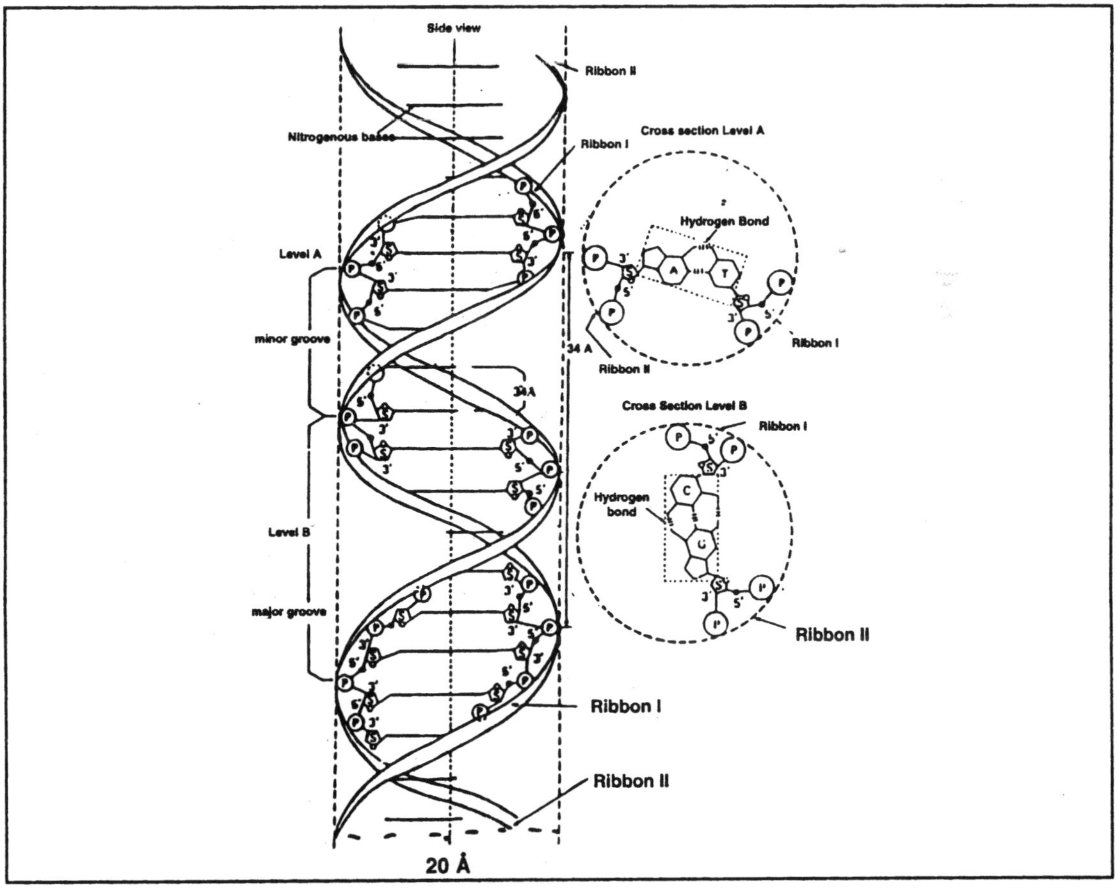

Fig. 2 : The Watson-Crick model for DNA as redrawn by Etkin
(Reprinted, with permission from the 1974 *Bio Science* 24:622 redrawn to maintain 3'-5' polarity).

Bacterial cells previously irradiated with UV are reactivated when additionally irradiated with visible light. The UV resistant strains have the ability to excise thymine dimers from their DNA. This is known as photoreactivation.

Environmental issues such as UV – B (2800-3200 Å) infiltration due to depletion of our planet's protective layer of ozone in the stratosphere, have brought about an awareness of the fragility of our atmosphere, as well as, the vulnerability of the living world.

Bioconcentration of Radio-Isotopes

A nuclear power plant in the U.S.A. released insignificant amounts of radioactivity in its cooling water. An examination of algae in the water, however, showed a concentration upto 6000 times the level in the water itself. These algae serve as food for the bluegill fish; the bones of this fish had a concentration 200 times that in the water.

The great outcry raised by scientist and the public all over the world has to a certain extent caused the cessation of nuclear tests in the atmosphere, due to which environmental contamination has markedly abated. But this is far from achieving the objective of banning nuclear proliferation other than for peaceful uses.

References

Charlton, J.S. Ed., Radioisotope Techniques in Problem Solving in Industrial Process Plants.

Feinehdegen, L.E., Biological damage from radioactive nuclei incorporated into DNA of cells; implications for radiation biology and radiation protection. Proc. 6, Symp. on Microdosimetry; Harwood Acad. Publ. London.

International Nuclear Fuel Cycle Evaluation (INFCE) (1979). Release consequence analysis for a hypothetical geologic radioactive waste repository in hard rock. INFCE/DEP/WG7/21.

Keeling, C.D., Whorf, T.P., Wahlen, M. and van der Pilcht. J. (1995). International extremes in the rate of rise of atmospheric carbon dioxide since 1980. Nature, 375, 666-670.

Mookerjee, A. and Bhattacharjee, S.B., (1984). Aspects of Radiation Biophysics, Interprint, New Delhi, India.

Mookerjee, A., (1985). "Environment, Nursery of Life", Publications Division, Ministry of Information and Broadcasting, Govt. of India, New Delhi-110001. Proceedings of Seventh International Congress of Radiation Research held in July 1983. Martinus Nijhoff, (1984).

Peak, M.J. and Peak, J.G., The role of solar ultra violet radiation in marine ecosystems (ed. by John Calkins) Pleflum Publishing Corporation, 1982, 325-336.

Siddons, R.A., (1974). "Production of Radioactivity in Peaceful Nuclear Explosions" (Proc. Panel Vienna, 1972), IAEA, Vienna 354.

Turner, J.E., Atoms, radiation and radiation protection.

Whicker, F.W. and Schultz, V. Radioecology : Nuclear Energy and the Environment, Vol. 1, 7, CRC Press, Inc. Boca Raton, Florida.

Proceedings of the National Seminar on Environmental Biology, April 03-05, 1998
***Edited By :* A.K. Aditya & P. Haldar**
DAYA PUBLISHING HOUSE, DELHI

Responses of Antioxidase and Detoxifying Enzymes of *Balanus Balanoides* and *Saccostrea Cucullata* Collected from Polluted and Non-polluted Zones of Hoogly Estuary

S. Biswas, S. Niyogi & A.G. Dutta
Department of Life Science and Biotechnology
Jadavpur University, Calcutta-700032

Abstract

A comparative study on the response of antioxidase enzymes (Catalase and DT-diaphorase) and a detoxifying enzyme (Glutathione-S-transferase) has been carried out in a barnacle, *Balanus balanoides* and a bivalve, *Saccostrea cucullata*. Both these filter feeding, sedentary, marine invertebrates were collected from non-polluted as well as polluted zones, *Balanus balanoides* was collected from Lighthouse, Sagar Island (remote, non-polluted zone inside Sunderban). Haldia oil jetty (heavily polluted zone due to waste disposal from oil refinery and oil disembarkment activities of numerous oil tankers) and Sankarpur fishing harbour (another highly polluted zone due to activities of numerous mechanised fishing trawlers and vessels). The bivalve mollusc, *Saccostrea cucullata* was collected from two stations. Lighthouse (Sagar Island) and Sankarpur fishing harbour. Significant changes were recorded in the activities of the above mentioned enzymes in both the animals collected from non-polluted as well as polluted zones. The results indicate that antioxidase and detoxifying enzymes may be useful in detecting the effects of pollutants on marine biota and also they may be considered as complementary parameters in environmental monitoring.

Introduction

Pollution in the marine environment affects not only aquatic life as such but poses a serious threat to all life on the planet due to the continuous degradation of the environment. The multifarious activities of mankind have led to a continual influx of various toxic xenobiotics into the oceans, particularly polycyclic aromatic hydrocarbons (PAH), polychlorinated biphenyls (PCB), heavy metals (Cd, Zn, Cu, Pb, Hg,......................) etc. which may upset the fragile balance of life in the seas. Estuaries and coastal water are particularly at risk from anthropogenic pollution.

Xenobiotic biotransformation of lipophilic organics are mostly dependent on a oxidative step catalyzed by the microsomal cytochrome P-450 dependent mixed function oxidase (MFO) system [Phase I metabolism]. It has been shown that MFO mediated PAH metabolism may result in the formation of potentially carcinogenic, mutagenic and/or cytotoxic intermediates which result in DNA damage and cell death (De Pierre & Ernster, '78; Sole *et al.* '94)

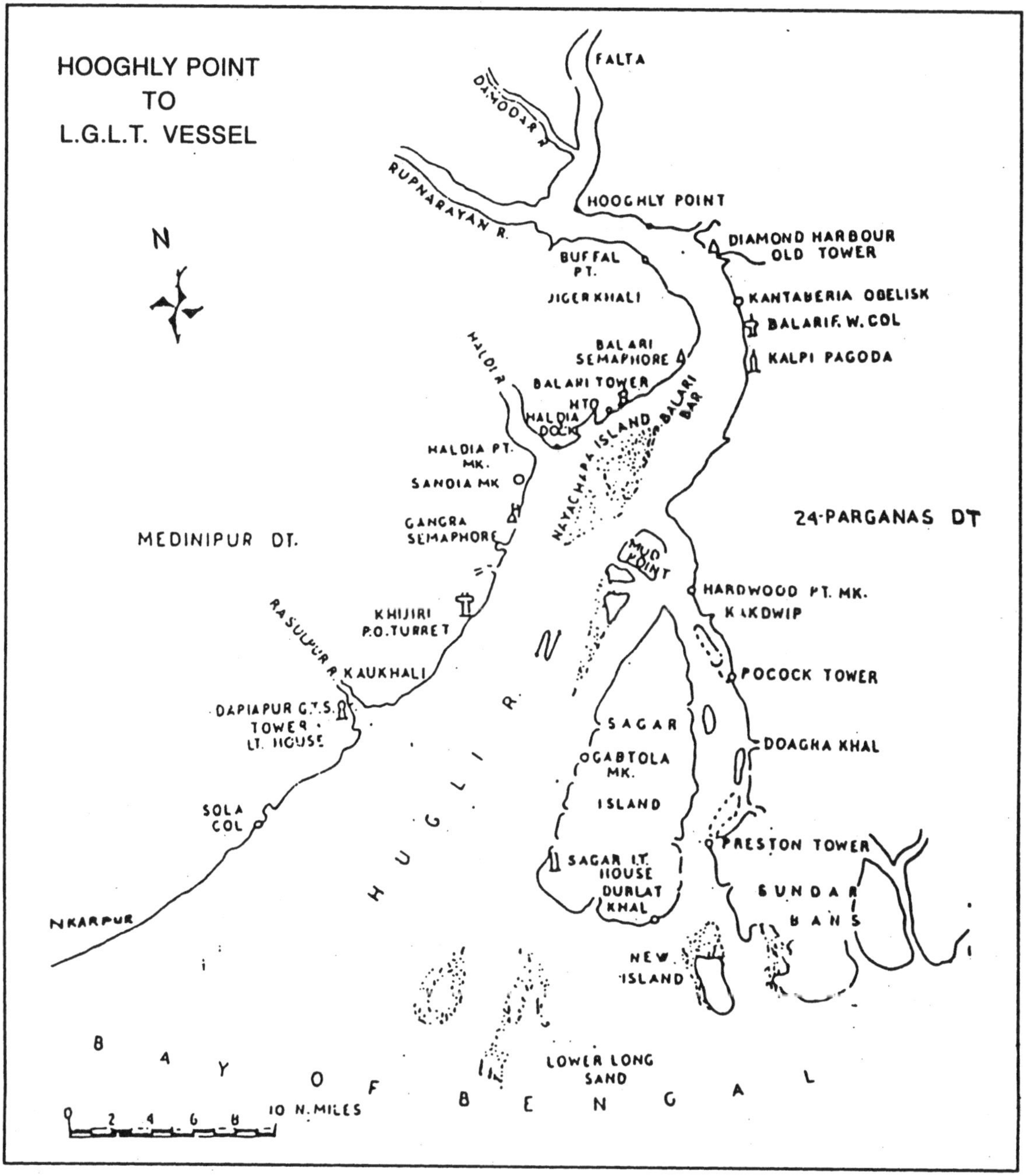

MAP OF THE STUDY AREA

Phase I biotransformation is usually followed by Phase II metabolism or conjugation reaction which detoxify the reactive intermediates of Phase I. Glutahione (L-γ-glutamy-ζ-cysteinyl glycine), an ubiquitous tripeptide and usually the most prevalent intracellular thiol, plays a most important role in this sort of metabolism by being conjugated with electrophilic xenobiotics (either native xenobiotic or Phase I metabolite) mediated by the Glutathione-S-transferase enzymes resulting in a water soluble and readily excretable and product Redox cycling of various xenobiotics also result in the formation of reactive oxygen species, the superoxide anion (O^-_2))] the hydorxyl radical (OH) and hydrogen peroxide (H_2O_2) and lipid peroxidation (Sole *et al*, '94; Livingstone *et al* '89. Winston & Di Giulio. '91 and Morgenstren *et al*, '81).

In the marine environment the cytochrome P-450 dependent MFO system has been well characterized in molluses, fishes and various arthropods (Porte *et al.* '91; Schlenk & Buhler, '89. Stegeman & Binder, '79 and Stegeman & Kaplan, '81) but there is very little information regarding Phase II and Antioxidant enzymes in these animals. The purpose of our research is to examine the detoxication enzyme Glutathione-S-transferase (G-S-T) and the antioxidant enzymes DT-diaphorase and catalase in the barnacle *Balanus balanoides* and oyster *Saccostrea cucullata* collected from polluted and non-polluted zones of the Hooghly estuary and to determine their efficacy as indicators of pollution.

The cirriped crustacean *Balanus* and the mollusc *Saccostrea* are both sessile animals. Among molluscs, mussles have been used widely to monitor pollution (Goldberg *et al*, '78) but very little work has been done with crustaceans, especially *Balanus*. As both barnacles and oysters are able to survive and reproduce in extremely polluted waters, it was assumed that they may have some protective enzymatic machinery against toxic xenobiotics. Various previous studies have indicated this fact in molluscs and to some extent in crustaceans.

Of the sampling stations, Haldia oil jetty complex and Sankarpur fishing harbour have been designated as polluted zones: the former due to heavy industrial activity and movement of shipping including oil tankers, while in the latter pollution is largely due to the activities of large number of mechanized fishing trawlers. PAHs are probably the most prevalent xenobiotics in these waters (water analysis of these stations are being carried out at present). The control station at the old Lighthouse. South Sagar Island is away from the industrial zones, also there is very little movement of mechanized fishing boats. It is assumed the pollution load will be the lowest here (*see map*).

Material and Methods

Sample collection

Barnacles were collected every month during August '97 to February '98 from the old Light-house at Sagar Island and oil jetty complex at Haldia. Oysters were collected from same site at Sagar Island and Sankarpur fishing harbour during January and February '98. The animals were transported to the laboratory in ambient esturaine water.

Preparation of Cytosol

The soft body parts of barnacles were removed from their shells and washed with ice cold 0.065 M Na-Phosphate buffer ph 7.0 containing 1.15% KCL and 3 mM $MgCl_2$. The tissues were mixed and homogenized in 4 volumes of this buffer incorporating 0.01 mM Phenyl methyl sulphonyl fluoride (PMSF) with a glass-tefleon Potter-Elvehjem homogenizer till no particulate matter could be discerned. Post mitochondrial supernatant (PMS) was prepared by centrifuging the homogenate at 10,000 x g for 30 minutes. Cytosol was prepared by centrifuging the PMS at 10,000 x g for 90 minutes. The supernatant cytosol from this step was aliquoted and stored at– –20°C (Stegeman & Kaplan, '81)

The digestive tissue of oysters were dissected out and washed with ice cold 20mM tris-HCI buffer, pH 7.6, incorporating 0.5 M sucrose and 0.15 M KCI. The tissues were mixed and homogenized in 4 volumes of this buffer containing 0.001 mM PMSF with an all glass Potter-Elvehjem homogenizer till no particulate matter could discerned Cytosol was prepared as described above. The cytosol was aliquoted and stored – 20°C (Schlenk & Buhler, '89)

Biochemical determinations

DT-diaphorase : Cytosilic NADH-dependent DT-diaphorase activity was measured as NADH-dichlorophenol indophenol (DCPIP) reductase activity. The reduction of DCPIP was measured at 600 nm (Ernster, Methods in Enzymology)

Catalase : Cytosolic catalase activity was determined by following the decomposition of H_2O_2 at ph 7 and 240 nm (Beers & Sizer, '52)

Glutahione-S-transferase : G-S-T activity was measured at pH 6.5 by monitoring the increase in absorbance of GSH-I-chloro-2, 4-dintrobenzene adduct at 340 nm (Gibson & Skett. '94) Protein was measured according to Bradford (1976) using Bovine Serum Albumin as standard.

Results and Discussion

The results indicate that there is a significant difference in the activities of all three enzymes studied between the polluted and non-polluted zones (Tables 1-6). The cytosolic catalase activity in barnacles show a definite seasonal pattern–increasing during monsoon and then decreasing in the post monsoon period–at both Sagar Island (control) and Haldia (polluted zone). The per cent stimulation of the enzyme in polluted zone over control however shows a decrease only till December and then registers an increase during January-February (Table 1, Fig. I).

Table 1 : Variation in activities of Cytosolic Catalase of *Balanus balanoides* collected from three different places along Hooghly Estuary

Specific activity (S.A.) is expressed as ΔOD/min/mg protein

Month/ Year	Lighthouse, Sagar Island (Control)	Oil Jetty complex, Haldia	% stimulation compared to control area (Sagar)	Fishing Port, Sankarpur	% stimulation compared to control area (Sagar)
August, 97	0.17	0.23	35.3%	—	—
September, 97	0.19	0.26	36.8%	—	—
November, 97	0.145	0.19	31%	—	—
December, 97	0.105	0.13	23.8%	—	—
January, 98	0.055	0.080	45.5%	0.07	27.3%
February, 98	0.55	0.075	50%	0.063	36%

Note : 1. Sampling from Sankarpur was started from January, 98 onwards.
2. Results depicted are average of at least three experiments.
3. ΔOD represents change in absorbance (arbitrary units).

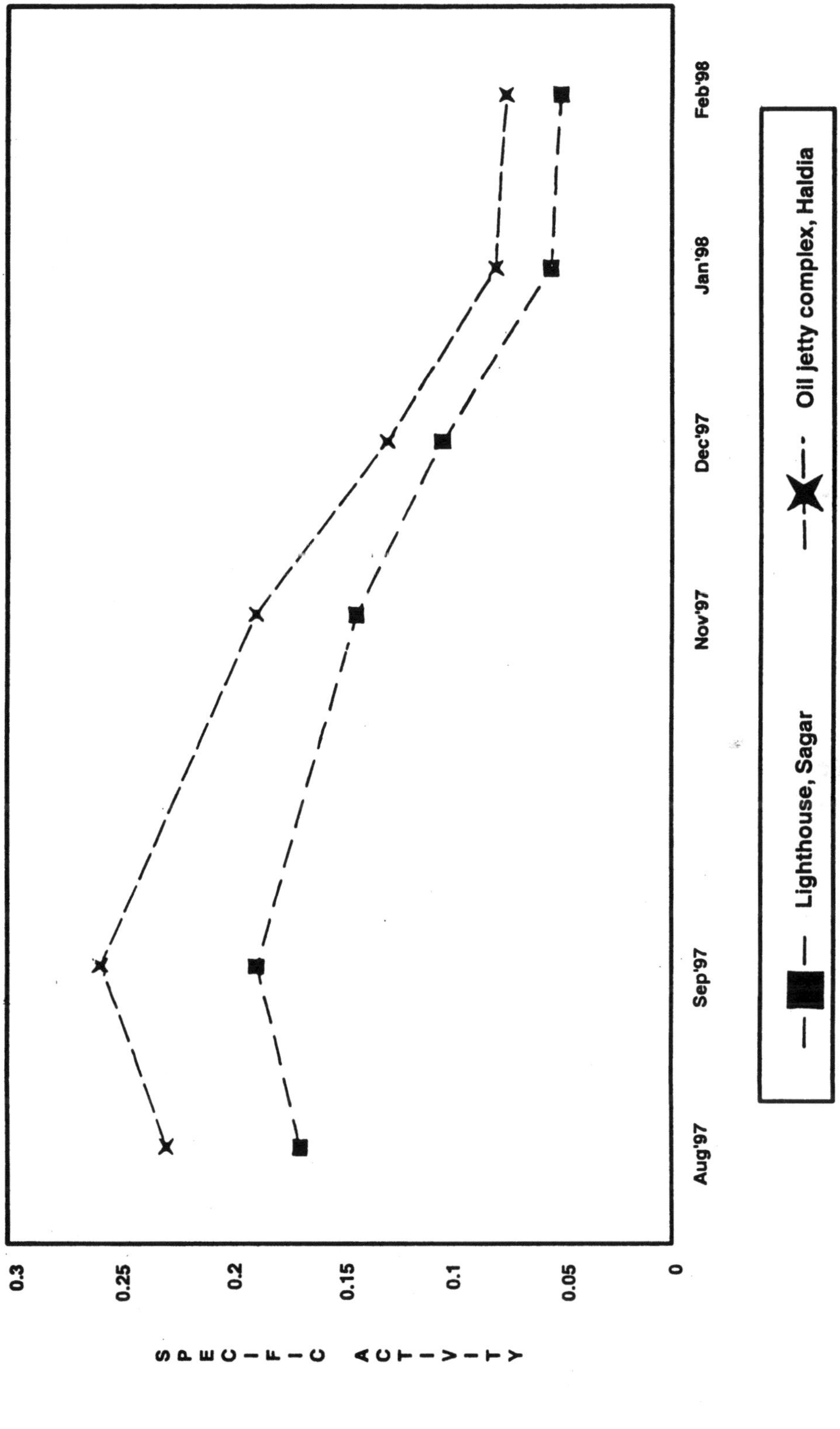

Fig. II : Graphical representation of Monthly variation of cytosolic catalase activity in *Balanus balanoides* collected from Lighthouse, Sagar Island and Oil Jetty Complex, Haldia.

The other antioxidant enzyme studied in barnacle, cytosolic NADH dependant DT-diaphorase, also shows a similar seasonal pattern of activity–increasing during monsoon and then decreasing in the post monsoon period. However, the specific activity of this enzyme is higher than that of catalase. The per cent stimulation of this enzyme in barnacles from polluted zone is greater than catalase but exhibits a similar pattern (Table 2, Fig. II)

Table 2 : Variation in activities of Cytosolic NADH-dependent DT-diaphorase in *Balanus balanoides* collected from three different places along Hooghly Estuary

Specific activity (S.A.) is expressed as ΔOD/min/mg protein

Month/ Year	Lighthouse, Sagar Island (Control)	Oil Jetty complex, Haldia	% stimulation compared to control area (Sagar)	Fishing Port, Sankarpur	% stimulation compared to control area (Sagar)
August, 97	0.42	0.83	97.6	—	—
September, 97	0.50	0.90	80	—	—
November, 97	0.40	0.65	62.5	—	—
December, 97	0.23	0.37	60.8	—	—
January, 98	0.19	0.32	68.4	0.43	126
February, 98	0.18	0.30	66.6	0.39	116

Note : 1. Sampling from Sankarpur was started from January, 98 onwards.
2. Results depicted are average of at least three experiments.
3. ΔOD represents change in absorbance (arbitrary units).

The Phase II enzyme, cytosolic Glutathione-S-transferase in barnacle, does not show any apparent seasonal activity in the two stations studied. However, the percent stimulation of the enzyme in the polluted zone over control area shows an increase to a peak level in December before reaching a steady state during January-February (Table 3, Fig. III)

Table 3 : Variation in activities of Cytosolic Glutathione-S-transferase of *Balanus balanoides* collected from two stations along Hooghly Estuary

Specific activity (S.A.) is expressed as ΔOD/min/mg protein

Month/ Year	Lighthouse, Sagar Island (Control)	Oil Jetty Complex, Haldia	% stimulation compared to control area
September, 97	1.5625	2.0753	32.82
November, 97	0.782	1.2121	55
December, 97	1.211	2.668	138
January, 98	0.857*	1.768*	106.3
February, 98	0.75	1.53	104

** Assay carried out with Post mitochondrial supernatant (10,000 x g supernatant).*

Note : 1. Sampling from Sankarpur was started from January, 98 onwards
2. Results depicted are average of at least three experiments.
3. ΔOD represents change in absorbance (arbitrary units).

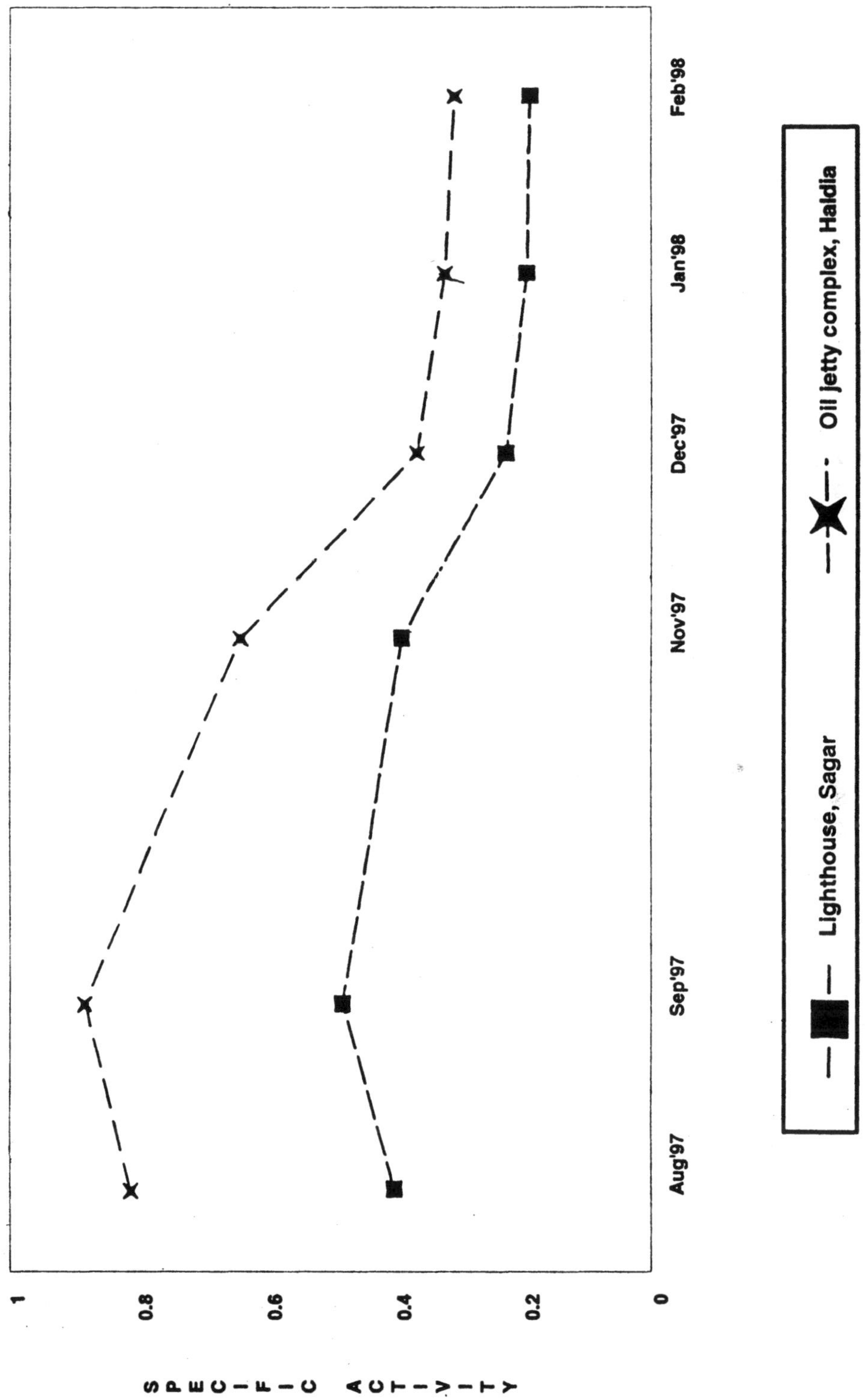

Fig. II : Graphical representation of Monthly variation of cytosolic NADH-dependent DT-diaphorase activity in *Balanus balanoides* collected from Lighthouse, Sagar Island and Oil jetty complex, Haldia.

Sampling at Sankarpur was done only during January February, 98. Catalase levels in barnacles from here show a lower percent stimulation over control than those in Haldia (Table 1), while DT-diaphorase stimulation is greater (Table 2).

In oysters, the specific activity of catalase and G-S-T are considerably higher than that in barnachles, but cytosolic NADH dependant DT-diaphorase does not show such a high difference. All three enzymes show greater activity in the polluted zones than in the control area (Table 4-6).

Taule 4 : Variation in activities of Cytosolic catalase in digestive tissue of *Saaccostrea cucullata* collected from two different places along Hooghly Estuary

Specific activity (S.A.) is expressed as ΔOD/min^{-1} / mg^{-1} protein

Month/ Year	Lighthouse, Sagar Island (Control)	Fishing Port, Sankarpur	% stimulation compared to control area (Sagar)
January, 98	6.80	11.40	67.6
February, 98	6.1	9.3	52.4

Note : 1. Sampling from Sankarpur was started from January, 98 onwards.
2. Results depicted are average of at least three experiments.
3. ΔOD represents change in absorbance (arbitrary units).

Table 5 : Variation in activities of Cytosolic NADH-dependent DT-diaphorase in digestive tissue of *Saccostrea cucullata* collected from two different places along Hooghly Estuary

Specific activity (S.A.) is expressed as ΔOD/min^{-1} / mg^{-1} protein

Month/ Year	Lighthouse, Sagar Island (Control)	Fishing Port, Sankarpur	% stimulation compared to control area (Sagar)
January, 98	0.26	0.75	188
February, 98	0.24	0.55	129

Note : 1. Sampling from Sankarpur was started from January, 98 onwards.
2. Results depicted are average of at least three experiments.
3. ΔOD represents change in absorbance (arbitrary units).

Table 6 : Variation in activities of Cytosolic Glutathione-S-transferase in digestive tissue of *Saccostrea cucullata* collected from two different places along Hooghly Estuary

Specific activity (S.A.) is expressed as ΔOD/min / mg protein

Month/ Year	Lighthouse, Sagar Island (Control)	Fishing Port, Sankarpur	% stimulation compared to control area (Sagar)
January, 98	3.96	6.72	69.7
February, 98	3.92	5.18	32.14

Note : No detectable level of enzymatic activity (all three enzymes) was found in the cytosolic fraction of gill and mantle of *Saccostrea cucullata* collected from both the stations.

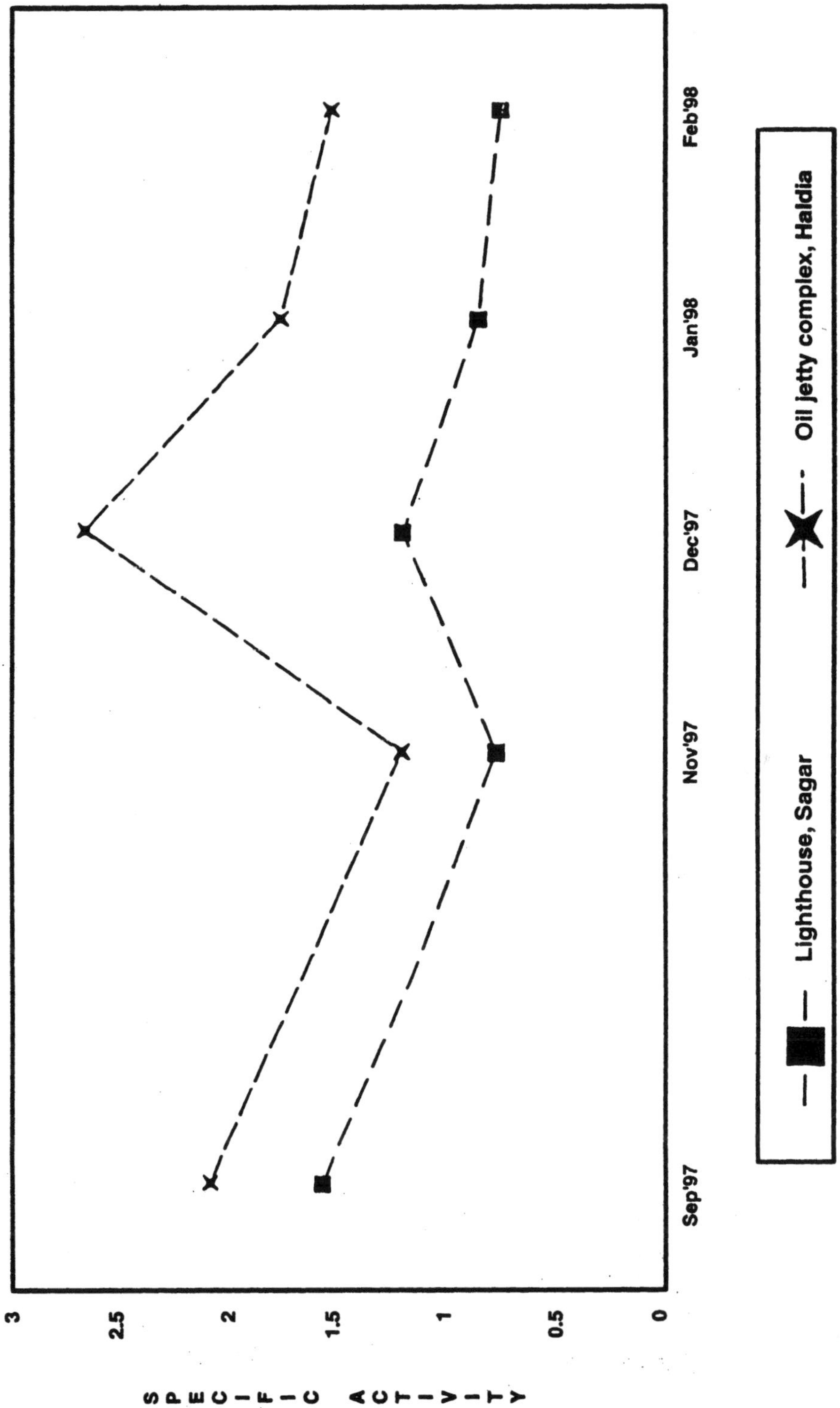

Fig. III : Graphical representation of Monthly variation of cytosolic Glutathione-S-transferase activity in *Balanus balanoides* collected from Lighthouse, Sagar Island and Oil jetty complex, Haldia.

Increased oxyradical production due to pollution may occur by several mechanisms, including uptake of redox cycling metals and organic xenobiotics and metabolism of xenobiotics to redox cycling derivatives (Livingstone *et al*, '89; Porte *et al*, '91 & Livingstone, '88) It is reported that DT-diaphorases stimulate the redox cycling of quinones and semiquinones which may be produced by the MFO mediated metabolism of PAHs (Sole *et al*, '94 & Porte *et al*, '91). It has been reported that G-S-T enzymes play a role in the detoxication of these PAH metabolites (De Pierre & Ernster, '78 & Morgenstern *et al*, '81).

From our result it is clear that at least one of the above factors is responsible for the induction of the enzymes studied in polluted zones, but a definite conclusion cannot be reached yet. The elevated specific activities of the enzymes in oyster when compared to barnacles is probably due to more efficient xenobiotic metabolism in the former organism which is also a well known bioaccumulator of pollutants (Sole *et at*, '94 & Goldberg, '78). While it is clear that these enzymes in both organisms can be used as indicators of pollution, further research is needed to correlate pollutant type and levels of enzyme induction.

Acknowledgement

This work was supported by Ministry of Enviornment and Forests, Govt. of India.

We would like to express our heartfelt gratitude to Dr. R. N. Hati, Scientist E-II, Indian Institute of Chemical Biology, for his valuable advice and the infrastructural support he has extended to us, without which we would have been unable to do this work.

References

Beers R.F. and Sizer I W (1952). A spectrophotometic method for measuring the breakdown of Hydrogen Peroxide by Catalase *J. Biol. Chem.* 195 : 133-140.

Bradford M.M. (1976). A rapid and sensitive method for the quantitation of microgram quantities of protein utilizing the principle of protein dye binding. *Anal. Biochem.* 72 : 248-254.

De Pierre J.W. and Ernster L (1978). The metabolism of polycyclic hydrocarbons and its relation to cancer *Biochim Biophys Acta.* 473 : 149-186.

Ernster L., DT Diaphorase Method in Enzymology. Vol. X : 309-317.

Gibson G.G. and Skett. P. (1994). Introduction to Drug Metabolism. 2nd Edition, Blackie Academic and Professional. Glasgow pp 238.

Goldberg E.D., Bowen V T, Farrmgton J W, Harvey G, Martin. J.H., Parker. P.L., Risebrough R.W, Robertson W, Schneider W. and Gamble. E. (1978). The Mussel Watch, *Environ Conserv* 5 : 101-125.

Livingstone D R, Garcia-Martinez P, Stegeman J. J and Winston. G. W. (1988b) : Benzo(a) pyrene metabolism and aspects of oxygen radical generation in the common mussel. *Mytilus edulus* l. *Biochemical Society Transactions.* 16: 779.

Livingstone. D R, Garcia-Martinez P. and Winston G.W. (1989a). Menadione stimulated oxyradical formation in digestive gland microsomes of the common mussel. *Mytilus edulis* L. *Aquat. Toxicol.* 15 : 213-236.

Morgenstem R., De Pierre J W. Lind C, Guthenberg C., Mannervik. B. and Ernster. L. (1981). Benzo(a) pyrene Quinones can be generated by Lipid Peroxidation and are conjugated with Glutathione-S-Transferase B from rat liver. *Biocchem. Biophys. Res. Commun* 99. No 2 : 682-690.

Porte C., Sole M., Albages J. and Livingstone D. R. (1991). Responses of mixed function oxygenase and antioxidase enzyme system of *Mytilus* sp. to organic pollution. *Comp. Biochem Physiol* Vol 100 C. No 1/2; 183-186.

Schlenk D and Buhler D.R. (1989). Xenobiotic biotransformation in the western oyster, *Crassostrea gigas* *Comp Biochem Physiol* 94C : 469-475.

Sole M., Porte C and Albaiges J (1994). Mixed function oxygenase system components and antioxidant enzymes in different marine bivalves its relation with contaminant body burdens. *Aquat. Toxicol.* Vol. 30 : 271-283

Stegeman J. J and Binder R. L. (1979) : High Benzo (a) pyrene hydorxylase activity in the marine fish. *Sienotomus versicolor Biochem Pharmacol.* 28 : 1686-88.

Stegeman J. and Kaplan H.B. (1981). Mixed function oxygenase activity and benzo[a] pyrene metabolism in the Barnacle *Balanus eburneus* (Crustacea : Cirripedia) *Comp. Biochem Physiol.* Vol 68C : 51-61.

Winston G. W. and Dr. Grulio R. T. (1991). Pro-oxidant and antioxidant mechanisms in aquatic organisms *Aquat Toxicol.* 19: 137-161.

Proceedings of the National Seminar on Environmental Biology, April 03-05, 1998
***Edited By* : A.K. Aditya & P. Haldar**
DAYA PUBLISHING HOUSE, DELHI

Inorganic Mercury Specifically Binds to Rat Platelet Membrane and Initiates a Cascade of Actions

Shelley Bhattacharya, S. Vinaya Kumar & Rakhi Bose
Environmental Toxicology Laboratory
Department of Zoology, Visva Bharati University
Santiniketan 731235

Inorganic mercury has gained attention since the episode of Minamata Bay. This extremely toxic Group II B metal is ubiquitous in its distribution in the environment which has encouraged a large number of research groups to focus their attention to unraveling the mechanism of action of this metal. Various physiological and biochemical effects of inorganic mercury are on record (Imura *et al.*, 1980; Ghosh and Bhattacharya, 1992; Bose *et al.*, 1993). It has been shown that when inorganic mercury enters into the blood it first binds with α-globulins and erythrocytes with a half time of about 15 days and the systemic effect ensues within a few hours of exposure (Harvey, 1975). Later authors have also demonstrated that inorganic mercury is maximally retained by blood (Aihara and Sharma, 1986) and it has a cooperative binding with erythrocyte membrane (Bose *et al*, 1994). Prostaglandin thromboxane receptors in mammalian platelets have essential sulfhydryl groups (Dorn, 1990) and therefore it is expected that inorgainc mercury would bind to platelet membrane and induce changes in its function, namely, platelet aggregation.

It is well known that platelets are primarily involved in the processes of thrombosis and hemostasis and it has also been proposed that it has a significant role to play in eliciting the responses of the blood to injury (Packham *et al.*, 1968). Later Mustard and Packham (1970) have shown that even xenobiotics are able to elicit a platelet response but no conclusive evidence was provided. Extensive reports are available on the action of physiological agents on platelet aggregation (Jamaluddin and Krishnan, 1987a, b; Jamaluddin *et al.*, 1988) but very few on the role of xenobiotics in altered platelet function (Chaudhury *et al.*, 1987, 1988). More recently, Jamaluddin and Thomas (1992) demonstrated inhibition of platelet aggregation at 10mM Hg^{2+} in the presence of Ca^{2+}. These results point to a specific role of inorganic mercury in platelet aggregation which is likely to be mediated via the plasma membrane. *In vivo* treatment with Hg^{2+} causes both time and dose dependent inhibition of agonist induced platelet aggregation. Among the four (ADP, arachidonic acid, collagen and epinephrine) agonists tested alteration in the ADP induced platelet aggregation was more drastic (Chaudhury *et al.*, 1996). These results suggested that Hg^{2+} inhibits Ca^{2+} entry into platelets or the rate of mobilization of Ca^{2+} from intraplatelet sources as observed earlier with crude petroleum oil (Chaudhury *et al.*, 1988). Platelet aggregation is known to be a membrane associated phenomenon (Berlin *et al.*, 1984). Eaton *et al.*, (1992) demonstrated the entry of Cd^{2+}, another Group II B metal through a non-selective cation channel which probably blocks the influx of extracellular Ca^{2+} as reported by Mukhopadhyay *et al.*, (1994). It may thus be opined that Hg^{2+} being another member of the Group II B metal family inhibits the rate of Ca^{2+} influx thereby reducing adrenaline-mediated platelet aggregation.

Interestingly, low doses of Hg^{2+} can induce aggregation even in the absence of Ca^{2+} (Chaudhury *et al.*, 1996) as seen with Mg^{2+} (Garvin, 1961; Haslam, 1964; Des Prez 1967). Contrastingly, with higher doses of Hg^{2+}, with or without Ca^{2+} (or the inophore, 0.5mM A23187, or the calcium blocker, 3mM verapamil) the pattern of inhibition of platelet aggregation is very similar to that elicited by excess of Mg^{2+} in pig and rabbit platelet (Born and Cross, 1964). It has long been established that thiol groups on platelet membrane are essential components to maintain the integrity of platelet structure and other enzymatic and biological functions of the platelet (Aledort *et al.*, 1968; Mitchell and Sharp, 1964; O'Brien, 1962). Some very interesting observations on the effect of methylmercuric nitrate (MMN) on washed dog platelet were reported by Robinson *et al.*, (1965). It was found that MMN inhibited the ability of the washed platelet to aggregate with ADP in the presence of Ca^{2+} while at low concentrations MMN induced platelet aggregation with a concomitant increase in ATPase activity. Thus it may be concluded that being an avid SH blocker Hg^{2+} at higher concentrations inhibit platelet aggregation by specific binding to the thiol groups of membrane proteins.

Stimulation of platelet aggregation by inorganic chemicals has not been addressed by many workers. Considering the role of cAMP as a second messenger in the signal transduction mechanisms studied in various types of cells it was thought worthwhile to proceed to the assessment of cAMP, adenylate cyclase and cAMP-dependent phosphodiesterase. Marcus and Zucker (1965) has reported the inhibition of platelet aggregation caused by cAMP which was further corroborated by Ardlie *et al.*, (1967). Any agent which increases the concentration of intracellular cAMP is liable to inhibit platelet aggregation as evidenced by prostaglandins (Smith *et al.*, 1974). It was opined by Kaser-Glanzmann *et al.*, (1977) that platelet aggregation is regulated by the cAMP adenylate cyclase-phosphodiesterase system. That this pathway is actually involved was proved by Chaudhury *et al.*, (1996). In their experimental system a dose and time dependent inhibition of Ca^{2+} and ADP induced aggregation was noticed by cAMP addition which effect was further enhanced by theophylline, a strong inhibitor of cAMP-dependent phosphodiesterase present in the system.

The action of inorganic mercury on ADP-induced platelet aggregation could not be reversed by either cAMP alone or a combination with theophylline. This refractoriness of cAMP or theophylline indicates a more sustained action of Hg^{2+} which stimulates influx of Ca^{2+} blocking adenylate cyclase to synthesise cAMP. Inhibition of this enzyme tends to mobilize Ca^{2+} from internal stores thereby enhancing the rate of platelet aggregation. Phosphodiesterase activity declines to increase the cytosolar cAMP concentration to the physiological level during the recovery phase of the cell from tissue injury. Thus it became apparent that the lower doses of Hg^{2+} counteracts the cAMP-intraplatelet phosphodiesterase activity (Chaudhury *et al.*, 1996). Aspirin is known to acetylate platelets (Woodbury and Fingle, 1975) and inhibit cyclooxygenase (Jamaluddin and Thomas, 1992). Chaudhury *et al.*, (1996) demonstrated that aspirin stimulates ADP induced aggregation when it is added after the addition of Hg^{2+}. Jamaluddin and Thomas (1992) recorded a similar action of aspirin on H_2O_2– induced calf platelet aggregation in the absence of Ca^{2+}. These events are due to the balance between cyclooxygenase inhibition and catalase inhibition. Chaudhury *et al.*, (1996) further demonstrated that preincubation of platelets with aspirin completely abolished the stimulatory action of Hg^{2+} on platelet aggregation. It was logically surmised that once cyclooxygenase inhibition sets in Hg^{2+} is unable to exert its effect and vice versa. And at low concentrations therefore, Hg^{2+} stimulates the activity of cylooxygenase.

Platelet activation is a prerequisite of platelet aggregation. The agonists must occupy a sufficient number of membrane receptors to induce the activation of a guanine nucleotide binding protein or G-protein linked to phospholipase C (PLC) at the cytoplasmic side of the plasma membrane. As soon as PLC is activated it catalyses the hydrolysis phosphatidylinositol 4, 5-biphosphate (PIP2) at a resting intracellular calcium ion concentration of about 100 nM. This results in two key second messenger molecules – the hydrophilic inositol 1,4,5-triphosphate (IP3) and the lipophilic diacyglycerol (DAG). IP3 releases the calcium ions from the intracellular stores which raises the intracellular calcium ion concentration transiently. This event may be regulated by cAMP levels and protein kinase C (PKC) activities which are known to change during platelet activation. Receptor activation also opens the so-called receptor operated calcium channels which in the presence of extracellular calcium ions (~ 1 mM) add to the intracellular calcium thereby maintaining the concentration at ~ 1-15 mM. Thus calcium is found to have many direct and indirect effects on several key enzymes and proteins that control platelet activation (Jamaluddin, 1991).

Secretion of the platelet is also mediated by intracellular calcium through phospholipase A2 and PKC. Once activated PLA2 sets arachidonate free from phospholipids, mainly phosphatidylcholine, Interestingly PKC does also have a negative control over intracellular calcium, cAMP levels and receptor-effector coupling. Free arachidonic acid is quickly converted by cyclooxygenase to prostaglandins G2 ad H2 (PGG2 and PGH2). Only PGH2 accumulates which is converted to thromboxane A2 by thromboxane synthetase. Both PGH2 and TXA2 are well known agonists for human platelets inducing platelet activation through the PLC-PKC pathway. The enzyme cyclooxgenase is irreversibly inactivated by aspirin which explains the antagonistic effect of this drug on platelet aggregation. Free arachidonic acid is also partly acted upon by lipooxygenase producing hydroperoxy and hydroxyenoic acids (Jamaluddin, 1991).

It has been observed that rat platelet aggregation was enhanced by the agonist, collagen in presence of 2mM Ca^{2+} with all the xenobiotics. ADP and arachidonic acid inhibits aggregation with cadmium, mercury and arsenic while epinephrine accelerates it under Hg and Cd treatment and inhibited under As treatment. It appears that under 1/10 LD50 *in vivo* xenbiotic treatments collagen is the most potent agonist. It may be possible that in such cases the transformation of the platelet from a discoid to spheroid shape is very rapid (Vinaya Kumar, unpublished observations).

The results of platelet aggregation *in vitro* with increased concentrations of xenobiotics (5, 10 and 20 pmoles) reveal that all the agonists accelerate the rate of aggregation except Cd^{2+} treatment which inhibits the relative aggregation dose dependently. Reports are available that Cd^{2+}, group II B metal enters liver through a non selective cation channel (Eaton *et al.*, 1992) and blocks the influx of extracellular Ca^{2+} (Mukhopadhyay *et al.*, 1994). Thus it can be concluded that Cd^{2+} perhaps in a similar manner inhibits Ca^{2+} influx and reduces the ADP mediated platelet aggregation.

As reported by various workers cAMP is considered as a regulatory molecule in the event of platelet aggregation. Interestingly we have observed that the cytosolar cAMP profile demonstrates a significant decline in the *in vivo* (1/10 LD 50) metal treated rats, but under *in vitro* treatment with 5, 10 and 20 pmoles of metal As and Hg treatment register a decline while Cd shows a 90-100% increase in the cytosolar cAMP content. The data encourage us to conclude that

Cd is a potent inhibitor of phosphodiesterase. Furthermore the rates of platelet aggregations were inversely correlated to the cAMP content in the metal treated platelet thus proving the mediatory role of this biomolecule in platelet aggregation. Since the concentration of cAMP is dependent on the cataytic rate of the membrane bound adenylate cyclase and the cytosolar phosphodiesterase, analyses of the activity of these two enzymes suggest a dynamicity of the cAMP-Adenylate cyclase-Phosphodiesterase system in the platelet.

It is surmised that all the activation reactions of the platelet leading to aggregation can be triggered by specific binding of the metals to the surface membrane. From the metal binding assay done with ^{203}Hg, it was clear that this heavy metal effects specific binding to the surface membrane. Scatchard analysis of the assay revealed a specific binding of inorganic mercury to the platelet membrane, The Bmax was calculated to be 7 nmoles/mg protein and a Ka of $1x10^7$ moles. Specific binding was demonstrated also in rat liver (Bhattacharya *et al.*, 1997) fish liver (Sarkar, 1997) and fish oocyte (Mondal *et al.*, 1997) plasma membrane. In each of these cases the Hg specifically bound to the membrane bound enzyme Na^+-K^+-ATPase initiating a cascade of events leading to the synthesis of a stress protein, metallothionein, in the liver or an enzyme 3β-hydroxysteriod dehydrogenase in the oocyte. It is proposed that in rat platelet Hg-binding to the membrane may also lead to the synthesis of a specific protein which warrants further investigation.

References

Aihara, M. and Sharma, R.P. (1986). Effects of endogenous and exogenous thiols on the distribution of mercurial compounds in mouse tissues. Arch. Environ. Contam. Toxicol. 15: 629-636.

Aledort, I.M., Troup, S.B., and Weed, R.I. (1968) Inhibition of sulfhydry1 dependent platelet functions by penetrating and non penetrationg analogues of parachloro mercuribenzene. Blood, 31: 471-475.

Ardlie, N.G., Glew, G., Schultz, B.G. and Schwartz, C.J. (1967). Thromb. Diath. Haemorrh. 18: 670-673. Cited by Le Breton et al., (1982). Relationship of cyclic nucleotides and calcium on platelet function. In Calcium Regulation by Calcium Antagonists (R.G. Rahwan and D.T. Witiak, Eds) pp. 153-174. American Chemical Society, Washington D.C.

Berlin, E., Shapiro, S.G. and Friedland, M. (1984) Platelet membrane fluidity and aggregation of rabbit platelets. Atherosclerosis, 51: 223-239.

Bose, S., Ghosh, P., Ghosh S., Chaudhury, S., and Bhattacharya, Shelley. (1993). Time dependent distribution of [^{203}Hg] mercuric nitrate in the subcelluler fractions of rat and fish liver. Biomed. Environm. Sc., 6: 195-206.

Bose, S., Chaudhury, S., Ghosh, P., Ghosh, S., and Bhattacharya, Shelley (1994). Mercury binding to the plasma membrane in rat erythrocytes. In Environmental Toxicology in Southeast Asia (B. Widianarko, K. Vink, and N.M. Van Straalen, eds.) pp215-219. VU University Press, Amstrdam.

Bhattacharys, Shelley, Bose, S., Mukhopadhyay, B., Sarkar, D., Das, D., Bandyopadhyay, Bose R., Majumdar, C., Mondal S., and Sen, S (1997). Specific binding of inorganic mercury to Na^+-K^+- ATPase in rat liver plasma membrane and signal transduction. Biometals, 10:157-162.

Chaudhury, S., Macko, S., and Rahimtula, A.D. (1987). Inhibition of rat platelet aggregation by a prudhoe bay crude oil and its aliphatic, aromatic and heterocyclic fractions. Toxicol. Appl. Pharmacol. 90: 347-356.

Chaudhury, S., Khan, S., and Rahimtula, A.D. (1988). Comparison of the inhibitory effects of some compounds present in crude oils on rat platelet aggregation: Role of intra-and extra-cellular calcium. Toxicology, 51: 35-46.

Cahudhury, S., Mukhopadhyay, B., Bose, S., and Bhattacharya, Shelley (1996). Involvement of mercury in platelet aggregation-probable mechanism of action. Bimed. Enviornm. Sci, 9: 26-36.

Des Prez, R.M., Bryant, R.E., Katz, J.A., and Brittingham, T.E. (1967). Platelet aggregation by magnesium ion. Thromb. Diath. Haemorrh., 17: 516-531.

Dorn, H.G.W. (1990). Cyclic oxidation-reduction regulate thromboxane A/prostaglandin H_2 receptor number and affinity in human platelet membranes. J.Biol. Chem., 265:4240-4246.

Eaton, D.C., Marunaka, Y., and Ling, B.N. (1992). Ion channels in epithelial tissue: Single channel properties. In Membrane Transport in Biology (J.A. Scafer, H.H.Ussing, P.Kristensen, and G.H. Giebisch, eds.), Vol. 5, pp 73-165., Springer Verlag, New York.

Garvin, J.E. (1961). Factors affecting the adhesiveness of human leucocytes and platelets *in vitro.* J Exp. Med., 114 : 51-73.

Ghosh, N., and Bhattacharya, Shelley (1992). Thyrotoxicity of the chlorides of cadmium and mercury in rabbit. Biomed. Environm. Sci., 5:236-240.

Harvey, S.C. (1975). Heavy Metals. In The Pharmacological Basis of Therapeutics (L.S. Goodman, and A. Gilman, Eds.) pp. 924-945. Macmillan Pub. Co. Inc. New York.

Haslam, R.J. (1964). Role of ADP in the aggregation of human blood platelets by thrombin and by fatty acids. Nature, 202: 765-768.

Imura, N. Miura, K., Inokawa, M., and Nakada, S. (1980). Mechanisms of mrthylmercury cytotoxicity: By biochemical and morphological experiments using cultured cells. Toxicology, 17 : 241-254.

Jamaluddin, M. (1991). New perspectives in blood platelet aggregation. Current Science, 61 : 526-533.

Jamaluddin, M., and Krishnan, L.K. (1987a) A spectrophotometric method for following initial rate kinetics of blood platelet aggregation. J Biochem. Biophys. Methods, 14 : 191-200.

Jamaluddin, M., and Krishnan, L.K. (1987 b). A rate equation for blood platelet aggregation. J Theor. Biol., 129:257-261.

Jamaluddin, M., Krishnan, L.K., and Thomas, A. (1988). Ajonene inhibition of platelet aggregation: possible mediation by a heamoprotein. Biochem. Biophys. Res. Commun., 153:479-486.

Jamaluddin, M., and Thomas, (1992). Competitive inhibition of hydrogen peroxide induced aggregation of calf platelets by prostaglandin H_2 /thromboxane A_2 receptor ligands. J Biosci., 17: 129-140.

Kaser-Glanzmann, R., Jakabova, M., George, J.N., and Luscher, E.F. (1977). Stimulation of calcium uptake in platelet membrane vesicles by adenosine 3',5'–cyclic monophosphate and protein kinase. Biochem. Biophys.Acta, 466:429-440.

Marcus, A.J., and Zucker, M. (1965). The Physiology of Blood Platelets. Grune and Stratton, Inc. New York.

Mitchell, J.R.A., and Sharp, A.A. (1964). Platelet clumping *in vitro.* Brit. J. Haematol., 10:78-93.

Mondal, S., Mukhopadhyay, B., and Bhattacharya, Shelley (1997). Inorganic mercury binding to fish oocyte plasma membrane induces steroidogenesis and translatable messenger RNA synthesis. Biometals, 10:285-290.

Mukhopadhyay, B., Bose, S., Bhattacharya, S., and Bhattacharya, Shelley (1994). Induction of metallothionein in rat liver by cadmium chloride : Probable mechanism of action. Biomed. Environm. Sci., 7: 232-240.

Mustard, J.B., and Packham, M.A. (1970). Factors influencing platelet function; Adhesion, release and aggregation. Pharmacol. Rev. 22: 97-187.

O'Brien, J.R. (1962). Platelet aggregation. I. Some effects of the adenosine phosphates, thrombin and cocaine upon platelet adhesiveness. J Clin. Pathol., 15: 446-452.

Packham, M.A., Nishizawa, E.E., and Mustard, J.F. (1968). Response of platelets to tissue injury. Biochem. Pharmacol. Suppl. : 171-184.

Robinson, C.W. Jr., Kress, S.L., Wagner, R.H., and Brinkhous, K.M. (1965). Platelet agglutination and deagglutination with a sulphydryl inhibitor, methyl mercuric nitrate : relationships to platelet ATPase. Exp. Mol. Pathol., 4: 457-464.

Sarkar, D. (1997). Role of reduced glutathione, glutathione-S-transferase, metallothionein and lipid peroxidation in the detoxication mechanism in *Anabas testudineus* (Bloch). PhD. Dissertation, Visva Bharati University, Santiniketan.

Smith, J.B., Silver, M.J., Ingerman, C.M., and Kocsisi, J.J. (1974). Prostaglandin D_2 inhibits the aggregation of human platelets. Thromb. Res., 5:291-299.

Woodbury, D.M., and Fingl, E. (1975). Analgesics, antipyretics, antiinflammatory agents and drugs employed in the therapy of gout. In The Pharmacological Basis of Therapeutics (L.S. Goldman and A. Gilman, Eds.) pp. 325-358. MacMillan Pub. Co. Inc., New York.

Proceedings of the National Seminar on Environmental Biology, April 03-05, 1998
Edited By : A.K. Aditya & P. Haldar
DAYA PUBLISHING HOUSE, DELHI

Impact of Roadside Soil Extracts on Germination and Growth of *Phaseolus Aureus* Roxb. and *Dolichos Biflorus* L.

V. Roopashree and R.K. Somashekar
Department of Environmental Science
Bangalore University, Bangalore-560056

Abstract

Aqueous extract of soil smaples collected during summer and rainy seasons along the three different road ways of Bangalore city, namely, Double Road, RV Road and Bangalore Bus Station having high traffic density were tested for their impact on seed germination and growth of two edible crops *Dolichos biflorus* L. and *Phaseolus aureus Roxb.* and compared with the control sample collected from the Jnanabharathi campus of Bangalore with negligible vehicular density. A negative correlation between the concentrations used and root and shoot growth was observed. The enzyme activities such as acid phosphotase and a-amylase, besides proteins, phenolics and chlorophyll contents in both the sets of plants were assessed.

Introduction

In recent years Automobile pollution in developing countries has assumed alarming heights. The degree of air pollution by automobiles depend on their density, traffic flow pattern and, more directly the commuting activities. India is likely to have over 3.8 million vehicles by the end of 1998, of which more than 65 per cent represent two-wheelers operationg on petrol (Khopakar, 1995).

From all the major cities of India about 800-1000 tons of pollutants are emitted into air daily of which 50 per cent comes form automobile exhaust. In the major metropolitan cities, vehicular exhaust accounts for 70 per cent of all carbon monoxide, 50 per cent of all hydrocarbons, 30-40 per cent of all oxides and 30 per cent of all suspended particulate matter.

Once introduced into the atmosphere, the pollutants enter biogeochemical cycles by different routes. The immediate absorbants of pollutants are the roadside soil and plants which disturb the mineral make up of the roadside soil, and are likely to affect plant growth.

Soils due to their capacity to complex with organic substances, oxides, carbonates and bicarbonates, carry high retention capacity for the pollutants. The concentration of the pollutants in the soil continues to increase as long as the source remains active and, ultimately, this soil is likely to interfere with plant growth.

In view of the above, the present study was taken up for examining the effect of roadside soil extracts on seed germination and growth of two edible crops, *Phaseolus aureus* Roxb. and *Dolichos biflorus* L.

Material and Methods

The sites selected for the study have no other polluting sources in the vicinity other than automobile exhaust. The locations selected were –

1. Janabharathi campus (Control)
2. Bangalore Bus Station (BBS)
3. Double Road
4. Rastriya Vidyalaya Road (RV Road)

The soil samples were collected from region of road divider, during summer as well as rainy season using soil augur.

Soil pH was determined by suspending 10 g of soil in 25 ml distilled water (Jackson, 1973) and the same suspension was used for determining Electrical Conductivity. Percentage organic carbon was determined titrimetrically according to walkey and Black's Wet Oxidation method (Black, 1965). The concentration of sodium and potassium were estimated flame photometrically using neutral normal ammonium acetate. Micro nutrients were determined by DTPA method according to Lindsay and Norwell (1978).

Different concentrations of the aqueous extracts (5, 10, 15, 20, 25%) of the soil samples were prepared by blending appropriate quantity of samples with distilled water and by using an electric shaker at 100 rpm for one hour and, filtered through Whatman No. 42 filter paper.

The seeds of *Phaseolus aureus* and *Dolichos bifloruls* procured from authorised stockists were surface sterilised with 0.2% mercuric chloride and soaked in aqueous extracts for 3 hours. Ten seeds in each case were placed on sterile blotting papers in sterile pertriplates and fed with different concentrations of the extracts. The set treated with distilled water served as control. After 48 hours, root and shoot length of germinated seeds were recorded daily for ten days. Triplicate sets wre maintained in all the cases.

Once hundred pre-soaked seeds were arranged serially on blotting paper and sprayed with extracts, rolled and wrapped with butter paper. After 48 hours percentage of germination was recorded.

Treated germinated seeds were used for the assay of proteins, a-amylase and acid phosphatase. Plants, 15-20 days old, were used for estimating phenolics and chlorophyll. Protein estimation was carried out according to Hawk *et al.* (1948). α-amylase and acid phosphatase were assayed according to the procedure of Sadashivam and Manikam (1992). Phenol estimation was done following the procedure of Malik and Singh (1980), and total Chlorophyll estimation was made according to Arnon (1949).

Results and Discussion

Soil is an important natural resource, containing both living, non-living components and their decayed products. Soil is a three dimensional biological system, providing support, water, nutrients and oxygen to the plants.

Most of the samples analysed were alkaline and only the summer samples from BBS were slightly acidic when compared with the rating chart for soil fertility test data (Bharat Prasad & Srivastava, 1978) (Table 1) This might be due to continuous vehicular emission, which increases the soil temperature, due to which, water tends to evaporate faster, allowing the salts to accumulate. The micronutrients like Zinc, Copper, Manganese, Iron and Boron occurred in very high con centrations (Table 1). These micronutrients are required by plants in small quantities and when present in large quantities, affect the plant system in several ways.

Table 1 : Physico-Chemical Characteristics of Roadside Soils (All values except pH and EC are in Mg/kg)

Parameters	Control	BBS	DR	RV	Critical Limits
pH	7.8	6.00 (–4)	8.,81 (7.8)	8.45 (8.4)	6.3–8.5
Electrical Conductivity (m mhos)	0.21	0.28 (0.8)	0.25 (0.1)	0.31 (0.4)	1.0–2.0
Carbonates	2.4	3.6 (2.4)	4.8 (3.6)	2.4 (4.8)	1.0–2.0
Bicarbonates	4.8	7.3 (4.8)	8.5 (4.8)	4.8 (5.8)	–
Organic Carbon	1.8	3.8 (5.4)	4.4 (4.6)	2.2 (3.0)	0.50–0.75
Calcium	729.4	1094 (437.6)	729.4 (255.2)	401.17 (875.3)	–
Magnesium	728.6	291.8 (932.9)	728.6 (918.3)	832.83 (328.2)	–
Available Nitrogen	210	201.6 (268.8)	156.8 (112)	201.6 (168)	112–224
Sodium	16	53 (66)	24 (19)	16 (28)	–
Potassium	21	26 (41)	38 (14)	25 (21)	50–120
Zn+	3.56	10.79 (9.92)	10.62 (9.36)	7.9 (7.9)	0.5–1.0
Cu++	1.39	7.23 (10.79)	8.13 (9.11)	4.09 (4.46)	0.20
Mn+	7.2	11.54 (14.18)	12.71 (6.24)	12.71 (4.48)	1.00
Fe++	4.64	17.56 (25.86)	16.80 (15.15)	6.13 (7.58)	2.50
Bo++	0.55	1.08 (0.93)	0.95 (0.46)	0.62 (0.47)	0.50

Values in paranthesis are for Rainy Season

The results of the plant growth studies have been summerised in Table 2. Though, there was not much impact on the rate of germination, a little damaging influence was imparted by RV road soil extracts followed by BBS and Double Road soil extracts during summer. However, during rainy season, more inhibition was seen in the case of Double Road followed by RV road and BBS. This may be attributed to the alkaline nature of the soil samples.

This in turn acted upon the enzymatic activity at the early stages of germination finally resulting in low germination. A similar observation was made by Vijayakumari *et al.* (1993) in certain millets and pulses and, Ray & Khaddar (1995) in wheat. Reduction in the percentage of germination at higher concentration of soil extracts also indicated the possible injury caused to the embryo due to higher osmotic pressure resulting from salt concentration as noted by Joshi and Iyengar (1985) and Sahai and Srivastava, (1986).

Root elongation test has been extensively used as a bioassay technique for detecting phytotoxic effects of various substances. (Behra and Mishra, 1982; Wong and Bradshaw, 1982: Wong, 1987). In the present investigation, there exists a negative correlation between the concentration of soil extracts and the root and shoot length (Table 2). The soil extracts from BBS produced a marked effect (Table 4). This might be due to the occurrence of micro nutrients (Table 1) such as Zinc, Copper, Manganese, Iron and Boron in quantities above the cirtical limits

Table 2 : Impact of Soil Extracts on Germination and Growth of the Plants

Conc.	Germin. %		Shoot/Root Length (Cm)				Shoot/Root Ratio		Fresh Weight (g)		Dry Weight (g)	
	P	D	P	D	P	D	P	D	P	D	P	D
Bangalore Bus Station												
	100	100	13.2	12.0	5.4	5.3	2.44	2.26	1.28	1.50	0.11	0.12
Control	(100)	(100)	(14.5)	(10.7)	(5.4)	(5.5)	(2.68)	(1.94)	(1.79)	(1.95)	(0.13)	(0.13)
5%	100	100	12.5	11.3	5.2	4.7	2.40	2.40	1.22	1.26	0.10	(0.13)
	(100)	(80)	(13.8)	(9.0)	(5.2)	(5.0)	(2.65)	(1.80)	(1.78)	(1.31)	(0.11)	(0.12)
10%	80	100	12.0	11.0	5.2	4.2	2.30	2.61	1.27	1.16	0.08	0.10
	(100)	(100)	(13.0)	(8.9)	(5.0)	(3.7)	(2.60)	(2.40)	(1.61)	(1.24)	(0.11)	(0.12)
15%	90	90	11.3	12.2	4.8	4.6	2.35	2.21	1.29	1.16	0.10	0.19
	(100)	(70)	(12.0)	(8.5)	(5.0)	(3.7)	(2.40)	(2.29)	(1.67)	(1.17)	(0.10)	(0.12)
20%	80	80	11.0	8.5	4.3	4.1	3.55	2.07	1.22	1.13	0.08	0.11
	(90)	(90)	(13.0)	(8.7)	(3.8)	(4.3)	(3.42)	(2.02)	(1.52)	(1.17)	(0.10)	(0.09)
25%	70	80	9.9	6.7	3.0	4.0	3.30	1.67	0.99	1.04	0.07	0.10
	(80)	(80)	(10.0)	(6.6)	(3.5)	(3.3)	(2.85)	(2.00)	(1.04)	(0.74)	(0.09)	(0.08)
Double Road												
	100	100	12.0	13.7	5.9	6.0	2.03	2.28	1.52	1.80	0.12	0.15
Control	(100)	(100)	(13.5)	(11.0)	(6.0)	(5.5)	(2.25)	(2.01)	(1.77)	(1.61)	(0.11)	(0.10)
5%	100	100	11.8	13.0	5.6	5.6	2.10	2.32	1.20	1.55	0.11	0.12
	(100)	(100)	(12.3)	(10.5)	(5.6)	(5.2)	(2.19)	(2.02)	(1.53)	(1.50)	(1.10)	(0.09)
10%	100	100	10.4	12.3	5.2	5.3	2.00	2.32	1.03	1.53	0.11	0.13
	(90)	(80)	(11.0)	(10.0)	(5.4)	(4.8)	(2.03)	(2.08)	(1.17)	(1.50)	(0.09)	(0.09)
15%	90	100	10.1	11.7	4.6	4.8	2.19	2.43	1.65	1.19	0.13	0.11
	(90)	(90)	(9.5)	(9.3)	(5.0)	(4.4)	(1.90)	(2.11)	(1.10)	(1.37)	(0.08)	(0.08)
20%	100	80	9.3	11.0	3.9	4.3	2.38	2.55	1.12	1.15	0.10	0.10
	(100)	(90)	(8.6)	(8.5)	(4.9)	(3.8)	(1.75)	(2.23)	(0.99)	(1.25)	(0.08)	(0.08)
25%	90	90	6.8	7.9	3.7	3.5	1.83	2.25	1.1	1.10	0.10	0.10
	(90)	(90)	(7.5)	(7.3)	(4.5)	(3.5)	(1.66)	(2.08)	(0.82)	(1.14)	(0.07)	(0.07)

Contd. . .

Conc.	Germin. %		Shoot/Root Length (Cm)				Shoot/Root Ratio		Fresh Weight (g)		Dry Weight (g)	
	P	D	P	D	P	D	P	D	P	D	P	D
R.V. Road												
	100	100	14.2	12.5	6.2	6.1	2.29	2.04	1.74	1.78	0.14	0.22
Control	(100)	(100)	(10.0)	(12.0)	(5.6)	(6.0)	(1.78)	(2.00)	(2.60)	(2.22)	(0.29)	(0.14)
5%	100	90	13.5	11.2	5.4	5.8	2.50	1.93	1.94	1.49	0.16	0.20
	(90)	(100)	(8.9)	(10.9)	(4.8)	(5.6)	(1.85)	(1.94)	(2.46)	(2.11)	(0.22)	(0.09)
10%	90	100	12.8	10.0	4.9	5.2	2.61	1.92	1.70	1.64	0.14	0.17
	(90)	(80)	(9.0)	(9.8)	(4.2)	(5.2)	(2.14)	(1.88)	(2.58)	(2.12)	(0.22)	(0.16)
15%	90	90	12.2	8.3	4.1	4.7	2.97	1.76	1.64	1.21	0.13	0.11
	(80)	(100)	(8.3)	(10.2)	(4.0)	(4.7)	(2.07)	(2.17)	2.50)	(1.30)	(0.25)	(0.10)
20%	80	100	10.0	8.9	3.7	4.3	2.97	2.06	1.63	1.10	0.12	0.10
	(80)	(80)	(8.0)	8.2)	3.9	(4.0)	(2.05)	(2.05)	(2.46)	(1.75)	(0.21)	(0.13)
25%	100	90	8.3	6.5	3.2	4.0	2.59	1.62	1.49	0.95	0.11	0.08
	(90)	(70)	(5.5)	(7.0)	(3.3)	(4.0)	(1.83)	(1.75)	(1.88)	(1.99)	(0.17)	(0.15)

P = Phaseolus aureus *D = Dolichos biflorus;* Values of Brackets are for summer.

as opined by Bharath Prasad and Srivastasava (1978) which are potentially toxic to plants (Lou *et al.* 1995). The present data are in agreement with the earlier investigations involving inhibiton of growth by Zinc (Lata, 1983) in Legumes and *Cajanus cajan* (Singh and Singh, 1997). The growth inhibition might also be due to high soil pH, which influences the solubility and availability of soil elements as observed by O'Born *et al.* (1995).

Table 3 : Data on Certain Biological Parameters

	Total Chlorophyll (μ g/g)	Proteins (μ g/g)	α - Amylase (μ g/g)	Acid Phosphatase (mili moles/g)	Phenolics (μ g/g)
Rainy Season					
C P	1.19	40	2400	0.12	235
C D	1.15	38	2250	0.13	230
B1 5% P	1.07	24	1830	0.11	225
B1 5% D	1.01	21	1700	0.11	230
B1 25% P	1.00	27	2100	0.12	230
B1 25% D	1.05	25	1780	0.11	210
D1 5% P	1.12	36	1860	0.13	190
D1 5% D	1.07	30	2200	0.11	210
D1 25% P	1.10	33	2100	0.15	200
D1 25% D	1.02	36	2400	0.13	225
R1 5% P	1.18	30	1980	0.11	195
R1 5% D	10.14	21	2200	0.12	225
R1 25% P	1.07	24	2000	0.13	195
R1 25% D	1.04	33	1860	0.12	230
Summer Season					
C P	1.12	42	2500	0.15	240
C D	1.10	39	2300	0.13	255
B1 5% P	0.83	24	1500	0.10	225
B1 5% D	0.90	19	1700	0.08	225
B1 25% P	0.08	21	1800	0.08	200
B1 25% D	0.87	17	1700	0.07	225
D1 5% P	0.96	30	2000	0.11	200
D1 5% D	0.99	21	1830	0.10	200
D1 25% P	0.92	30	1780	0.11	210
D1 25% D	0.90	21	2100	0.13	230
R1 5% P	0.96	38	1830	0.07	200
R1 5% D	0.87	30	1800	0.08	225
R1 25% P	0.98	30	1800	0.08	195
R1 25% D	0.82	24	1700	0.10	230

Index :

P = Phaseolus aureus — *D = Dolichos biflorus*

B1 = Bangalore Bus Station — D1 = Double Road

R1 = Rastriya Vidhayalaya Road — C = Control

Table 4 : Correlation Co-efficient (r) between root lenght, shoot length, and soil extracts

	Summer		Rainy Season	
	Root Length	Shoot Length	Root Length	Shoot Length
BBS				
P. aureus	–0.926	–0.820	–0.880	–0.794
D. biflorus	–0.671	–0.795	–0.750	–0.963
Double Road				
P. aureus	–0.995	–0.982	–0.985	–0.936
D. biflorus	–0.995	–0.847	–0.985	–0.978
RV Road				
P. aureus	–0.627	–0.996	–0.934	–0.997
D. biflorus	–0.892	–0.877	–0.993	–0.993

A dose dependent relationship was observed between the fresh weight and the dry weight of the plants and the concentration of the extracts used. This reduction might be due to the existence of higher amount of Copper and Zinc above the critical limits (Table 1). This damaged the plant cells as suggested by Singh and Singh (1997) resulting in poor growth. Balashouri and Prameela Devi (1995), made similar observations in the case of *Vigna radiata* and *Sorghum bicolor.*

Acid phosphatase activity was more in the seedlings treated with the extracts of Double Road, and least in respect of BBS extracts during the rainy season (Table 3). However, of the summer extracts, the activity was high in respect of Double Road. This could be the other parameter involved in inhibition of germination rate as acid phosphatase helps in mobilizing nutrient reserves (Flinn and Smith, 1967).

α-amylase (Table 3) remained scanty among germinating seeds treated with the extracts of BBS of both the seasons and higher with the extracts of RV Road than with those Double road during rainy reason. A reverse trend existed during the summer. This might be due to alkaline pH, as the activity of α-amylase is controlled by pH and temperature (Sadashivam and Manikam, 1992).

Plants grown with the extracts of BBS contained more phenolics and those with Double road soil contained lesser amount. The protein content was more in the plants treated with the extracts of Double road and minimum amount was found in the case of BBS (Table 3). This could be because, phenols cross link proteins and denature them (Salisburry and Ross, 1986). Consequently, plants exhibited a reduction in protein content and thus resulting in a stunted growth. Similar observations have been made by Mesmar and Jaber (1991) in Wheat.

The extracts of the summer samples imparted significant impact on the chlorophyll content during the rainy season. It was observed that the samples from BBS exerted more impact (Table 3). This might be due to higher Zinc, Copper and magnesium (Moustakas *et al.* 1994), and their resultant toxicity on photosynthetic activity (Singh and Singh, 1997: Sharma *et al.* 1997).

References

Arnon, D.I. (1949). Copper enzyme in isolated chloroplasts, polyphenol oxidase in *Beta vulgaris. Plant Physiol,* **24**:1-25.

Balashouri and Prameela Devi (1995). Effect of Zinc on germination, growth, pigment content and phytomass of *Vigna radiata* and *Sorghum bicolor. J. Eco. Bio,* **7(2)**109-114.

Behra, B.K. and Mishra, B.N, (1982). Analysis of the effect of industrial effluent on growth and development of rice seedlings. *Envtl. Res.,* **28:** 10-12.

Black, C.A, (1965). Methods of soil Analysis. *Am. Soc. Agron. Inc. Madeson, Wisconsin, USA.*

Bharat Prasad and Srivastava, K.P. (1978). *Academics Dictionary of Agricultrue.* Academic (India) Publishers New Delhi 110060.

Flinn, A.M and Smith, D.L. (1967). the localisation of enzymes in cotyledons of *Pisum satium* during germination. *Planta,* **755:** 10-22.

Hawk, P.B., Oser, B.L. and Summerson, W.H. (1948). *Physiol Chem.,* Blakistan Company USA.

Jackson, M.L. (1973). *Soil Chemical Analysis.* Prentice Hall of India, New Delhi

Joshi, A.J and Iyengar, E.R.R. (1985). Salinity effects on seed germination in *Saud nudiflora Moq. J. Indian Bot. Soc.,* **64:** 362-364.

Khopkar, S.M. (1995). *Environmental Pollution analysis.* Centre for Environmental Science Engineering. IIT, Bombay.

Lata, K. (1983). Effects of Heavy metal Zinc on seedling growth of some legumes, J. *Indian Bot. Soc.,* **62** (Suppl.): 82-86.

Lindsay and Norvel, (1978). *Development of micro nutrients soil test,* Agronomy abstract : **84.**

Lou, Rimmer, Youngming, David, L. (1995). Zinc- Copper interactiosn affecting plant growth on a metal contaiminated soil. *Environ. Poll.,* **88:** 78-83.

Malik, C.P. and Sigh, M.B. (1980). *Plant Enzymology and Histology.* Kalyani Publishers, New Delhi.

Moustakas, M., Lanaras, T., Symeonidis, L. and Karatagalis, S. (1994). Growth and photosynthetic Characters of some field grown *Avena sativa* under Cu and Pb stress. *Photosynthetica (Prague).* **30(3):** 389-396.

Mesmar, M.N. and Jaber, K. (1991). The toxic effect of Lead on seed germination, growth, Chlorophyll and protein contents of Wheat and Lens. *Acta Bilogica Hungarica,* **42(4):** 331-344.

Ndiokwere, C.L. (1984). A study of heavy metal pollution from motor vehicle emission and its effect on road side soil. Vegetation and crops in Nigeria. *Environ. Poll.,* **B7:** 247-254.

O Born, I., Jonson, G and Johnson, L. (1995). A field study on the influence of soil. pH on trace elemet levels in Spring Wheat *(Triticum aestivum),* Potato and Carrot. *Water, Air, Soil Polln.* **82(2):** 835-840.

Ray, N. and Khaddar, V.K. (1995). A study on the effects of soil salinity, sodicity and their combination on early seedling growth in Wheat. *J. Environ., Biol.,* **16(3):** 193-199.

Sadashivam, S. and Manickam, A. (1992). *Biochemical methods for Agricultural Science.* Willey Eastern Limited, New Delhi.

Sahai, R., and Srivastava, N (1986). Effect of distillery waste on seed germination, seedling growth and pigment content of *Cajanas cajan. J. Indian Bot. Soc.,* **65:** 208-211.

Salisbury, F.B. and Ross, C.W. (1986). *Plant Physiology.* CBS publishers, New Delhi, India.

Sharma, S.K., Srivastava, A., Singh, V.P. (1997). Effect of rubber factory effluent on growth in *Vigna mungo. J. Envtl. Pol.,* **4(3):** 175-177.

Singh. P., and Singh V.P. (1997). Phytotoxic effect of city waste on growth and chlorophyll content in Pigeon pea. *J. Envtl. Poll.,* **4(4):** 265-269.

Vijayakumari K., Kumuda, P., Siddhuraju, P. Janardhana, K. (1993). Effect of Soap factory effluent on seed germination and early seedling growth of certain Millets and Pulse crops. *J. Envtl.. Bitol.*, **14(4):** 275-281.

Wong, W. and Bradshaw. A.P. (1982). A comparison of the toxicity of heavy metals using elongation of Rye grass, *Lolium perenne. New phytol.*, **91:** 255-261.

Wong, W. (1987). Root elongation method for toxicity testing of organic and inorganic pollutants. *Envtl. Toxicol. Chem.*, **6:** 409-414.

Proceedings of the National Seminar on Environmental Biology, April 03-05, 1998
***Edited By :* A.K. Aditya & P. Haldar**
DAYA PUBLISHING HOUSE, DELHI

Heavy Metal Accumulation in Fish : An Assessment in Sewage Fed Aquafarm of East Calcutta, India

S.C. Santra & Nasima Bano
Department of Environmental Science
University of Kalyani, Nadia
West Bengal

Abstract

The accumulation of toxic metals in higher tropic orders in an ecosystem is well known. Sewage of city and municipal township contained a considerable amount of toxic metals along with rich organic and inorganic nutrients. Such waste when used as feed in aquafarming system as in sewage fed fisheries, apparently the fish yield is enhanced very much but the toxic metal accumulation in the fish took place very much with time. The present paper deals with the accumulation toxic metals like chronium, cadmium and lead in fishes cultivated in city's sewage fed system in the east Calcutta region. The metal level in fish body parts are not only accumulates but also magnified manyfold with age of the fish.

Introduction

City sewage is extensively used for fish culture in Calcutta. The sewage is diverted through canal system to the fishery bheries after dilution with fresh water usually in the ratio of 1:4. The fingerlings of Rohu, Catla and Telapia were stocked in the appropriate ratio. Sewage was fed at a regular interval at least once in a month. The yield was reported to enhanced form 4 tons/hect. to 8 tons/hect. on fedding the bheries with diluted waste water.

There were a number of reports of existence of toxic heavy metal contamination in city sewage (Bhawal *et al.*, 1987; David 1959; Adhikary *et al.*, 1994).

Even on dilution of such waste water was fed to fishery bheries, there are chances of contamination of toxic metals like Cr, Cd, Zn, Cu, Hg & Pb in particular. The toxic metals which are present at a lower concentration in water and sediment part of the sewage fed bheries, but with time each metal gets accumulated more and more in fishes that are cultured on such bheries by following the principles of "biomagnification".

In the present study, an attempt has been made to assess the status of metal accumulation in clutureable fishes in sewage fed bheries in the eastern part of Calcutta where traditional sewage fed fisheries were promoted as useful eco-friendly technology. In fact, earlier no serious attempt has been made for toxicological risk assessment of sewage fed fisheries.

Materials and Method

Sampling

Three common culturable fishes viz., Catla (*Catla catla*), Ruhu (*Labeo rohita*) and Telapia (*Telapia mosumbica*) wre harvested from the bheries after 3 months growth of fingerling in sewage fed system. They were thoroughly washed in clean water thrice and then dissected for isolation of body segments like liver, kidney and muscles for toxic metals like chromium, cadmium and lead content analyses. These tissues were dissected out carefully in uncontaminated form and then digested with acids by conventional protocol. Separate samples were prepared from each fish types for different metal assay. Three replica of each samples were prepared. Identically water and sludge samples were also collected from sewage fed fisheries for toxic metal assay by following identical protocol.

Sample Preparation

Separate organs of fish samples were dissected out and then washed in distilled water. Finally blotted the tissues and digested in glass made digestion flask in a micro oven using HNO_3 – perchloric acid mixture. Clear digested filtrate is then collected in volumetric flask and finally desired volume was made before assay of metal contents by Perkin Elmer-AAS against standard reference metal solution (APHA, 1992).

Identically sediment and water samples were collected separately from the sewage fed fish bheries, and then digested flask using conc. HNO_3 for extraction metals in acid solution using micro-oven. The clear digested solution is then filtered and a desired volume was made with volumetric flask. Finally metal content of the sample solutions were determined by using AAS as stated earlier.

Table 1 : Metal accumulation status in sewage fed fishes

Sample(s)	*Metal content (in ppm) [Mean value ± SD]*		
	Lead	***Chromium***	***Cadmium***
1. Water	0.4 ± 0.01	1.45 ± 0.05	0.04 ± 0.02
2. Sediment	1.5 ± 0.1	4.65 ± 0.05	0.03 ± 0.01
3. Fish – *Catla*			
(a) Liver	0.67 ± 0.3	1.62 ± 0.02	0.15 ± 0.05
(b) Kidney	0.83 ± 0.1	1.87 ± 0.05	0.25 ± 0.01
(c) Muscle	0.42 ± 0.3	1.50 ± 0.02	0.06 ± 0.01
4. Fish – *Ruhu*			
(a) Liver	0.30 ± 0.1	0.2 ± 0.01	0.50 ± 0.01
(b) Kidney	0.40 ± 0.02	0.2 ± 0.01	0.70 ± 0.02
(c) Muscle	0.25 ± 0.01	0.1 ± 0.01	0.45 ± 0.01
5. Fish – *Telapia*			
(a) Liver	1.95 ± 0.5	0.37 ± 0.02	0.65 ± 0.01
(b) Kidney	1.6 ± 0.2	0.40 ± 0.05	0.75 ± 0.02
(c) Muscle	1.8 ± 0.2	0.57 ± 0.05	0.55 ± 0.01

Result and Discussion

The details of metal level in water, sediment and fish organs as accumulated on sewage fed fishery system at east Calcutta region is given in the Table-I. From the result it appears that considerable amount toxic metals like Pb, Cr and Cd are accumulated in fish organs particularly in liver and kidney region. This indicate the fact that there is a potential health risk of raw sewage used in aquaculture. Similar observations were also made earlier by Dev and Santra (1997) and Sadhukhan *et al.* (1996) from Calcutta region.

Bioaccumulation of toxicants like metals and pesticides in aquatic ecosystem have a great significance. The pollutants level build up a higher order in various trophic level through food chain. As fish is a very common commercial species cultured in different aquatic bodies, thus the chances of pollutant level build up in such fishes in fairly high. There is rise of persistent pollutant contamination in aquatic systems also. If such fishes which are cultured in pollutant loaded aquatic system, are regularly consumed by the communities for over decade, there are a number of possible potential health risks that might crop up in the long run. A good number of examples of similar bioaccumulation in aquatic ecosystems was reported earlier (Anand Kumar, 1989; Deb and Santra, 1997; Jhingran and Joshi, 1989; Sadhukhan *et. al.*, 1996). Thus it is better to use at least partially treated sewage for aquafarm use.

References

Adhikary, S.; Mitra, A.; Gupta, S.K. and Banerjee, S.K. (1994) : Possibilities of using Calcutta sewage effluents and sludges for irrigation and manurial purposes; Part-I - physico-chemical characteristics and impact of heavy metal contents of soil and water. Proc. Indian National Acad, B60, (6), 541-552.

Anonymous (1992) : Standard method for the examination of water and waste-water, APHA, 19th Ed. (Mc Graw Hill Co., NY).

Anand Kumar, S. (1989) : Bioaccumulation of copper in *Heteropneustes fossilis*. Poll. Res. 8, 45-48.

Bhowal, S.K.; Chakravorty, A.K.; Dhar, B. (1987) : Heavy metal contamination in sewage sludge of Calcutta Metropolitan Area. Indian J. Environ. Hlth. 29(1), 66-71.

David, A. (1959) : Effects of Calcutta Sewage upon the fisheries of the Kulti estuary and connected culturable fisheries. J. Asiatic Soc. New Ser. I(4). 339-363.

Dev, S.C. and Santra, S.C. (1997) : Bioaccumulation of metals in sewage fed aquatic systems - a case study from Calcutta (India). Intern. J. Environ. Studies. 52, 117-126.

Jhingram, A.G. and Joshi, H.G. 1989. Heavy metals in water, sediments, and fish in the river Jamuna (India). J. Int. Fish Soc. India, 19(1), 13-23.

Kabiraj, A. (1989) : Heavy metal concentration of fish and prawn collected from Hoogly Estuary, Science & Culture, 55, 257-260.

Ray, D.; Banerjee, S.K. and Chatterjee, M. 1989. Bioaccumulation of nickel and vanadium in tissues of the Cat fish *Clarius batrachus*. J. Inorg, Biochem. 5(1), 1-7.

Sadhukhan, P.C.; Ghosh, S; Ghosh, D.K.; Choudhuri, J. and Mandal, A. (1996) : Accumulation of Mercury in edible fish from wetlands of Calcutta. Indian *J. Environ.* Hlth. 38, (4), 261-268.

Proceedings of the National Seminar on Environmental Biology, April 03-05, 1998
***Edited By :* A.K. Aditya & P. Haldar**
DAYA PUBLISHING HOUSE, DELHI

Bioindicators for Freshwater Mussel, *Lamellidens marginalis* (Lamarck) Farming

A. Biswas & S.K. Raut
Ecology and Ethology Laboratory
Department of Zoology, University of Calcutta
35, Ballygunge Circular Road, Calcutta-700 019

Abstract

The freshwater mussels *Lamellidens marginalis* (Lamarck) are consumed by a large section of people. The increasing demand of these mussels necessitates farming of the same at an early date. Since, of the many factors, success of mussel farming is much dependent on the food organisms an attempt was made to record the same in different mussel populations occurring in southern part of West Bengal. Examinations of gut contents revealed the presence of 32 types diatoms, 107 types green alga and 3 types zooplankton. Irrespective of locality a marked difference in food selection has been noted in juvenile and aged individuals. Of the many types of microorganisms of the waterbodies as the mussels are selective for certain types only, it is likely that these food organisms be considered as bioindicator for mussel farming.

Introduction

The freshwater mussels *Lamellidens marginalis* (Lamarck) are widely distributed in India, Bangladesh, Burma and Sri Lanka (Subba Rao, 1989). In addition to pearl formation (Raut and Biswas, 1989) these mussels are economically important because of a source of animal protein. They are also palatable food item to a large section of people. For the said reason these mussels are sold regularly in different markets of eastern India including Calcutta. Accordingly, they are being considered as a good source of income by the village people. Since the demand of these mussels is increasing day by day, in near future, the village people may not be able to get these molluscs from the nature for their trade. So, the only alternative to meet up the demand is to develop mussel industry at an early date. As such an attempt depends on the knowledge gained on the factors influencing the growth and multiplication of *L. marginalis* the present authors have taken the liberty to study the same. In the mean time, data on the role of water, soil and the larval hosts in regulation of mussel population have been collected (Raut and Biswas, 1990, 1991). The present communication deals with the foods of mussels and their possible role as bioindicators for mussel farming.

Materials and Methods

In course of surveys the mussels *L. marginalis* belonged to different size groups were collected from ten ponds – one from each of the districts : Howrah (Dasnagar), Burdwan (Golapbag) and Murshidabad (Balirghat), two from Nadia (Chuadanga, pond A and B) and five from 24 Parganas district (Chandpara, pond A, B, C, Fulsara, Thakurnagar) during June 1987 to July

Table 1 : List of food items isolated from the gut of *L. marginalis* collected from the nature

	Varieties		Varieties
Diatoms		*Kirchneriella obesa*	1
Amphora sp.	1	*K. lunaris*	1
Cocconeis disculus cleve	1	*Kirchneriella* sp.	1
Cyclotella sp.	1	*Merismopedia glauca*	1
Cymbella ventricosa	1	*Microcystis* sp.	1
Cymbella sp.	2	*Nephrocytium obesum*	1
Gomphonema sp.	1	*Oocystis borgei*	1
Melosira sp.	2	*Oocystis borgei*	1
Navicula crucicula	1	*Oscillatoria irrigua*	1
Navicula bacilloides	1	*Oscillatoria curvicepa*	1
Navicula sp.	14	*Pandorina morum*	1
Nitzschia umblicata hustedt	1	*Pediastrum duplex*	2
Nitzschia sp.	1	*Phacus tortus*	1
Pinnularia sp.	1	*Phacus acuminatus*	1
Stauroneis wislouchii	1	*Phacus caudatus*	1
Staruoneis eluta hustedt	1	*Phacus longicauda*	1
Surirella splendid	1	*P. stokessil*	1
Synedra ulna var acqualis	1	*Phacus* sp.	7
		Schroederia planctonica	1
Green algae		*Schroederia* sp.	2
Ankistrodesmus convolutus	1	*Scenedesmus acuminatus*	3
Ankistrodesmus sp.	4	*Scenedesmus dimorphus*	5
Chlamydomonus sp.	1	*Scenedesmus quadircuda var bicaudatus*	5
Chlorella vulgaris	1	*Scenedesmus bigugatus*	2
Chlorococcum infusionum	1	*S. smithii*	1
Chlorococcum sp.	1	*S. quadricuda*	7
Chroococcus minor	1	*S. perforatus*	1
Chroococcus minutus	1	*S. armatus*	1
Coelastrum microporum	1	*Scenedesmus* sp.	6
Coelastrum sp.	1	*Selenestrum gracile*	1
Cosmarium perfissum	1	*Staurastrum* sp.	1
Crucigenia tetrapedia	1	*Tetraedrom trigonum*	1
Crucigenia rectangularis	1	*T. minimum*	1
Crucigenia crucifera	1	*T. muticum*	1
Crucigenia sp.	1	*T. pusillum*	1
Eudorina elegans	1	*Tetraedron* sp.	3
Euglena polymorpha	1	*Tetrastrum stannogeniaeforme*	1
Euglena acus	1	*Trachelomonas*	1
Euglena cyclopicola			
Euglena sp.	14	**Zooplankton**	
Gleocapsa granosa	1	*Brachyonus* sp.	3

1988. The mussels were collected from 50 cm^2 quadrat at different sites of a pond irrespective of water depths and soil conditions. The quadrats were selected at random. For sampling, a mussel sampler made of wood, 2.5 m in height was used. The sampler was pushed gently inside the water down to the bottom and still further inside the soil to a maximum of 60 cm below the surface soil. Then the mussels present in the quadrat were collected by hand, and were brought to the laboratory for subsequent analysis. Of the collected mussels 20, ten from the juvenile (up to 25mm shell length) and ten from the aged (sexually mature) size groups were taken at random for dissection to examine the gut contents. The gut contents were collected and fixed in 4 per cent formalin in a watch glass individualwise at the collection spot and examined under microscope in laboratory. The organisms recovered from the gut contents were figured, identified and recorded.

Table 2 : Food chart of juvenile *L. marginalis*

Age of *L. marginalis*	Foods
1 day	*Chlorella* sp., *Chlorococcum infusionum, Chroococcus minor, Kirchneriella lunaris, Tetraedron muticum, Trachelomonas* sp., Cyst of *Trachelomonas* sp.
15 days	*Ankistrodesmus convolutus, Ankistrodesmus* sp., *Chlorella vulgaris, Chlorococcum infusionum, Cosmarium perfissum, Chroococcus minor, Chlorococcum* sp., *Kirchneriella lunaris, Schroederia planctonica, Staurastrum* sp., *Tetraedron muticum, Tetraedron* sp., *Trachelomonas* sp., Cyst of *Trachelomonas* sp.
1 month	*Chlorococcum* sp., *Chroococcus minutus, Chroococcus minor, Crucigenia crucifera, Crucigenia rectangularis, Eudorina* sp., *Oocystes borgei, Scenedesmus quadricud* var *bicaudatus, Scenedesmas bigugatus, Trachelomonus* sp., Cyst of *Trachelomonas* sp.
2 months	*Chlorococcum* sp., *Chroococcus minor, Chroococcus minutus, Crucigenia rectangularis, Crucigenia crucifera, Eudorina* sp., Oocystes sp., *Scenedesmus* sp. *Scenedesmus quadricuda* var *bicaudatus, Cymbella* sp., *Trachelomonas* sp., Cyst of *Trachelomonas, Tetraedron* sp., *Tetraedron minimum, Tetraedron muticum, Kirchneriella obesa, Melosira* sp.
3 months	*Chlorella* sp., *Chlorococcum infusionum, Chroococcus minor, Coelastrum* sp., *Cyclotella* sp., *Eudorina elegans, Gleocapsa granosa, Merismepodia* sp., *Staurastram* sp., *Tetraedron* sp., *T. muticum, Trachelomonas,* Cyst of *Trachelomonas.*
4 months	*Chlorococcum infusionum, Chlorella* sp., *Chroococcus minor, c. minutus, Crucigenia tetrapodia, Cyclotella* sp., *Eudorina* sp., *Gleocapsa* sp., *Kirchneriella obesa, K. lunaris, Melosira* sp., *Merismepodia* sp., *Oocystis borgei, Schroederia planctonica, Staurastrum* sp., *Tetraedron muticum, T. minimum, T. incus, Trachelomonas* sp., Cyst of *Trachelomonas.*
5 months	*Chlorella* sp., *Chroococcus minutus, Chroococcus minor, Coelastrum microporum, Cyclotella* sp., *Eudorina elegans, Gleocapsa* sp., *Kirchneriella obesa, K. lunaris, Merismopedia* sp., *Navicula* sp., *Oocystis borgei, Phacus* sp., *Phacus acuminatus, Trachelomonas,* Cyst of *Trachelomonas.*
6 months	*Ankistrodesmus* sp., *Chlorella* sp., *Celastrum microporum, Chroococcus minor, C. minutus, Cyclotella* sp., *Eudorina elegans, Euglena cyclopicola, Gleocapsa granosa, Microcystis* sp., *Melosira* sp., *Merismopedea* sp., *Navicula* sp., *Phacus acuminatus, Scharcoderia planctonica, Scenedesmus quadricuda* var *bicaudatus, Scenedesmus* spp., *Trachelomonus* sp.

In laboratory the mussels both juvenile and aged individuals were reared in a large sized plastic container (25 cm in depth and 29 cm in diameter) containing soils (10 cm in thickness) and water (10 cm in column height). The water of the container was changed regularly by fresh pond water. From these mussels ten specimens, five from each size group were sacrificed in each week and the gut contents were collected and fixed as per above mentioned method. The organisms present in the gut content were noted individualwise.

In all cases water samples were examined and the organisms in respect to the food-organisms revealed from the gut content were noted.

Results

Field : A total of 142 varieties food organisms have been noted from the gut of *L. marginalis* irrespective of size groups. Of these, diatoms were of 32 varieties, green alge were of 107 varieties and zooplanktons were of 3 varieties (Table-1, Figs, a_1-a_{32}, b_1-b_{107}). Though all the 142 varieties of food items were revealed from the gut contents of aged individuals a marked restriction in food selection in juvenile mussels is highlighted from the data recorded on young individuals (Table 2).

Laboratory : Except the green alge, *Crucigenia tetrapedia, Euglena* sp., *Scenedesmus dimorphus, S. smithii, S. perforatus* and *Tetraedron trigonum* as recorded in field specimens all other varieties are found in the aged mussels. Interestingly, in case of juveniles 8 varieties *Navicula* (Fig. a_{33}-a_{40}), in addition to those noted in field specimens were recorded.

Examinations of water samples revealed the occurrence of many kinds of diatoms, green algae and zooplanktons.

Discussion

From the results it is evident that the mussels *L. marginalis* are much selective in trapping the microorganisms as their food. Of the many varieties of diatoms, green algae and zooplanktons they are selective to a few only. Moreover, it is important to note the facts that the juveniles are too much specific for certain food organisms are mostly algae feeders. It is clear that the newly born mussels feed only on 6 varieties of green algae untill they became 15 days old. Thereafter, 8 more varieties algae are included in their menu and such food selection was continued for a period of next 15 days. Though many organisms are common food to the mussels, the 1-2, 2-3, 3-4, 4-5 and 5-6 months old individuals consumed 11, 18, 13, 21, 16 and 18 varieties of food organisms (green algae and diatoms) respectively. Interestingly, the green algae *Chroococcus minor* and *Thrachelomonas* sp. are fed by the juvenile mussels irrespective of age. The green algae *Chlorella* sp., *Chlorococcum infusionum* and *Tetraedron muticum* are next in order of preference. As the mussels switched over the new variety of food organisms with the increase of age it is apparent that higher the age diverge the food habits. This is well judged from the fact of trapping 142 types of food organisms by the adults in respect to only 42 varieties trapped by the juveniles. Though 32 types of diatoms are consumed by the adult mussels only 4 varieties are acceptable to the young mussels. Thus, of the 42 types of food organisms of the juveniles only 4 belonged to diatom group. Moreover, the diatoms are trapped by the juveniles only when they became 2 months old. While explanations regarding low frequency of diatoms trapping are wanting it can safely be concluded that the growth and survival of mussels, especially the juveniles are strictly governed by the green algae. Since juvenile mussels have selective choice for particular food organisms and the aged mussels are in need of varieties of food organisms it is sure that the mussel (*L. marginalis*) population is largely governed by these microorganisms. So, occurrence of these organisms in the waterbodies of any locality be considered as indicators for mussel farming.

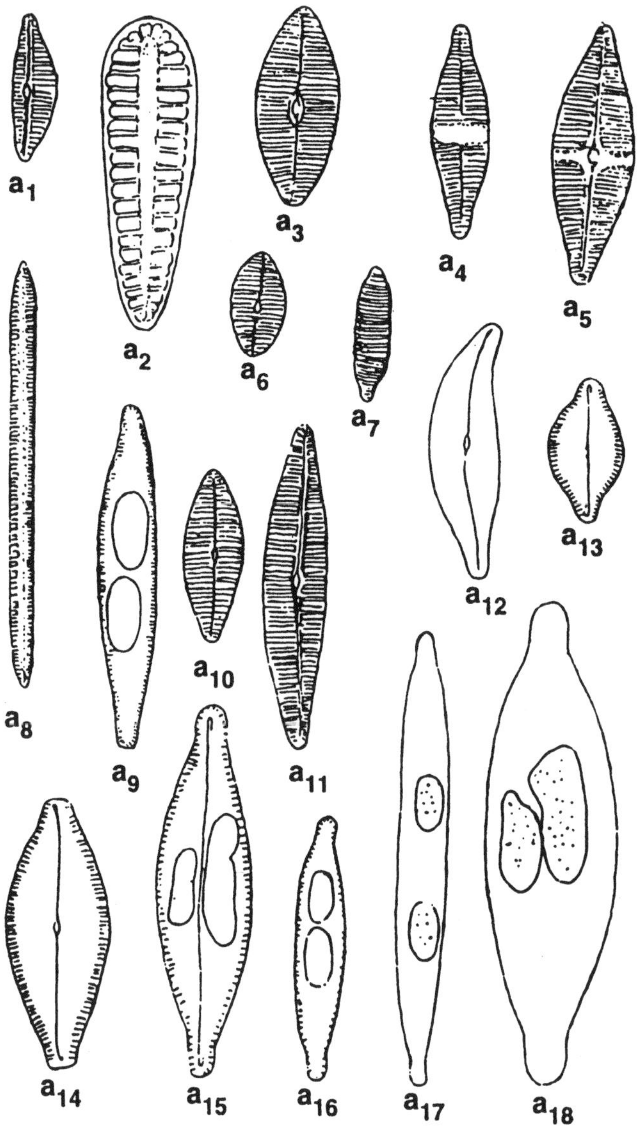

Plate I

Figs. a_1-a-$_{18}$: Different species of diatoms isolated from the gut contents of *L. marginalis* collected from the nature.

a_1 *Cymbella ventricosa* (Size : 10x100)
a_2 *Surirella splendida* (Size : 10x100)
a_3 *Navicula crucicula* (Size : 10x100)
a_4 *Stauroneis wislouchii* (Size : 10x100)
a_5 *Stauroneis eluta hustedt* (Size : 10x100)
a_6 *Cocconeis disculus cleve* (Size : 10x100)
a_7 *Nitzschia umbilicata hustedt* (Size : 10x100)
a_8 *Synedra ulna var acqualis* (Size : 10x100)
a_9 *Navicula* sp. (Size : 15x100)
a_{10} *Navicula bacilloides* (Size : 10x100)
a_{11} *Navicula* sp. (Size : 10x100)
a_{12} *Cymbella* sp. (Size : 10x100)
a_{13} *Navicula* sp. (Size : 10x100)
a_{14} *Navicula* sp. (Size : 10x100)
a_{15} *Navicula* sp. (Size : 10x100)
a_{16} *Navicula* sp. (Size : 15x100)
a_{17} *Nitzschia* sp. (Size : 10x100)
a_{18} *Navicula* sp. (Size : 15x100)

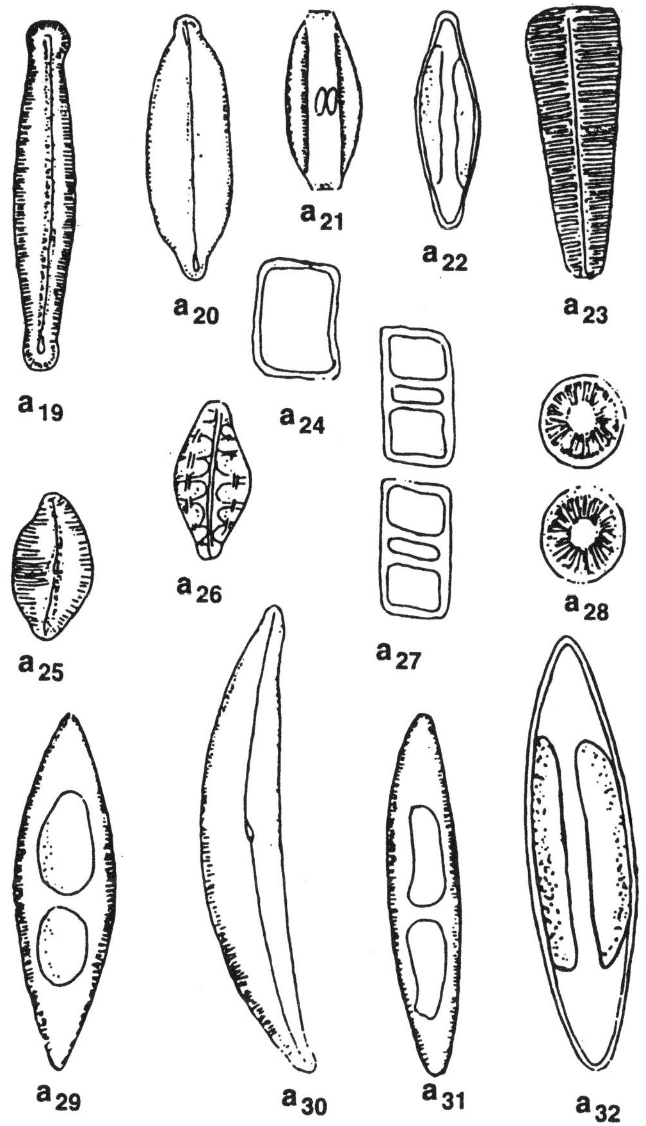

Plate II

Figs. a_{19}-a_{32} : Different species of diatoms isolated from the gut contents of *L. marginalis* collected from the nature.

a_{19} *Pinnularia* sp. (Size : 10x100)
a_{20} *Navicula* sp. (Size : 15x100)
a_{21} *Amphora* sp. (Size : 10x100)
a_{22} *Navicula* sp. (Size : 10x100)
a_{23} *Gomphonema* sp. (Size : 10x100)
a_{24} *Melosira* sp. (Size : 10x100)
a_{25} *Navicula* sp. (Size : 15x100)
a_{26} *Navicula* sp. (Size : 10x100)
a_{27} *Melosira* sp. (Size : 15x100)
a_{28} *Cyclotella* sp. (Size : 10x100)
a_{29} *Navicula* sp. (Size : 15x100)
a_{30} *Cymbella* sp. (Size : 5x100)
a_{31} *Navicula* sp. (Size : 10x100)
a_{32} *Navicula* sp. (Size : 15x100)

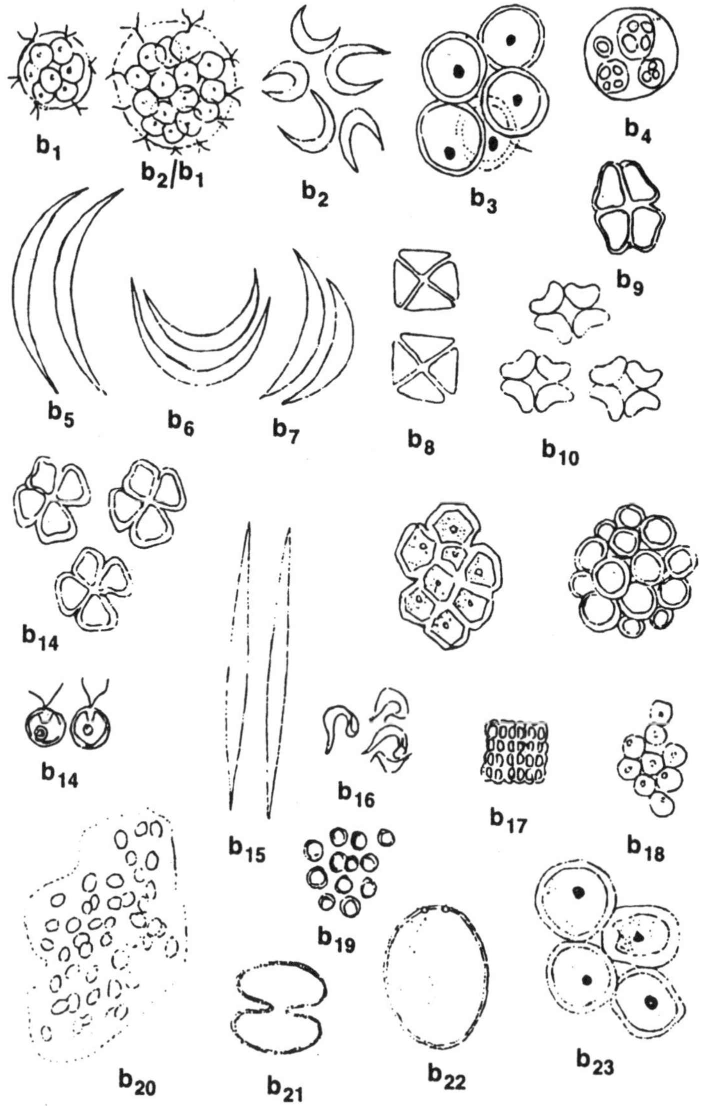

Plate III

Figs. b_1-b_{23} : Different species of green algae isolated from the gut contents of *L. marginalis* collected from the nature.

b_1 *Eudorina elegans* (Size : 10x100)
b_2 *Selenastrum gracile* (Size : 10x100)
b_3 *Chlorococcum* sp. (Size : 15x100)
b_4 *Gleocapsa granosa* (Size : 10x100)
b_5 *Ankistrodesmus* sp. (Size : 10x100)
b_6 *Ankistrodesmus* sp. (Size : 10x100)
b_7 *Ankistrodesmus* sp. (Size : 10x100)
b_8 *Crucigenia tetrapodia* (Size : 15x100)
b_9 *Crucigenia rectangularis* (Size : 10x100)
b_{10} *Crucigenia* sp. (Size : 10x100)
b_{11} *Crucigenia* sp. (Size : 15x100)
b_{12} *Coelastrum* sp. (Size : 10x100)
b_{13} *Coelastrum microporum* (Size : 10x100)
b_{14} *Chlamydomonas* sp. (Size : 10x100)
b_{15} *Ankistrodesmus* sp. (Size : 10x100)
b_{16} *Ankistrodesmus convolutus* (Size : 10x100)
b_{17} *Merismopedia glauca* (Size : 10x100)
b_{18} *Chlorococcum infusionum* (Size : 10x100)
b_{19} *Chroococcus minor* (Size : 10x100)
b_{20} *Microcystis* sp. (Size : 10x100)
b_{21} *Cosmarium perfissum* (Size : 10x100)
b_{22} *Tracheomonus* sp. (Size : 15x100)
b_{23} *Chlorella vulgaris* (Size : 10x100)

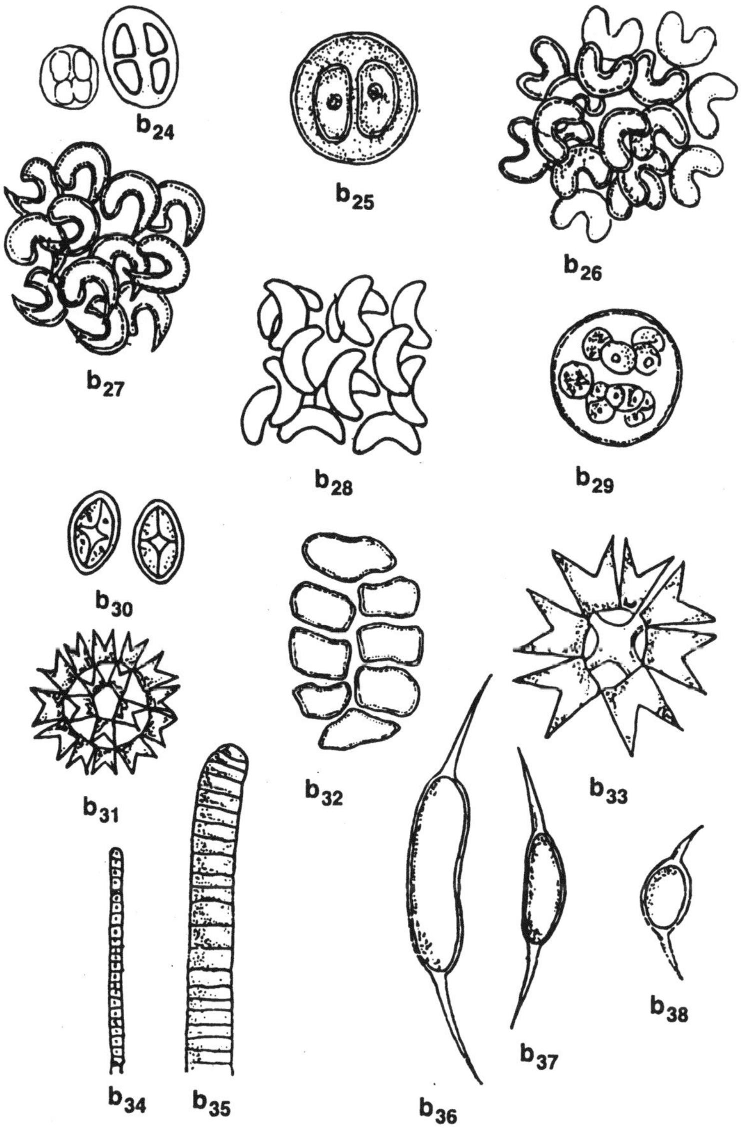

Plate IV

Figs. b_{24}-b_{38} : Different species of green algae isolated from the gut contents of *L. marginalis* collected from the nature.

b_{24} *Chroococcus minutus* (Size : 10x100)
b_{25} *Nephrocytium obesum* (Size : 10x100)
b_{26} *Kircheneriella obesa* (Size : 10x100)
b_{27} *Kircheneriella lunaris* (Size : 15x100)
b_{28} *Kircheneriella* sp. (Size : 15x100)
b_{29} *Oocystis borgei* (Size : 10x100)
b_{30} *Oocystis borgei* (Size : 10x100)
b_{31} *Pediastrum duplex* (Size : 15x100)
b_{32} *Pandorina morum* (Size : 15x100)
b_{33} *Pediastrum duplex* (Size : 15x100)
b_{34} *Oscillatoria irrigua* (Size : 10x100)
b_{35} *Oscillatoria curviceps* (Size : 10x100)
b_{36} *Schroederia* sp. (Size : 15x100)
b_{37} *Schroederia* sp. (Size : 15x100)
b_{38} *Schroederia planctonica* (Size : 15x100)

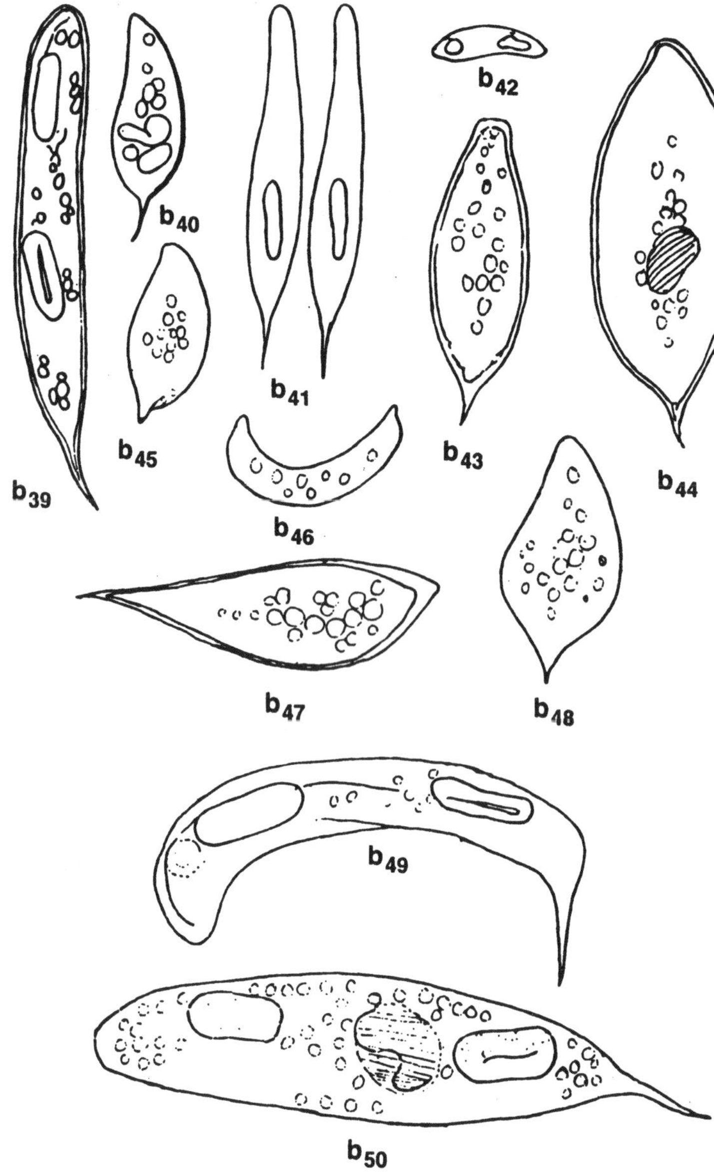

Plate V

Figs. b_{39}-b_{50} : Different species of green algae isolated from the gut contents of *L. marginalis* collected from the nature.

b_{39} *Euglena* sp. (Size : 10x100)
b_{40} *Euglena polymorpha* (Size : 10x100)
b_{41} *Euglena acus* (Size : 10x100)
b_{42} *Euglena cyclopicola* (Size : 10x100)
b_{43} *Euglena* sp. (Size : 10x100)
b_{44} *Euglena* sp. (Size : 15x100)
b_{45} *Euglena* sp. (Size : 10x100)
b_{46} *Euglena* sp. (Size : 15x100)
b_{47} *Euglena* sp. (Size : 10x100)
b_{48} *Euglena* sp. (Size : 10x100)
b_{49} *Euglena* sp. (Size : 10x100)
b_{50} *Euglena* sp. (Size : 10x100)

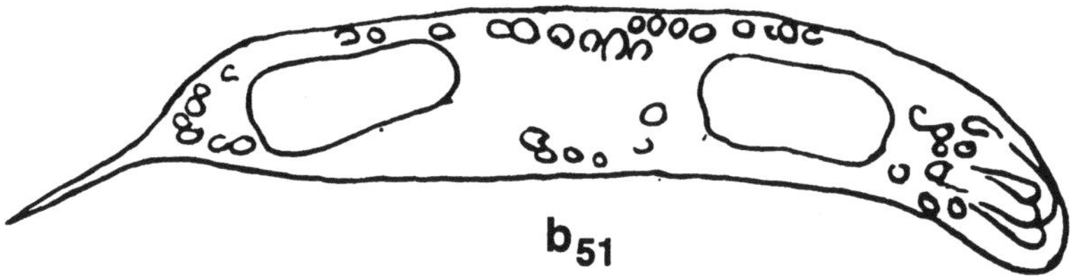

b_{51}

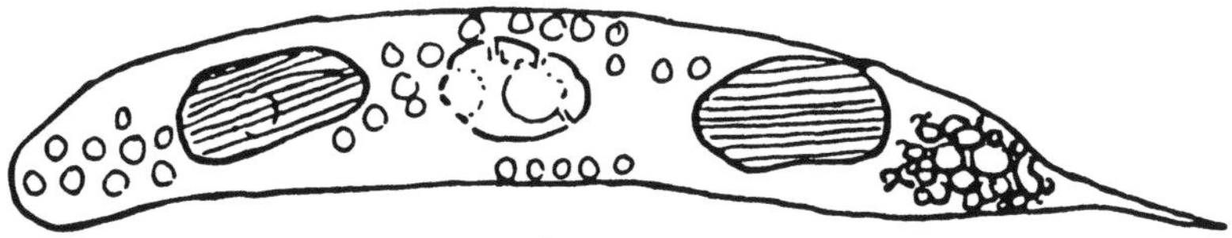

b_{52}

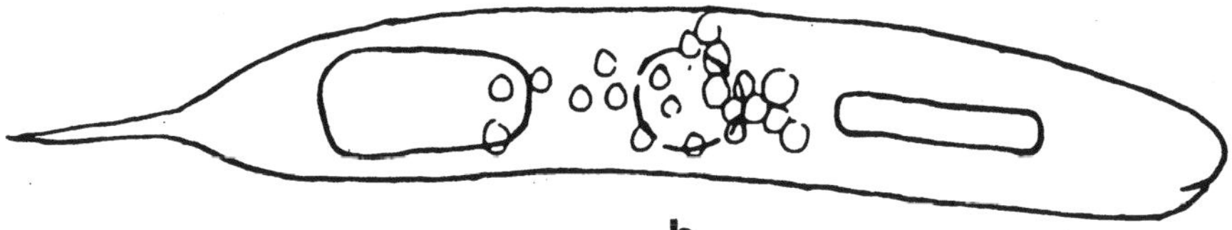

b_{53}

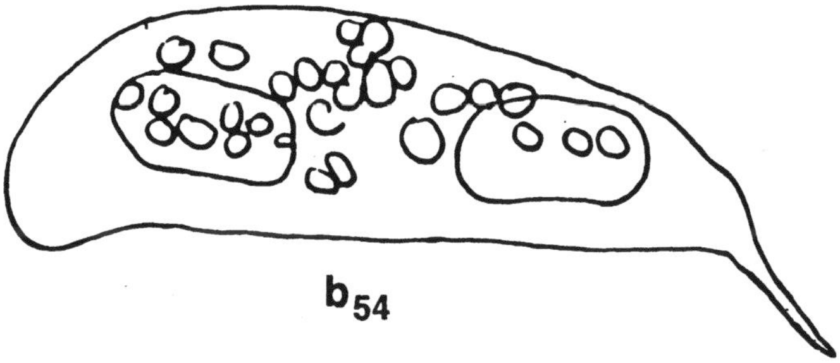

b_{54}

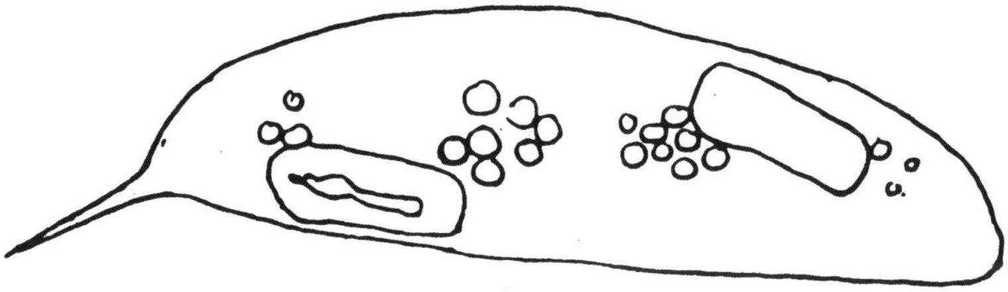

b_{55}

Plate VI

Figs. b_{51}-b_{55} : Different species of green algae isolated from the gut contents of *L. marginalis* collected from the nature.

b_{51} *Euglena* sp. (Size : 10x100)

b_{52} *Euglena* sp. (Size : 10x100)

b_{53} *Euglena* sp. (Size : 10x100)

b_{54} *Euglena* sp. (Size : 10x100)

b_{55} *Euglena* sp. (Size : 10x100)

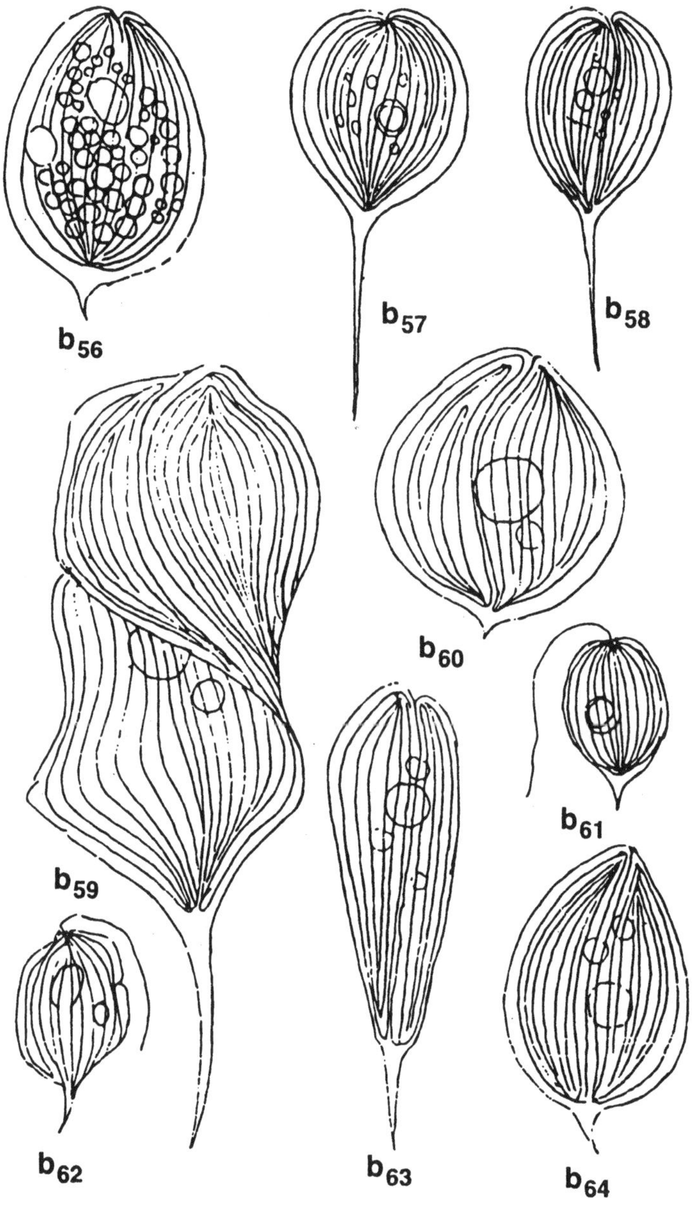

Plate VII

Figs. b_{56}-b_{64} : Different species of green algae isolated from the gut contens of *L. marginalis* collected from the nature.

b_{56} *Phacus* sp. (Size : 10x100)
b_{57} *Phacus* sp. (Size : 10x100)
b_{58} *Phacus* sp. (Size : 10x100)
b_{59} *Phacus tortus* (Size : 10x100)
b_{60} *Phacus* sp. (Size : 10x100)
b_{61} *Phacus acuminatus* (Size : 10x100)
b_{62} *Phacus caudatus* (Size : 10x100)
b_{63} *Phacus* sp. (Size : 10x100)
b_{64} *Phacus* sp. (Size : 10x100)

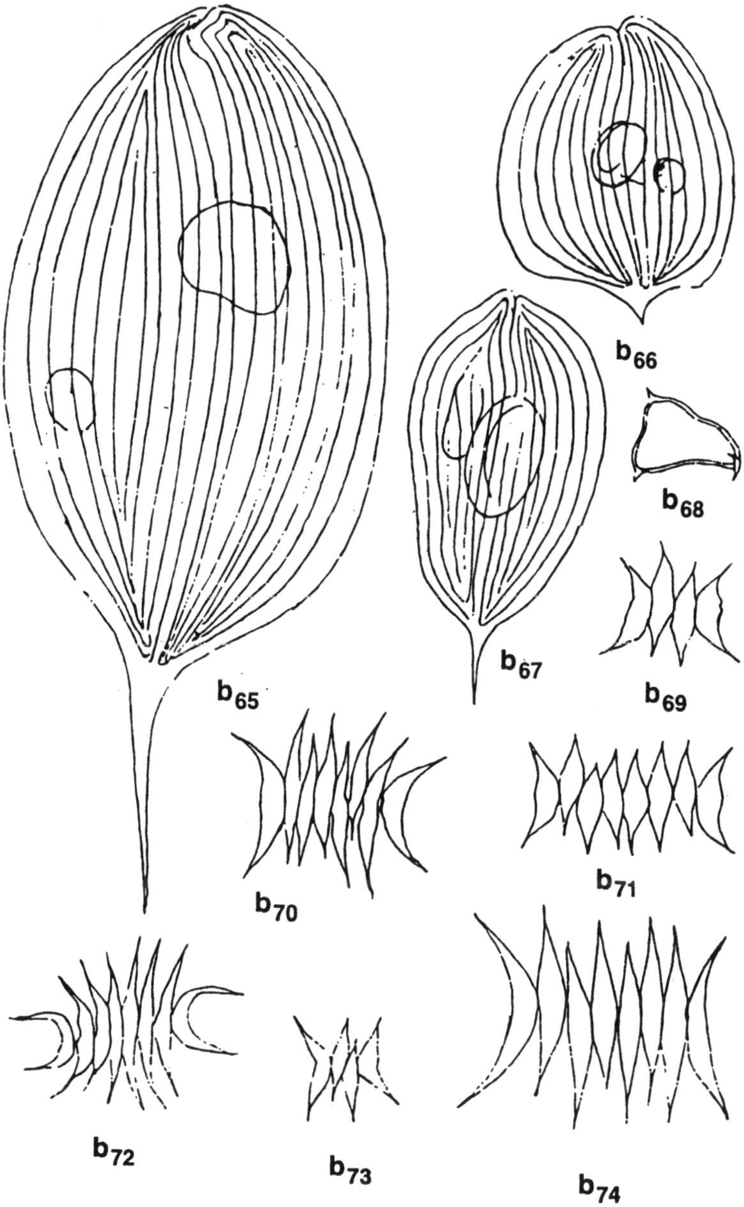

Plate VIII

Figs. b_{65}-b_{74} : Different species of green algae isolated from the gut contents of *L. marginalis* collected from the nature.

b_{65} *Phacus longicauda* (Size : 10 x 100)
b_{66} *Phacus stockessil* (Size : 10 x 100)
b_{67} *Phacus* sp. (Size : 10 x 100)
b_{68} *Staurastrum* sp. (Size : 10 x 100)
b_{69} *Scenedesmus acuminatus* (Size : 10 × 100)
b_{70} *Scenedesmus dimorphus* (Size : 10 × 100)
b_{71} *Scenedesmus acuminatus* (Size : 10 × 100)
b_{72} *Scenedesmus dimorphus* (Size : 10 × 100)
b_{73} *Scenedesmus acuminatus* (Size : 10 × 100)
b_{74} *Scenedesmus dimorphus* (Size : 10 × 100)

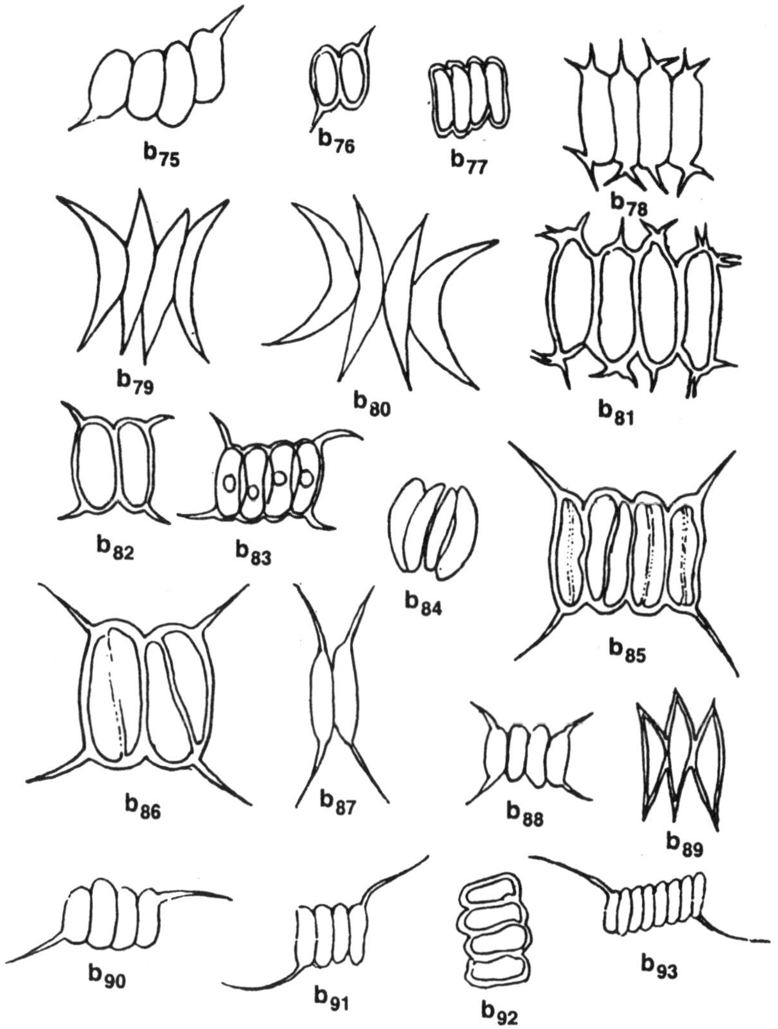

Plate IX

Figs. b_{75}-b_{93} : Different species of green algae isolated from the gut contents of *L. marginalis* collected from the nature.

b_{75} *Scenedesmus quadricuda* var *bicaudatus* (Size : 15 × 100)

b_{76} *Scenedesmus quadricuda* var *bicaudatus* (Size : 10 × 100)

b_{77} *Scenedesmus bigugatus* (Size : 10 × 100)

b_{78} *Scenedesmus* sp. (Size : 10 × 100)

b_{79} *Scenedesmus dimorphus* (Size : 10 × 100)

b_{80} *Scenedesmus dimorphus* (Size : 10 × 100)

b_{81} *Scenedesmus smithii* (Size : 15 × 100)

b_{82} *Scenedesmus quadricuda* (Size : 15 × 100)

b_{83} *Scenedesmus perforatus* (Size : 10 × 100)

b_{84} *Scenedesmus* sp. (Size : 10 × 100)

b_{85} *Scenedesmus armatus* (Size : 15 × 100)

b_{86} *Scenedesmus quadricuda* (Size : 15 × 100)

b_{87} *Scenedesmus quadricuda* (Size : 15 × 100)

b_{88} *Scenedesmus quadricuda* (Size : 15 × 100)

b_{89} *Scenedesmus* sp. (Size : 10 × 100)

b_{90} *Scenedesmus quadricuda* var *bicaudatus* (Size : 10 × 100)

b_{91} *Scenedesmus quadricuda* var *bicaudatus* (Size : 10 × 100)

b_{92} *Scenedesmus bigugatus* (Size : 10 × 100)

b_{93} *Scenedesmus quadricuda* var *bicaudatus* (Size : 10 × 100)

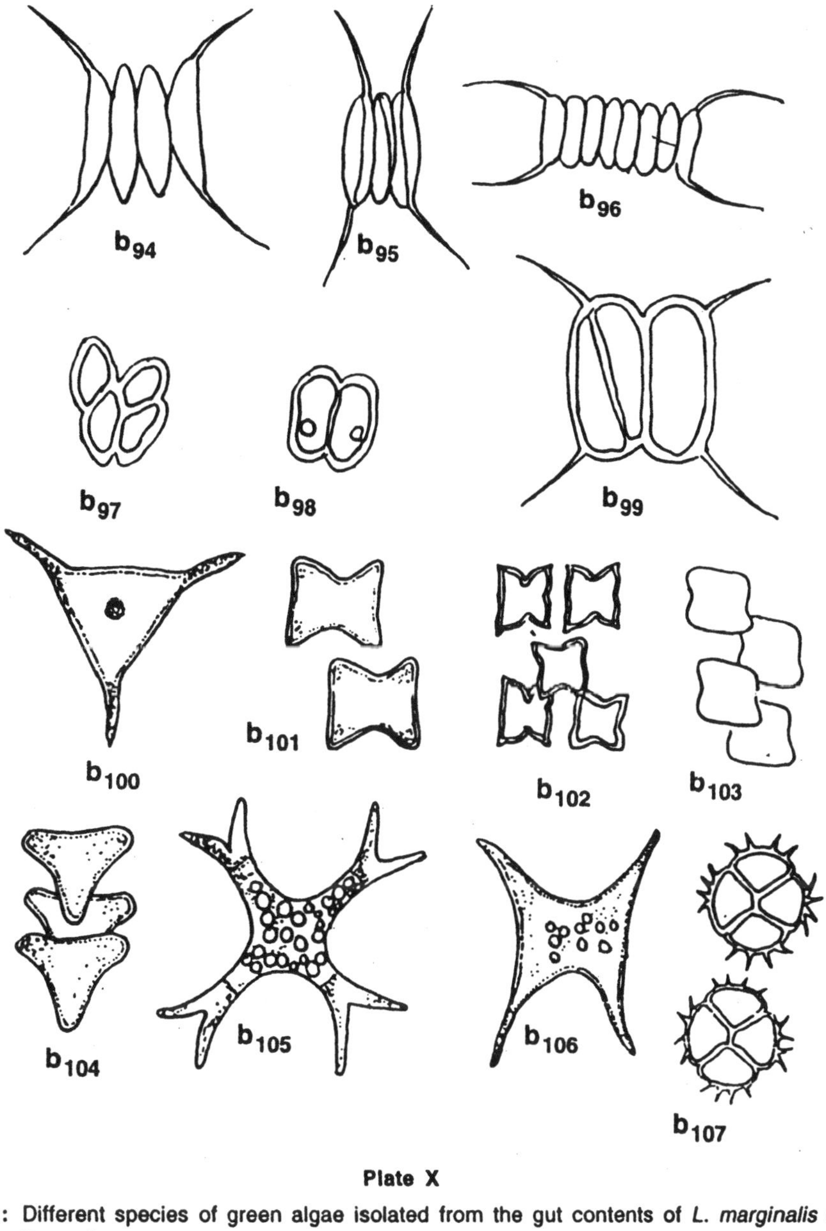

Plate X

Figs. b_{94}-b_{107} : Different species of green algae isolated from the gut contents of *L. marginalis* collected from the nature.

b_{94} *Scenedesmus quadricuda* (Size : 10 × 100)
b_{95} *Scenedesmus* sp. (Size : 10 × 100)
b_{96} *Scenedesmus quadricuda* (Size : 10 × 100)
b_{97} *Scenedesmus* sp. (Size : 10 × 100)
b_{98} *Scenedesmus* sp. (Size : 10 × 100)
b_{99} *Scenedesmus quadricuda* (Size : 15 × 100)
b_{100} *Tetraedron trigonum* (Size : 15 × 100)
b_{101} *Tetraedron* sp. (Size 10 × 100)
b_{102} *Tetraedron minimum* (Size : 10 × 100)
b_{103} *Tetraedron* sp. (Size : 10 × 100)
b_{104} *Tetraedron muticum* (Size : 15 × 100)
b_{105} *Tetraedron pusillum* (Size : 15 × 100)
b_{106} *Tetraedron* sp. (Size : 15 × 100)
b_{107} *Tetraestrum stannogeniaeforme* (Size : 15 × 100)

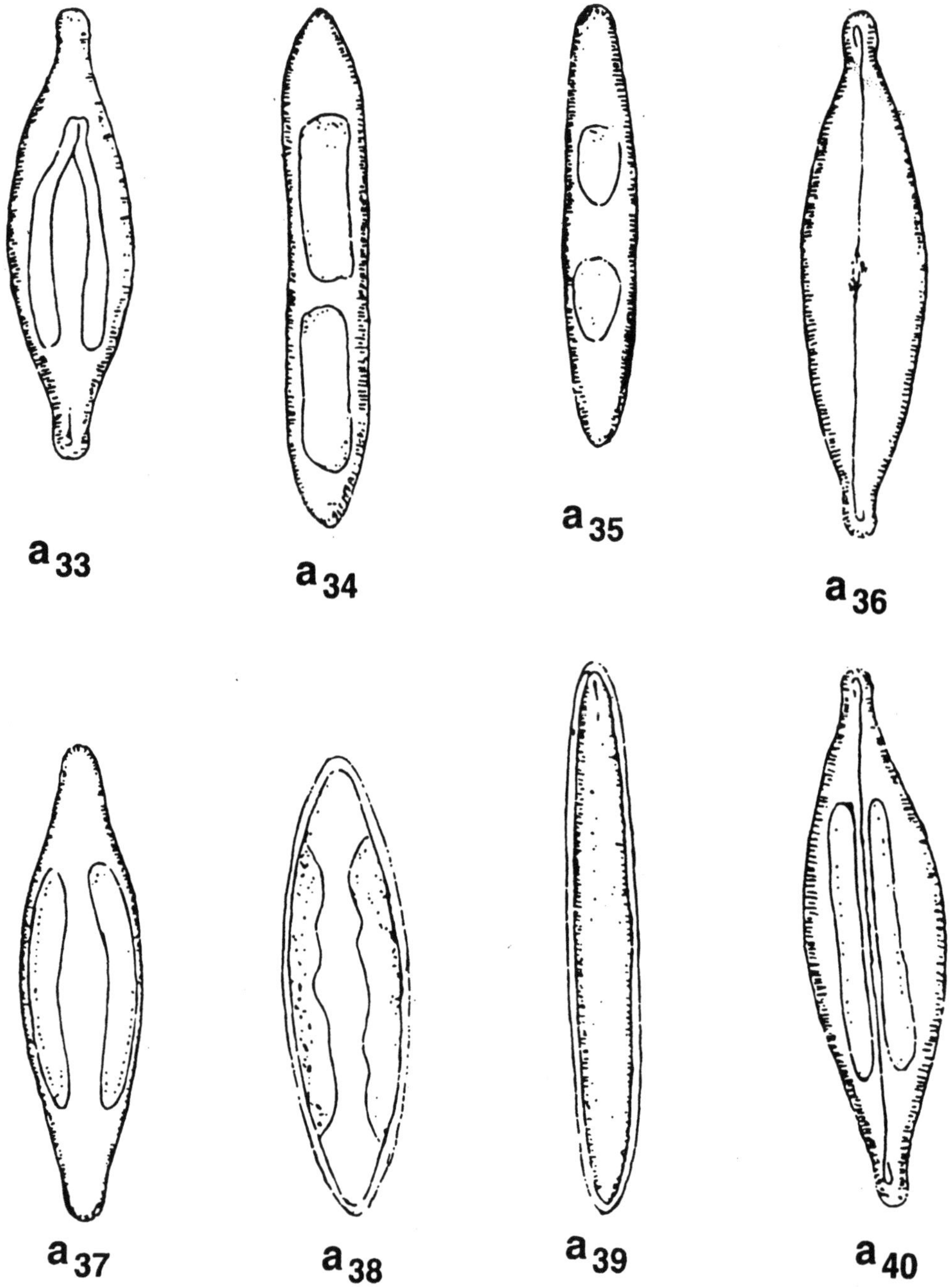

Plate XI

Figs. a_{33}-a_{40} : Different types of diatoms (*Navicula* spp.) (Size : 15 × 100) isolated from the gut of laboratory reared *L. marginalis.*

Acknowledgements

The authors are thankful to the Head of the Department of Zoology, Calcutta University for the facilities provided. Special thanks are due to Late Dr. Prabir Chatterjee, Department of Botany, Calcutta University for his kind help in identification of algae and diatoms. The financial assistance of the University Grants Commission, New Delhi is thankfully acknowledged.

References

Raut, S.K. and A. Biswas (1989) : Pearls in fresh water mussels *lamellidens marginalis* in India (Bivalvia : Unionidae). *Basteria*, 53; 105-109.

Raut, S.K. and A. Biswas (1990) : Fish hosts for larval freshwater mussel *Lamellidens marginalis* (Lamarck). *Apex*, 5 : 31-34.

Raut, S.K. and A. Biswas (1991) : Soils limiting the distribution of the freshwater mussel, *Lamellidens marginalis*. In : G.K. Veeresh, D. Rajagopal and C.A. Viraktamath eds. "Advances in Management and Conservation of Soil Fauna", pp. 859-864. Oxford & IBH Publishing Co. Pvt. Ltd., New Delhi, Bombay Calcutta.

Subba Rao, N.V. (1989) : Hand book : Freshwater molluscs of India. Zoological Survey of India, Calcutta, Calcutta, XXIII+289 pp.

Proceedings of the National Seminar on Environmental Biology, April 03-05, 1998
Edited By : A.K. Aditya & P. Haldar
DAYA PUBLISHING HOUSE, DELHI

Effect of Industrial Effluent on Germination and Growth of *Phaseolus aureus* L.

A.G. Malini Shetty & R.K. Somashekar
Department of Environmental Science
Bangalore University, Bangalore - 560 056

Abstract

The treated effluent of Bharat Heavy Electricals Limited was determined for its physico-chemical properties. A comparative effect of different concentrations of the effluent (10, 25, 50, 75 and 100%) on germination, growth and chlorophyll content of *Phaseolus aureus* was studied. The percentage of germination and growth showed significant variations between control and treated plants. The diluted effluent produced growth promoting effects which could be seen in terms of better percentage of germination, growth and chlorophyll content over the control. At higher concentrations (75 to 100%) a drastic reduction was noticed. A significant positive correlation existed between the growth and chlorophyll versus effluent used at lower concentration. The effluent contained several plant nutrients and therefore, could be recommended for irrigation at lower concentrations.

Introduction

The second major source of water pollution are the by products of rapid but unplanned industrialization. The waste water from industries not only contains harmful materials but also have a considerable amount of essential nutrients and therefore, at times can be used as an additional potential source of macro and micro-nutrients. Recycling and reuse of waste water in agriculture is one of the means of conserving water and nutrients for irrigation, but also offers low cost technology for the treatment and disposal of waste water. Indeed many farmers who live in the vicinity of industries use a mixture of treated effluents and fresh water and, also sewage for irrigation. It has been shown that seed germination test provides a method for evaluating the impact of contaminants present in waste water (Neelam and Sahai, 1988). The present study was carried out to determine the appropriate concentration of effluent beneficial for the germination and growth of *Phaseolus aureus*.

Material and Methods

Bharat Heavy Electricals Limited (BHEL) is a Central Government undertaking located in the outskirts of Bangalore City and is mainly concerned with the production of electrical equipments like insulators, transformers etc. The BHEL discharges about 280 m^3 effluent/day into a drain which later joins the nearby stream. The effluent was collected from the industry at the point of discharge from the treatment plant, at an interval of one month for a period of one year and analysed for physico-chemical parameters using the standard methods (APHA, 1985).

Germination Studies (*in vitro*)

The germination studies were carried out using different dilution of effluents prepared using distilled water viz., 10, 25, 50, 75 and also 100% effluent.

Phaseolus aureus commonly grown by the farmers inhabiting the surrounding areas of the factory, has been used for the study. The selected seeds were dressed with 0.2% mercuric chloride for fifteen minutes and thoroughly washed with distilled water. The seeds were then soaked in effluent for 3 hours.

Ten ml of effluent was used for soaking the filter paper in sterilized corning petriplates containing ten seeds per plate. A control set with distilled water was maintained in each case. A total of one hundred seeds were used in each case and four replicates were maintained for each experiment. The petriplates were kept in indoor laboratory conditions under diffused light at 28±2° C. Germination counts were taken after 48 hours and the shoot and root lenghts were measured on alternate days for a period of ten days.

Plant Growth Studies (*in vivo*)

This was conducted by employing pot culture technique. Each pot was filled with 3 kg of soil. Three sets of pots were maintained of which one set was irrigated with 50 ml of different dilutions of effluents daily and used to grow crop plants. Another set with distilled water served as control. One set without plants was irrigated with 100 per cent effluent to discern the impact of effluent on soil.

The seeds of *P. aureus* after treatment were soaked in different concentration of effluent and sown at the rate of 10 seeds per pot. Following maturity stage (9 weeks), they were harvested and analysed for biochemical parameters like chlorophyll (Arnon, 1949), proteins and phenolics by using prescribed methods (Sadashivam and Manikam, 1992).

The growth parameters like Shoot and root length, leaf area, fresh and dry weight of the plant and, yield characteristics were also estimated (Reddy and Rao, 1981). The soil samples were analysed for various physico-chemical characteristics using the methods of Piper (1950) and Jackson (1973).

Results and Discussion

Of the different physico-chemical parameters analysed (Table 1) the concentration of chlorides, sulphates and total hardness were high in tested sample.

Germination Studies (*in vitro*)

The data (Table 2) revealed a negative impact of the effluent on germination at higher concentrations. This inhibition at higher concentration is perhaps due to higher level of dissolved solids, which prevent the germination by contributing to salinity of the solute absorbed by seeds before germination. The higher concentration of nutrients inhibited seedling growth by altering the osmotic relation (Sahai and Srivastava, 1986). These results are in agreement with the observations of Somashekar *et al.*, (1984, 1993) Singh and Mishra (1987) and Vijayakumari *et al.* (1993). The lower concentration of the effluent on the other hand had a favourable influence on germination and growth of seedlings. The enhancement in germination percentage, shoot and root length may be due to the presence of plant nutrients such as chlorides, calcium and bicarbonates as reported by Manoharan and Lakshmanan (1988) in the case of tannery waste.

Table 1 : Physico-Chemical Parameters of Treated Effluent of Bharat Heavy Electricals Limited (All values except pH and EC are in mg/l)

1.	pH	7.3
2.	Electrical Conductivity (μ mhos)	3400.0
3.	Total Solids	3160.0
4.	Total dissolved solids	2260.0
5.	Total suspended solids	1320.0
6.	Turbidity	8.0
7.	Ammonical nitrogen	0.2
8.	Sulphates	132.8
9.	Total hardness	216.0
10.	Calcium	40.0
11.	Magnesium	40.2
12.	Phosphates	0.9
13.	Alkalinity	542.4
14.	Chlorides	390.0
15.	Dissolved Oxygen	6.5
16.	Biological Oxygen Demand	90.0
17.	Chemical Oxygen Demand	200.0
18.	Nitrate nitrogen	2.0
19.	Nitrite nitrogen	0.9
20.	Sodium	395.0
21.	Potassium	12.0
22.	Iron	0.8
23.	Silica	7.9
24.	Zinc	1.1
25.	Chromium	0.4
26.	Nickel	1.7
27.	Lead	0.2

Average Value of Ten Samples

Table 2 : Effect of Effluent on the Germination of *Phaseolus aureus L.*

Concentration	% of Germ	Shoot Length (Cm)	Root Length (Cm)	SL/RL	Vigour Index
100%	95	8.2	14.6	0.56	2166.0
25%	93	9.3	15.8	0.59	2334.3
50%	86	7.2	12.5	0.58	1694.2
75%	78	6.9	12.1	0.57	1482.0
100%	75	6.7	10.4	1.55	1282.5
Control	98	8.1	14.0	0.58	2165.8

Plant Growth Studies

There existed a dose dependent decrease in plant growth with the increase in the effluent concentration (Table 3). The effluent upto 25 per cent had a favourable impact, expressed in terms of better chlorophyll and protein content over the control. The vigour index also showed increase over control at lower concentrations. The overall percentage of germination, shoot and root growth of plant treated with higher concentration of the effluent remained lower than the control.

Table 3 : Effect of Effluent on the Growth and Yield of *Phaseolus aureus L*

Concentration	Root Length (Cm)	Shoot Length (Cm)	Plant weight (gms.)		No. of fruits	Leaf area (cm^2)
			Fresh	Dry		
10%	16.9	24.9	11.25	3.56	2	6.48
25%	18.6	26.0	10.96	3.66	3	6.16
50%	18.9	25.8	9.04	2.50	2	5.60
75%	15.2	23.3	8.2	2.20	1	4.80
100%	14.1	22.2	7.62	1.92	1	4.85
Control	17.2	25.4	11.12	3.26	2	8.00

The protein and chlorophyll content appear to have been affected following treatment (Table 4). Raza and Vijayakumari (1989) observed significant reduction in chlorophyll in distillery effluent treated fenugreek (*Trigonella foenum* graecum). Nanda *et al.* (1986) observed a decrease in chlorophyll content with the increase in waste concentration. Behra and Mishra (1983) reported an increase in total chlorophyll content in Paddy at lower concentration of distillery effluent.

Table 4 : Effect of Effluent on Biochemical Parameters of *Phaseolus aureus L.*
(All values except proline are in mg/g fresh wt.)

Concentration	Proteins	Proline (μ moles)	Phenolics	Carbohydrates	Chlorophyll
10%	7.32	0.47	6.40	3.8	1.9
25%	10.30	0.52	7.20	4.9	2.6
50%	6.82	0.59	7.80	4.9	1.7
75%	5.70	1.14	9.20	5.3	1.5
100%	5.47	4.15	10.80	5.6	1.4
Control	5.90	0.50	5.90	3.4	3.0

The Phenolic and carbohydrate content also showed dose dependent response. The lowest amount of phenol and carbohydrates were recorded in the control set, followed by plants treated with 10 per cent and 25 per cent effluent.

The higher concentration of the effluent had a negative effect on shoot growth. These results are in agreement with the observations of Vijayakumari *et al.* (1993) in some crop plants, Gautam and Bishnoi (1990) in *Avena sativa*, Elcey and Tiwari (1991) in Millet, Kannabiran and Pragasam (1993) in *Vigna mungo*, Madhappan (1993) in Grams, Vijayarengan and Lakshmanachary (1993) in *Vigna radiata*.

In general, a decrease in various growth parameters among the treated plants could be due to synergistic impact of pollutants on certain metabolic processes. Higher concentration of sodium, the strongest monovalent cataion might have changed the mineral composition of soil, culminating in adverse impact on the growth of the test crops. Tripathi and Ambasht (1981) also made similar observations.

The results of the soil analysis (Table 5), showed alkaline nature of effluent treated soil. A significant change between the composition of basic nutrients of the soil treated with different concentration of effluent and control was apparent. At higher concentration the progressive decrease in plant growth might be due to the accumulation of nutrients in the soil.

Table 5 : Effect of Different Concentration of Effluent on the Soil Characteristics (All values except pH and EC are in mg/kg soil)

Concentration	pH	E.C.	Organic Carbon (%)	Sodium	Potassium	Calcium	Magnesium
10%	8.1	0.2	0.12	975	70	3406.8	4142.01
25%	8.1	0.23	0.22	1200	78	3607.2	4629.30
50%	8.3	0.25	0.23	1375	88	3607.2	4994.77
75%	8.4	0.28	0.24	1413	100	3807.6	4994.77
100%	8.5	0.32	0.30	1438	103	3807.6	6822.12
Control	8.0	0.20	0.11	140	40	1856.2	2108.20

The data gathered from sand culture also revealed a dose dependent response (Table 6). In fact at 100 per cent concentration most of the plants died at seedling stage.

Table 6 : Effect of Effluent on Sand Culture of *Phaseolus aureus* L

Concentration	Root Length (Cm)	Shoot Length (Cm)	Plant weight (gms.)		Leaf area (cm^2)
			Fresh	Dry	
10%	8.3	18.8	3.40	1.10	1.90
25%	7.7	16.7	2.20	0.82	1.80
50%	7.6	17.8	2.31	0.82	2.20
75%	7.4	17.0	2.15	0.78	1.70
100%	5.4	17.1	2.20	0.76	1.70
Control	10.2	20.2	2.50	0.99	2.15

The study revealed that the raw effluent of BHEL contained excess of nutrients which affected the plant growth. But the diluted effluent at 25% concentration produced favourable impact. Hence the effluent in the range of 10 to 25 per cent dilution could be recommended for irrigation.

Acknowledgement

The senior author (A.G.M. Shetty) is grateful to the CSIR for providing financial assistance to conduct the study.

References

APHA (1985). *Standard Methods for the examinations of water and waste water*, 15th ed., American Public Health Association, American Water Works Association/Water Pollution Control Federation, Washington D.C.

Arnon, D.I. (1949). Copper enzymes in isolated Chloroplast, Polyphenol oxidase in *Beta vulgaris Plant Physiol.*, 24 : 1-15.

Behra, B.K. and Mishra B.N. (1983). Effect of the distillery effluents on Rice seedlings : Analysis of pigment, protein, nucleic acids, and 2, 6 dichlorophenol hill reaction of Rice seedlings. Envt. Res.; 31 : 381-389.

Elcey, C. and Tiwari, D. (1991). Effect of distillery effluent on seed germination, early growth and chlorophyll content of millet crop. *Geobios*, 18 : 134-136.

Gautam, D. and Bishnoi, S. (1990). Effect of Urmul dairy effluent on soil characteristics and plant growth in *Avena sativa L. Geobios*, 17 : 91-94.

Jackson, M.L. (1973). *Soil Chemical Analysis*. Prentice Hall of India, New Delhi.

Kannabiran, B. and Pragasam, A. (1993). Effect of distillery effluent on seed germination, seedling growth and pigment content of *Vigna mungo* (L)., Kepper (C.V.T.G.), *Geobios*. 20 : 108-112.

Madhappan, K. (1993). Impact of tannery effluent on seed germination, morphological characters and pigment concentration of *Phaseolus mungo L.* and *Phaseolus aureus. Poll. Res.*, 12 : 159-163.

Manoharan, M. and Lakshmanan, K.K. (1988). Effect of Tannery effluent on seed germination and early seedling growth in *Glycine max (L) Meer. J.Pl. Res.* 4 : 59-61.

Nanda, D.R., Mishra, B.B. and Mishra, B.N. (1986). Toxic effect of solid waste extract of chloro-alkali factory on pigment concentration of a crop plant, *Phaseolus aureus Roxb. J. Envt. Biol.* 7(2) : 95-99.

Neelam, S. and Sahai, R. (1988). Effect of fertilizer factory effluent on seed germination, seedling growth, pigment content and biomass of *Sesamum indicum L., J. Envt. Biol.*, 9 : 45-50.

Piper, C.S. (1950). *Soil and plant Analysis*, Inter Science Publishers, Univ. Adelaide, Australia/New York.

Raza, S.H. and Vijayakumari, M. (1989). Effect of liquor factory effluent on germination and growth of Paddy seedlings. *Indian J. Bot.*, 12 : 11-18.

Reddy, S.S. and Rao, G.M. (1981). Cytogenetic effects of agricultural chemicals. Effects of Insecticides B.H.C. and Novacron on chromosomal mechanism in relation to yield components in Chilli (*Capsicum annuum*). *Cytologia*, 46 : 699-707.

Sadasivam and Manikam A., (1992). *Biochemical Methods for Agricultural Sciences*, Willey Eastern Ltd., New Delhi.

Sahai, R. and Srivastava, N. (1986). Effect of distillery waste on seed germination, seedling growth and pigment content of *Cajans cajan* L., *J. Indian Bot. Soc.*, 65 : 208-211.

Singh, K.K. and Mishra, L.C. (1987). Effect of fertilizer factory effluent on soil and crop productivity, *Water Air and Soil Pollution*, 33 : 309-320.

Somashekar, R.K., Gowda, M.T.G. Shettigar, S.L.N. and Srinath, K.P. (1984). Effect of industrial effluents on crop plants. *Indian. J. Envtl. Hlth.* 26 (2) : 136-146.

Somashekar, R.K. and Siddaramaiah, (1993). Effect of soaps and detergent factory effluents on soil and crop plants. *J. Ind. Poll. Cont.*, 9(2) : 77-84.

Subramani, A., Sundaramurthy, P. and Lakshmanachary, A.S. (1995). Effect of Biologically treated distillary effluent on seed germination and seedling growth of Green gram. *Poll. Res.*, 14(1) : 37-41.

Tripathi, B.D. and Ambasht, (1981). Changes in the morphological behaviour of Wheat plants grown on polluted habitats (Varanasi, India). *Geo. Eco. Trop.*, 5 : 211-218.

Vijayakumari, K., Kumudha, P., Siddaraju, P. and Janardhana K. (1993). Effect of soap factory effluent on seed germination and early seedling growth of certain Millet and Pulse crops, J. Environ. Biol., 14 (4) : 275-281.

Vijayarengan, O. and Lakshmanachary, A.S. (1993). Effect of textile mill effluent on growth and development of Green gram seedlings. *Adv. Plant Sci.*, 6 : 359-365.

Section II

Human Impact on Ecosystem

Proceedings of the National Seminar on Environmental Biology, April 03-05, 1998
***Edited By :* A.K. Aditya & P. Haldar**
DAYA PUBLISHING HOUSE, DELHI

Heavy Metal Pollution of Subernarekha River : Its Ecological Impact on Water Quality and Biota

J.S. Datta Munshi, A.N. Mishra, J. Datta Munshi
Post Graduate Departments of Zoology, Chemistry and Botany
T.M. Bhagalpur University, Bhagalpur-812 007

Abstract

The Subernarekha river originates from a series of fissures of artesian wells situated in a flat field of Nagri village of Ranchi, and passes through the mineral rich industrial belt of South Bihar and Orissa and finally meets the Bay of Bengal. Due to the effluent discharges from various factories the water of the river specially at Ghatsila gets highly polluted due to the presence of heavy metals which ranged from 0.022 to 0.21 mgl^{-1} Cu, 0.13 to 0.635 mgl^{-1} Zn, 0.14 to 0.36 mgl^{-1} Pb, 0.725 to 2.125 mgl^{-1} Fe, 0.047 to 0.067 mgl^{-1} Cd and many species have vanished from the ecosystem. Accumulation of heavy metals in high concentrations were encountered in the liver, kidney and gills of fishes and sensitive bioassay methods for the detection of heavy metal pollution of the river have been developed. However, algal blooms and macrophytes like *Valisnaria, Potamogeton, Hydrilla* and *Eichhornia crassipes* (water hyacinths) found in river Subernarekha in all the seasons act as scavengers of heavy metals.

1. Introduction

1.1 Physiography of the Subernarekha River

The Subernarekha river is the largest river of Singhbum, flowing through the district for about 70 miles and draining over 2,000 square miles. The river orginates near Lodame-Piska road, two miles south of Nagri village (Ranchi dist.) from a series of fissures of artesian wells situated in a flat field. The river flows near about 85 km. in Ranchi dist. and 112 km. in Dhalbhum dist. The river Subernarekha is situated at an altitude of 686 meter (2250 feet) in Chhotanagpur plateau and traverses a distance of 393 km. (247 miles) crossing three states of Bihar, Bengal and Orissa and finally meets the Bay of Bengal in Jaleswar district of Orissa, about 15 km. West of Digha in West Bengal.

Analytical reports on the river water at Ranchi, Jamshedpur, and Ghatsila have been made earlier by Dora and Roy (1987), Singh *et. al.* (1989) and Singh and Singh (1980). But there are no studies available on the industrial effluents discharged into the river in the Jamshedpur Ghatsila region.

The seasonal variation in heavy metal status of the river water has been studied. As the effluents are considered to be the major pollutants, these have also been analysed during all the season with a view to assessing their contribution to pollution (Varma *et. al.* 1991, 1992, Singh & Munshi, 1990, 1992, Munshi *et. al.* 1990, Mishra *et. al.* 1994 a, b, c).

1.2 *The Effluents from the following sources have been Selected for Heavy Metal (Cu, Zn, Pb, Fe, Cr and Cd) Analyses :*

(i) The Tata Iron & Steel Co. (TISCO) : Jamshedpur,
(ii) The Rakha Copper : Galudih,
(iii) The Hindustan Copper Ltd. (HCL) : Ghatsila,
(iv) The Uranium Corporation of India Ltd. (UCIL): Ghatsila, and
(v) The HCL Concentration Plant : Ghatsila.

2. Methodology

2.1 *Water and Sediment Analysis*

The samples of river water were collected in plastic bottles in different seasons at the selected sites during the course of investigation (1987 to 1989), and preserved with concentrated HNO_3. The samples of sediments were collected seasonally at all the selected sites during the course of investigation (1987 to 1989), dried, crushed and preserved. All preserved water and sediment samples were digested with conc. HNO_3 following the standard methods of APHA (1981) and finally subjected to analysis using atomic absorption spectrophotometry, through commercial services of the RSIC, Bose Institute, Calcutta. The chemicals of AR grade and triple glass-distilled water were used.

2.2 *Analysis of Effluent of Factories*

Subernarekha river at Ghatsila is constantly being polluted by the effluents discharged from Hindustan Copper Limited (HCL) and Uranium Corporation of India Limited (UCIL). The concentration of heavy metals in water and effluents of the factories were analysed at Atomic Absorption Spectrophotometer.

2.3 *Hydrophytes*

Levels of three heavy metals (Cu, Zn & Pb) were determined in sampled water, surface soil and some aquatic macrophytes (*Vallisnaris spiralis, Potamogeton pectinatus, Potamogeton pusillus* and *Hydrilla verticellata*) collected from three sites of river Subernarekha at Ghatsila.

2.4 *Fish*

A few species like *Labeo pungusia, Glossogobius giuris* and *Puntius ticto* were collected from the polluted zones of the river. The different tissues such as liver, kidney, gill and whole fish were dried and then digested with conc. nitric and perchloric acid (2 : 1 v/v) for the determination of heavy metals. The heavy metals were estimated with the help of Atomic Absorption Spectrophotometer (APHA, 1981).

2.5 *Impact Assessment of Heavy Metals of Fish*

Specimens of C. *punctata* were procured from the local fish market, and were acclimatised in tap water (pH 7.1 to 7.4 DO 6.8 to 7.0 ppm; alkalinity 120 to 126 ppm as $CaCO_3$; hardness 94 ppm as $CaCO_3$). Sulphuric acid was mixed directly into the water and stirred well to obtain test solutions of different pH values. Bioassay experiments were carried out in plexiglass aquaria and LC_{50} values calculated following the methodology given in APHA (1981). Fishes were divided into three goups and experiments were conducted at pH values of 5.3, 6.0 and 6.6. A group of fish kept in tap water was used as a control. The ambient waters of control and experimental tanks were renewed every 24 h. Fishes were removed at 24, 90 and 350 h and killed by a blow to the head. The gill samples were quickly removed and fixed in 2.5 per cent glutaraldehyde. After fixation of the gills for 72 h at 4°C, the gill filaments were repeatedly washed in 0.1 M phosphate buffer (pH 7.4), dehydrated in graded alcohol, alcohol : acetone mixture and finally in anhydrous acetone at room temperature. The tissues were critically point dried in POLARON, gold coated and then studied under a scanning electron microscope (PSEM - 500, Philips) at the Regional Sophisticated Instrumentation Centre, Bose Research Institute, Calcutta.

2.6 *Bioassay of Heavy Metals*

Bioassay of two heavy metals, copper and mercury were carried out for 96 h on *Channa punctata*. The 96 h LC_{50} values for copper and mercury were 2.13 ppm, and 0.76 ppm respectively. Live specimens of *Channa punctata* of body weight 36.0 ± 2.09 were collected and acclimatized to laboratory conditions for one month. Bioassay experiments were carried out in Plexiglass aquaria (APHA, 1981). In two separate aquaria, sub-lethal concentration of copper suplhate (0.76 ppm = 1/3rd of LC_{50} value) and mercuric chloride (0.26 ppm = 1/3rd of LC_{50} value) were taken and five fishes were exposed to such solutions for different periods. Another group of five fishes was kept in freshwater aquarium and was used as control. After exposure for 24 h, 96h and 360h in sublethal doses of copper and mercury, the fishes were anaesthetized in an aqueous solution of MS 222. Few untreated fishes were also excised. Small pieces of gill were carefully removed and fixed in 2.5 per cent glutaraldehyde in 0.1 m Phosphate buffer, pH 7.4 at 4°C for 72 hrs. The fixed gill materials were repeatedly washed in the same phosphate buffer, dehydrated in graded alcohols, alcohol-acetone mixture and finally in anhydrous actone at room temprature. The tissues were critically point dried, gold coated, then studied under scanning electron microscope PSEM-500 (Philips, Holland) at RSIC, Bose Institute, Calcutta.

2.7 *Effects of Sublethal Mercuric Chloride Exposure on Melanophores of a Catfish, Heteropneustes fossilis (Bioch)*

A group of five fish (wt. 16 ± 2g, length 11 ± 2cm) was exposed to a sublethal concentration of $HgCl_2$ (0.1 ppm = 1/3rd of LC_{50}) for 24h, 72h, 96h, 160h and 360h. The same number of fish were kept in tap water to be used as control. The experimental and control ambient water were renewed every 24 hours. The exposed and unexposed whole fish were fixed in aqueous Bouin's after due periods of exposure to mercuric chloride. Skin pieces (5 x 8 mm) were removed from the predesignated area of the dorsal part of the fish. The tissues were dehydrated in graded series of alcohols and then mounted in DPX. Five permanent whole mount slides of each fish skin sample were made for measurement. The density, dimension and distance between two melanophores (measure from the centrospheres of melanphores) were counted and measured from the figures drawn from the slides on tracing graph paper (mm^2) under a projection microscope (Ermascope EP 333) at 10 x magnification. In all cases measurements were made on at least five places randomly chosen from each slide. The readings from these slides were taken and then the average was calculated. The changes in the shape of melanophores after acute and subacute exposure to sublethal concentration of mercury was observed from photomicrographs taken at x 400. Statistical analysis of results was done by Student's t-test.

2.8 *Oxygen Consumption and Ventilation Rate in Channa punctata (Bioch) Following Exposure to Sublethal Levels of Concentration of Copper Sulphate*

Live specimens of *Channa punctata* (32.0 ± 1g) were collected from local ponds and allowed to acclimatize in laboratory conditions (oxygen level of 6.8 - 7.2 ppm, pH 7.2 - 7.3 at a temperature of 30 ± 1°C). The method employed in the determination of VO_2 through gills under experimental (surfacing prevented condition) was adopted, as given by Munshi & Dube (1973). The 96-h cl50 value for *Channa punctata* was found to be 2.13 ppm. In order to study the effects of $CuSO_4 . 5H_2O$ on oxygen uptake of the fish, the following three sets of experiments were conducted.

(i) The toxic solutions of different concentrations of copper, i.e., 0.5, 1.0, 1.5 and 2.0 ppm (all below the LC_{50}) were taken in separate plexi-glass aquaria and well acclimatized specimens (N = 5) for each concentration of *Channa punctata* were introduced in them for 24 h. The VO_2 of the fish were measured in the respirometer with continuously flowing toxic solution of the same concentration of which the fish were acclimatized.

(ii) The respiratory activity of the fish was studied by exposing them in different sublethal concentrations of copper (0.5, 2.0, 1.5 and 2.0) without acclimatizing the fish in toxic solution.

(iii) The fish were exposed to sublethal concentration (0.5 ppm) of copper for a period of 20 days. The VO_2 was measured on the 1st, 5th, 10th, 15th and 20th day of exposure.

2.9 *Skin Melonophores as Indictor of pH of Ambient Waters*

Heteropneustes fossilis were procured from the local fish market, and acclimatized in tap water pH 7.1 - 7.4, Dissolved oxygen 6.0 - 7.0 ppm, alkalinity 120 - 126 ppm). Sulphuric acid was mixed directly into the water and stirred well to obtain test solutions of different pH values. Bioassay experiments were carried out in plexi-glass aquaria of $\sqrt{10}$ l capacity and LC_{50} was calculated (APHA, 1981). A group of 10 fishes (wt. 16 ± 2g, length 11 ± 2 cm) was exposed at pH 6.0 for 3h, 6h, 18h, 24h, 48h, 72h, 96h, 168h and 360h. Another group of fishes were to be used as control. The experimental and controlled ambient water were renewed every 24 hours. The exposed and control fishes were fixed in aqueous Bouin's after due periods of exposure to acidic water. Skin pieces (5 x 8 mm) were removed from the predesignated area of the dorsal part of the fish. The tissues were dehydrated in graded series of alcohols and then mounted in DPX. Five permanent wholemount slides of each fish skin sample made for measurment. The density, dimension and distance between two melanophores (measured from the centrophere of melanophore) were counted and measured from figures drawn the slides on tracing graph paper (mm^2) under projection microscope (Ermascope *EP* 333) at 10 x magnification. In all cases, measurements were made on at least five places randomly chosen from each slide. The readings from these slides were taken and then the average was calculated. The change in the shape of melanophores after exposure to acidic water (ph 6.0) was observed from the photomicrographs taken at 400 x. Statistical analysis of results was done by Student's 't' test.

2.10 *Integumentary Chormatophores and Mucous Glands of Fish as Indicator of Heavy Metal Pollution : A Simple Scale Method for Evaluation of Quality of Freshwater*

The toxic effects of sublethal concentration of 0.4 ppm $CuSO_4$ on the chromatophores and mucous glands of skin of a freshwater major carp, *Labeo rohita* (Ham.) has been studied (Pandey, Kunwar & Munshi, 1990). Skin flap enveloped scales of control and treated specimens were taken from the body of anaesthetised fish near the base of dorsal fin and immediately fixed in cold Bouin's fluid and processed for study under light microscope. A decrease in the number and increase in the area of mucous glands have been observed during different hours of treatment.

Mucous glands and chromatophores were counted with mm^2 grid and area of the mucous glands and chromatophores were examined under Ermascope. Number and area of mucous glands and chromatophores were used as indices of time dependent heavy metal pollution of water.

2.11 *The Effect of Heavy Metals on the Dimensions of the Gills of the Fishes – A Morphometric Assessment*

Live specimens of *Channa punctata* in the weight range of 20-85 g were collected from local ponds and were acclimatized to laboratory conditions in aquarium. Fishes were fed on chopped goat liver.

1000 ppm of stock solutions were made by dissolving appropriate amount of $CuSO_4$ and $HgCI_2$ in deionized water.

Static biossay experiments were carried out in plexiglass aquarium of 20 L capacity for 96 h in the manner suggested by APHA (1981). The 96 h LC_{50} values of coppr sulphate and mercuric chloride were observed to be 2.13 ppm and 0.76 ppm respectively. In order to study the effects of sublethal dose of copper and mercury on the gill dimensions of fish, the following three sets of experiments were performed:

1. Fishes of known body weight (weight range 20 to 85 g) were acclimatized in tap water and then fixed in Bouin's solution. This group of fishes were taken as control.
2. Fishes of the same body weight range (20 to 85 g) were exposed to a sublethal dose of 0.7 ppm (1/3 rd of LC_{50} value) copper sulphate solution for 7 days. The fishes were fixed in Bouin's.
3. Fishes of the same body weight range (20 to 85 g) were exposed to a sublethal dose fo 0.26 ppm (1/3rd of LC_{50} value) mercuric chloride. After exposure of 7 days, fishes were fixed in Bouin's.

All the four arches were excised and the morphometric measurement were carried out by weight method as described by Muir & Hughes (1969). All the measurements were made under dissecting binocular microscope and improved variety of camera lucida (Ermascope).

For each set of experiment 5 fishes were used (wt. range, 20-85 g) and the data of average weight 54 g was taken into consideration.

Student's-t test was applied to check the significance of data.

2.12 *Cytology of Macrophages as Indicator of Mercury Toxicity in Fishes*

Methodology : Live specimens of *Channa punctata* (Bloch) (34-36 g) were acclimatized to laboratory conditions for 15 days, during which they were fed chopped goat liver. Groups of 10 fish were exposed separately to various concentrations of mercury in plexiglass aquaria (20 L). Ambient water was changed daily to restore the mercury level. The lethal concentration (LC_{50}) value was calculated. Twenty fish were then exposed to sublethal concentration of $HgCl_2$ (0.248 ppm) and five control and experimental fish were sacrificed on the 7th and 15th days for histochemical tests.

3. Results and Discussion

3.1 *Heavy Metal in Water and Sediments*

The physico-chemical analysis of water of river Subernarekha at different sites in winter, summer and rainy seasons have been conducted. The levels of Cu, Zn, Pb, Fe, Cr and Cd were monitored in water and soil collected from various sites of river Subernarekha from its origin at Nagri to Ghatsila. The concentration of various heavy metals varied from 0.022 to 0.21 mgL^{-1} for Cu, 0.13 to 0.635 mgL^{-1} for Zn, 0.14 to 0.36 mgL^{-1} for Pb, 0.725 to 2.125mgL^{-1} for Fe, 0.047 to 0.067mgL^{-1} for Cd, in water. Higher values were obtained at Ghatisila, where effluents of Hindustan Copper Limited and Uranium Corporation of India are directly discharged into the river. The zooplankton and phytoplankton populations were drastically reduced in this area as compared to upstream of the river (Tables 1 a-c; Table 2).

Table 1a : The Seasonal and Overall Average Heavy Metal Concentration (mg^{-1} kg) in Soil Sediment of the River Subernarekha at Different Study Sites (Winter; 1987-1989)

Heavy Metal	Site	Site-1 Nagri	Site-2 Namkum	Site-3 Chandil	Site-4 Sonari	Site-5 Mango	Site-6 Galudih	Site-7 Ratmuhan	Site-8 Dhaiguda	Site-9 Amainagar
	W	15.3	20.3	18.7	28.3	32.7	29.5	91.0	822.3	522.7
	S	21.0	20.5	30.5	41.0	65.5	242.5	283.5	1581.5	1166.5
Cu	R	7.0	9.3	9.7	18.0	21.0	26.0	45.0	456.7	67.3
	Av	14.4	16.7	19.6	29.1	39.7	99.3	139.8	953.5	585.5
	W	31.0	37.0	38.3	57.3	57.3	205.5	64.3	482.7	238.3
Zn	S	37.5	51.5	41.5	63.0	58.0	349.5	76.0	987.5	936.0
	R	27.6	35.7	37.7	55.0	52.3	54.0	46.7	257.0	106.3
	Av	32.0	41.3	39.2	58.4	55.9	203.0	62.3	575.7	426.9

Note : W = Winter, S = Summer, R = Rainy, Av = Overall Average

Table 1b : The Seasonal and Overall Average Heavy Metal Concentration (mg^{-1} kg) in Soil Sediment of the River Subernarekha at Different Study Sites (Summer; 1987-1989)

Heavy Metal	Site	Site-1 Nagri	Site-2 Namkum	Site-3 Chandil	Site-4 Sonari	Site-5 Mango	Site-6 Galudih	Site-7 Ratmuhan	Site-8 Dhaiguda	Site-9 Amainagar
	W	33.7	33.0	20.3	33.0	32.3	31.0	34.7	68.0	34.0
	S	46.5	52.0	50.5	48.5	57.5	60.0	58.0	85.5	62.5
Pb	R	24.0	26.0	15.7	27.7	24.7	18.3	22.3	51.7	23.3
	Av	34.7	37.0	28.8	36.4	38.2	36.4	38.3	68.4	39.9
	W	19569.0	22664.0	18720.0	29200.0	29838.0	26305.0	26165.0	32000.0	26783.3
Fe	S	23468.0	24623.0	24696.0	27696.0	28240.0	40109.0	31062.0	38261.5	31575.5
	R	17083.3	16744.0	14480.7	23223.7	25820.0	23276.0	19413.3	23563.7	16973.3
	Av	20040.1	21343.7	19120.6	26706.6	27966.0	29896.7	25546.8	31275.1	25104.7

Note : W = Winter, S = Summer, R = Rainy, Av = Overall Average

Table 1c : The Seasonal and Overall Average Heavy Metal Concentration (mg^{-1} kg) in Soil Sediment of the River Subernarekha at Different Study Sites (Rainy; 1987-1989)

Heavy Metal	Site	Site-1 Nagri	Site-2 Namkum	Site-3 Chandil	Site-4 Sonari	Site-5 Mango	Site-6 Galudih	Site-7 Ratmuhan	Site-8 Dhaiguda	Site-9 Amainagar
	W	24.7	27.7	28.7	57.7	59.7	53.5	43.3	55.3	40.3
	S	31.5	26.5	25.5	66.5	65.5	63.5	34.5	64.0	42.0
Cr	R	21.3	21.0	18.0	36.3	45.0	32.0	28.0	43.0	32.3
	Av	25.8	25.1	24.1	53.5	56.7	49.7	35.3	54.1	38.2
	W	3.0	3.3	3.3	5.0	6.3	2.5	2.7	5.0	3.3
Cd	S	3.0	3.5	4.0	6.5	7.0	3.5	2.0	6.0	4.0
	R	2.3	2.3	2.3	4.7	3.0	1.5	2.0	4.0	2.7
	Av	2.7	3.0	3.2	5.4	5.4	2.5	2.2	5.0	3.3

Note : W = Winter, S = Summer, R = Rainy, Av = Overall Average

3.2 *Source of Pollution - Industrial Effluents*

The samples of effluents were collected in winter summer and rainy seasons from 1987 to 1989 and were analysed. The heavy metal concentrations show no seasonal variation. Hence the average concentration data of the metals have been presented in Table - 1. The most alarming is the level of lead which has been found to be much higher than the permissible level (100 μgL^{-1}) in all the effluents. The lowest is about 300% higher than the standard, as in Rakha Copper Effluent and the highest is more than 1,000% higher than the standard,as in HCL Effluent. The next is copper which raises an alarm only in the HCL and HCL - Concentration plant Effluents, where it is even more than one and half times the ISI standard. The zinc and cadmium levels are much below the permissible limits of 5,000 and 2,000 μgL^{-1} , respectively. Chromium found only in UCIL and HCL Concentration plant Effluents and the level is again much below the permissible one (2,000 μgL^{-1}). In the absence of the ISI standard, it is difficult to comment on the level of iron, but is appears to be high (28,565 μgL^{-1}) in the HCL Concentration plant effluent (Table 3).

Table 2a : The Seasonal and Overall Average Heavy Metal Concentration ($\mu g\ L^{-1}$) in Water of the River Subernarekha at Different Study Sites (Winter; 1987-1989)

Heavy Metal	Site	Site-1 Nagri	Site-2 Namkum	Site-3 Chandil	Site-4 Sonari	Site-5 Mango	Site-6 Galudih	Site-7 Ratmuhan	Site-8 Dhaiguda	Site-9 Amai-nagar	ISI Std NEERI, 1986
	W	29.7	39.3	45.0	43.3	48.7	43.5	74.7	520.7	123.7	
Cu	S	31.5	46.5	51.5	67.0	56.5	71.0	65.0	1,639.0	200.5	50.0
	R	20.3	48.0	44.3	51.0	48.0	38.0	63.0	465.0	91.3	
	Av	27.2	44.6	46.9	53.8	51.1	50.8	67.6	874.9	138.5	
	W	113.3	151.3	190.7	214.3	217.3	208.0	195.7	575.7	273.3	
Zn	S	109.5	153.0	194.0	257.0	170.5	245.0	223.0	843.0	224.5	5,000.0
	R	73.3	119.3	196.7	162.7	168.0	178.0	239.7	601.3	210.0	
	Av	98.7	141.2	193.8	211.3	185.3	210.3	219.5	673.3	235.9	

Note : W = Winter, S = Summer, R = Rainy, Av = Overall Average

Table 2b : The Seasonal and Overall Average Heavy Metal Concentration ($\mu g\ L^{-1}$) in Water of the River Subernarekha at Different Study Sites (Winter; 1987-1989)

Heavy Metal	Site	Site-1 Nagri	Site-2 Namkum	Site-3 Chandil	Site-4 Sonari	Site-5 Mango	Site-6 Galudih	Site-7 Ratmuhan	Site-8 Dhaiguda	Site-9 Amai-nagar	ISI Std NEERI, 1986
	W	107.7	159.0	295.0	216.3	285.7	402.0	318.7	395.0	232.3	
Pb	S	175.5	320.0	383.5	620.0	613.0	702.5	666.5	1,057.5	467.5	100.0
	R	85.3	60.3	143.3	144.3	142.3	330.5	226.3	243.3	153.0	
	Av	122.8	179.8	273.9	326.9	347.0	478.3	403.7	565.3	284.3	
	W	551.7	599.3	917.3	1,433.3	1,661.6	1,690.0	1,201.7	1,770.0	971.3	
Fe	S	609.0	1,147.0	1,159.0	1,795.0	1,920.0	1,587.5	1,192.5	1,626.5	1,470.0	300.0
	R	647.7	682.7	1,384.7	1,101.0	979.0	1,166.0	1,080.3	1,219.3	1,055.3	
	Av	602.8	809.7	1,153.7	1,443.1	1,520.2	1,481.2	1,158..2	1,538.2	1,165.5	

Note : W = Winter, S = Summer, R = Rainy, Av = Overall Average

Table 2c : The Seasonal and Overall Average Heavy Metal Concentration ($\mu g\ L^{-1}$) in Water of the River Subernarekha at Different Study Sites (Winter; 1987-1989)

Heavy Metal	Site	Site-1 Nagri	Site-2 Namkum	Site-3 Chandil	Site-4 Sonari	Site-5 Mango	Site-6 Galudih	Site-7 Ratmuhan	Site-8 Dhaiguda	Site-9 Amai-nagar	ISI Std NEERI, 1986
	W	nil	nil	nil	83.3	83.3	nil	83.3	nil	nil	
Cr	S	6.5	6.5	nil	nil	nil	nil	nil	nil	nil	50.0
	R	nil	nil	nil	nil	nil	nil	nil	nil	nil	
	Av	2.2	2.2	nil	27.8	27.8	nil	27.8	nil	nil	
	W	10.3	22.3	19.0	31.3	28.7	26.0	30.0	38.7	35.0	
Cd	S	8.5	21.0	24.0	29.0	24.5	25.5	23.5	42.0	32.0	10.0
	R	20.3	29.0	26.3	15.7	13.3	22.5	33.0	44.3	33.7	
	Av	13.1	24.1	23.1	25.3	22.2	24.7	28.8	41.8	33.6	

Note : W = Winter, S = Summer, R = Rainy, Av = Overall Average

Table 3 : Heavy metal concentration in water & effluents discharged in the river Subernarekha at Ghatsila

Heavy metals	Water	HCL eff.	HCL Conc. Plant	UCIL eff.	ISI Standards (NEERI)
Cu (ppm)	1.648	4.020	5.423	0.098	0.05
Zn (ppm)	0.898	2.406	3.515	0.538	5.00
Pb (ppm)	1.020	1.432	0.620	0.420	0.1
Fe (ppm)	1.678	5.840	44.640	10.240	0.3
Cd (ppm)	0.049	0.028	0.027	0.058	0.01

An examination of the heavy metal loads of the effluents suggests the following gradation : HCL Concentration plant >> HCL > UCIL > TISCO > Rakha Copper. Place-wise, the above gradation will mean : Ghatsila >> Jamshedpur > Galudih.

It seems that the pre-disposal treatment plant of Hindustan Copper Limited, Ghatsila were not effective in removing heavy metals from the effluents discharged into the river Subernarekha during the period of this investigation.

3.3 *River Sediments*

The analytical seasonal data on soil sediments for the period 1987-1989 were obtained from each of the study sites and seasonal average.

Table 4 : Accumulation of Heavy metals in tissues and whole fish of river Subernarekha at Ghatsila

Heavy metals	*Labeo pungusia*			*Glossogobius giuria*			*Puntius ticto* Whole fish
	Gill	Liver	Kidney	Gill	Liver	Kidney	
Cu (mg/g)	0.038	0.074	0.048	0.027	0.065	0.035	0.058
Zn (mg/g)	0.026	0.068	0.058	0.024	0.059	0.038	0.042
Pb (mg/g)	0.014	0.048	0.022	0.019	0.034	0.024	0.034
Fe (mg/g)	0.884	1.324	0.902	0.624	0.985	0.854	0.006
Cd (mg/g)	0.012	0.018	0.014	0.010	0.017	0.015	0.029

As expected, the heavy meatal concentrations in the soil sediments show a trend of seasonal variation. This is highest in the summer and lowest in the rains for all the sites and for all metals selected due to the lowest and the highest volumes of the river water in those seasons, respectively. As the river advances, there is, in general, a gradual increase in the metal load of the sediment due to metallic run off from the catchment area, but an abrupt rise in load has been recorded at Ghatsila, due to high sedimentation rate. At Ghatsila, the pollution load on river bed is the highest due to the multi unit industry. Among all the six metals, iron has been found to be in much higher concentration than others. The soil of ther area otherwise also, contains much iron.

3.4 *River Water*

The seasonal analytical data on river for the year 1987-1989 were obtained from each of the study sites and seasonal average data were computed. The pollution level is highest in the summer and lowest in the rains due to low and high volume of water in the river. In gereral, there is a gradual increase in the metal concentration in the river water as the river passes through Ghatsila. However, there is a subsequent fall in metal concentration due to high sedimentation rate as the river further proceeds down. Among all the six metals selected iron has been found to be in much higher concentrations than others.

3.5 *Phytoplankton Community Structure*

The phytoplankton density ranged between 55 μL^{-1} and 406 μL^{-1} . The minimum density was recorded at Dahigora (55 μL^{-1}) (Munshi *et. al.*, 1990). Members of Myxophyceae dominated (54.54%) at this site which was followed by chlorophyceae (34.54%) and Bacillariophyceae (10.90%). *Microcystis aerugnose* formed the dominat member of Cyanophyceae. Other forms recorded were *Oscillatoria chlorina, Anabaena unispora,* and *Eunotia praeurapta.* Bacillariopyceae were mainly represented by *Fragilaria capucina, Synedra ulna, Pinnularia nobilis* and *Denticula* sp. Member of Chlorophyceae- viz., *Spirogyra*, sp. *Closterium setaceum, Consnarium moromazum, Eudorina elegans, Pandorina sp. Hydrodictyon reticulatum, Hymenomonas rosela, Gymnoiyga moniliformis, Rhizoclnius sp.* and *Chlorella vulgaris* were reported from the same site.

Table 5 : Ventilation rate and oxygen consumption (V_2O) : mlO_2/h and $mlO_2/kg/h$) in *Channa punctata* (Bloch), (Surfacing prevented) in relation to copper sulphate of different concentration for 24 h exposure at water temperature 30±1º C (n = 3; ± standard deviation)

	Ventilation rate/minute				Oxygen consumption (VO_2)		
Body weight g	Experimental conditions		Value	% Change	mlO_2	mlO_2/kg/h	% Change
32.0±1	Control (in tapwater)		52±1	–	2.931±0.144	90.66±5.773	–
	24 h exposure in relation	0.5 ppm	58±2*	+11.53	2.506±0.219	75.333±0.237*	–16.66
	to different concentrations	1.0 ppm	60±2**	+15.38	1.833±0.145	57.066±4.300**	–36.66
	of copper sulphate solution	1.5 ppm	64±4*	+23.07	1.473±0.046	46.2±0.558***	–48.88
		2.0 ppm	68±2**	+30.76	0.976±0.49	30.536±1.512***	–66.66

Table 6 : Concentration of Cu, Zn and Pb (µg/g dry plant) in some aquatic macrophytes of river Subernarekha

Plants	Plant parts	Metal	Sites		
			Concentration (µg/g)		
Vallisnaria spiralis	Whole	Cu	194.0 ± 2.052	2055.1 ± 1.252	1805.0 ± 3.501
Potamogeton pusillus	,,	,,	222.1 ± 0.405	661.0 ± 2.048	497.0 ± 0.726
P. pectinatus	,,	,,	452.3 ± 1.357	1194.0 ± 2.027	575.0 ± 1.322
Hydrilla verticellata	,,	,,	100.0 ± 2.746	560.5 ± 2.891	544.0 ± 1.555
Vallisnaria spiralis	Whole	Zn	804.2 ± 2.110	1751.2 ± 3.520	1518.5 ± 2.221
Potamogeton pusillus	,,	,,	41.2 ± 0.520	1065.0 ± 1.386	233.2 ± 1.821
P. pectinatus	,,	,,	63.0 ± 1.373	176.1 ± 0.806	20.2 ± 1.638
Hydrilla verticellata	,,	,,	108.0 ± 0.624	380.0 ± 1.824	375.2 ± 1.513
Vallisnaria spiralis	Whole	Pb	60.0 ± 2.255	82.1 ± 1.502	72.1 ± 2.221
Potamogeton pusillus	,,	,,	18.0 ± 1.201	29.0 ± 1.154	16.5 ± 0.986
P. pectinatus	,,	,,	18.2 ± 1.462	40.4 ± 1.789	18.0 ± 0.971
Hydrilla verticellata	,,	,,	18.2 ± 0.493	82.0 ± 3.218	76.0 ± 1.069

3.6 *Hydrophytes*

The concentration of metals were maxium in water (Cu, 0.074; Zn, 0.788; Pb, 0.864 ugml) and soil (Cu, 1112.10; Zn, 3228.00; Pb, 95.00 μg/g) at Dahigora, the effluents discharge zone. The concentration of metals was also maximum in plants (2055.1, 661.0, 1194.0, 560.5 μg/g of Cu; 1751.2, 1065.0, 176.1, 380.0 μg/g of Zn and 82.1, 29.0, 40.4, 82.0 μg/g of in *V. spiralis, P. pectinatus, P. pusillus* and *H. vertucellata* respectively) collected (Tables 6 & 7).

Table 7 : Quantity of Copper (Cu) in the experimental tank and Cu-accumulation by *Hydrilla verticellata*

Exposed Cu-concentration	Accumulated Cu-Concentration in whole plant (μg/g)
Control	16.0 ± 4.005
1 ppm	602.0 ± 0.178
5 ppm	2796.0 ± 0.09
10 ppm	5914.0 ± 0.287
15 ppm	8830.0 ± 2.250
20 ppm	12787.0 ± 0.527

3.7 *Zooplankton Community Structure*

The zooplankton community constituted 5.17 per cent to 19.40 of the total plankton hauled and its density ranged between 3 μL^{-1} and 42 μL^{-1} . The minimum density was recorded at Dahigora and maximum at Namkum. In all the sites the members of Protozoa dominated which was followed by Rotifera and Copepoda. The Protozoa members reported at different sites were *Arcella discoides, Difflugia corona, D. labostoma, Centropyxix aerophila, c. ecornis, C. constricta* and *C. arcellodes.* Rotifera memebrs were dominated by *Asplanchna prodonta.* Other forms recorded were *Lecane ohioensis prodonta. L. luna, L. depressa, Euchlanis dilatat, Brachionus calyciflorus* and *B. forficula.* Copepoda members were represented by Nauplius larva, *Cyclops. sp. Mesocyclops leuckarti* (Munshi *et. al.,* 1990).

3.8 *Accumulation of Heavy Metals in Fishes (Table 4)*

The concentration of heavy metals in some tissue of *Labeo pungusia, Glassogobius giuris and Puntius ticto* collected around the effluents discharge zone have been estimated by Atomic Absorption Spectrophotometer. Liver accumulated higher amounts of these metals followed by kidney and gills. The heavy metal concentration in water were in order of Fe > Cu > Pb > Zn > Cd. But the accumulation of these metals in tissues of *Labeo pungusia, Glossogobius giuris and Puntius ticto* (as whole) were in the order of Fe > Cu > Zn > Pb > and Cd. Concetration of Pb in water was high in comparison to Zn but in fish tissues the case was reversed. Bryan (1976) also reported that the average concentration of Pb in phytoplankton and seaweed was 4 mg kg^{-1} dry weight whereas the corresponding values for filterfeeders, omnivores and carnivores were 3.25, 3.5 and 4.25 mg kg^{-1} respectively. The concentration of Cu, Zn, Pb, Fe and Cd was high in liver in comparison to kidney and gill in *L. pungusia* and *G. giuris* is in accordance with the work of Establier (1970), Brown *et. al.* (1986) and Ruparelia *et. al.* (1987). Wharfe and Van den Brock (1977) reported Cu in the liver of eel, whiting and flounder from Medway Esturay (U.K.) 5-60 times greater than those in muscle. Brooks and Rumsay (1974) reported average Zn concentration 8, 76, 78, 24, 93, 73 & 22 mg kg^{-1} in muscle, liver, kidney, heart, gonad, spleen and gills of eight sps. of Newzealand seafish (Olson *et. al.*, 1973; Singh, Varma & Munshi, 1990).

3.9 *Scanning Electron Microscopic Evaluation of Effects of Low pH on Gills of Fishes (Figs. 1-3)*

On exposure of fish to acidic water of different pH values several changes take place in the organization of gills. (1) The lamellae swell up and fuse with each other decreasing interlamellar spaces (Fig. 3a-b). (2) The mucus galand openings, increases in numbers indicating hyperactivity, secreting copious amount of mucus. (3) At low pH of 5.3 the branchial epithlium ruptures to expose blood vessels (Fig.-2) (Munshi & Singh, 1992).

Similar histopathological changes have been reported in gills of fishes in response to different irritants (Eller, 1975; Roy *et. al.*, 1986). Leino *et. al.*, (1987) have observed the reduction in interlamellar and interfilamentar space of the gill of *Semotius margarta* (Cope) in an acidic Canadian lake. Further, they have observed fusion of gill filaments. McDonald (1983) reported ion losses due to acidification. Hyperplasia and reduction of filament surface lead to reduction of ion loss through the gills.

Hypertrophied mucous glands and excessive mucus production in gills have been observed by Daye and Garside (1976) in *Salvelnus fontinalis* (Mitchill) exposed to water pH 5.2. According to McDonald (1983) it is a highly appropriate physiological response of fishes to environmental conditions. Such polysnious, which trap and concentrate cations, help in ion exchange. Chevalier *et. al.* (1985) suggested that the combination of pH, low calcium, and increased quantum of metal ions might together contribute to a more severe gill pathology than only low pH.

3.10 *Integumentary Chromatophores and Mucous Glands of Fish as Indicator of Heavy Metal Pollution*

Exposure of *L. rohita* to 0.4 ppm $CuSO_4$ showed a change in the number and area (pigment dispersion or aggregation) of chromatophores and mucous gland during different hours of exposure. During early hours of exposure there is gradual increases in the number and decrease in the area of chromatophores. But during late hours, i.e., 24 and 48 hrs. there is increase in the number as well as area of chromatophores. From this study it appears that $CuSO_4$ has aggregating action on chromatophores. Such phenomenon involves nervous pathway to the central nervous systems, then either to endocrine glands affecting the chromatophores through blood borne hormones or by efferent nervous pathway directly to the chromatophores where chemical mediators are liberated from the nerve terminals. There is considerable variations among fishes to the normal mechanism of control. The great majority of teleost fishes possess dineuronic melanophores, having double innervation with separate dispersing and concentrating fibers (Singh, 1989, Singh & Munshi, 1992).

3.11 *Gill Morphometric Evaluation of Pollution*

The change in the gill dimensions of *Channa punctata* of average weight 54 g, exposed to a sublethal concentration of 0.7 ppm Copper sulphate and 0.26 ppm mercuric chloride for 7 days have been studied.

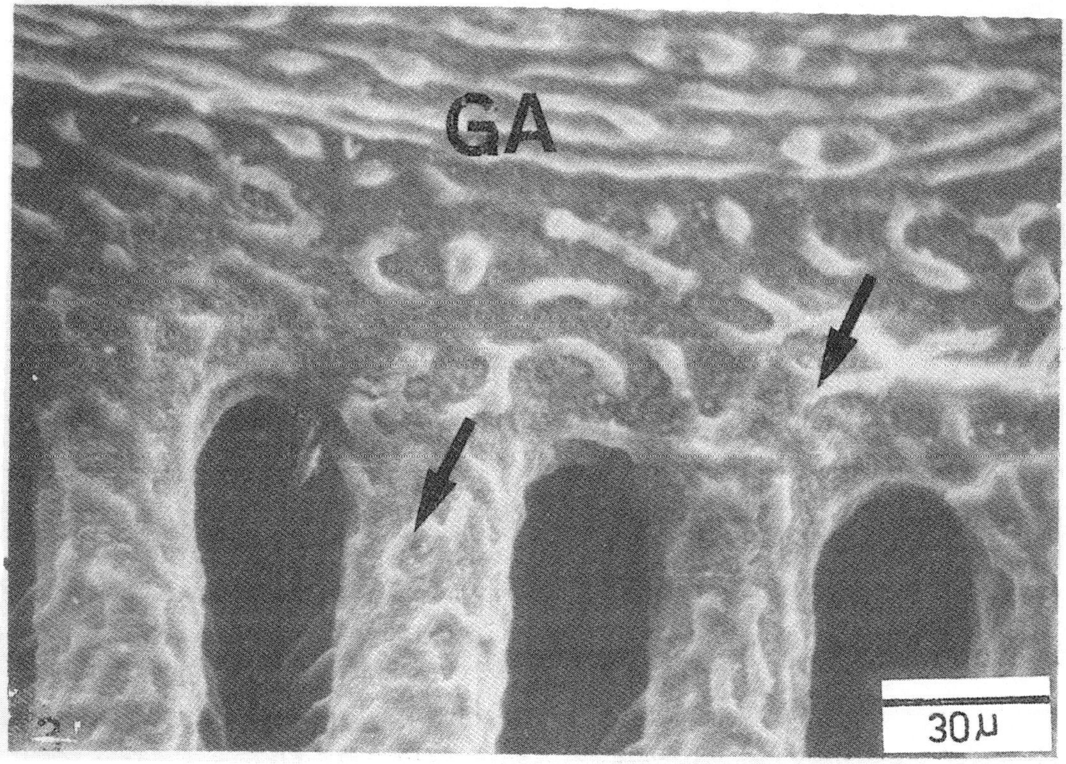

Fig. 1 : Scanning electron photomicrograph of gill head and filament surface of *C. punctata* showing thickened microridges of surface epithelium after 96 h exposure at pH 6.0
GA - Gill arch, GF - Gill filament, Arrow Microridge.

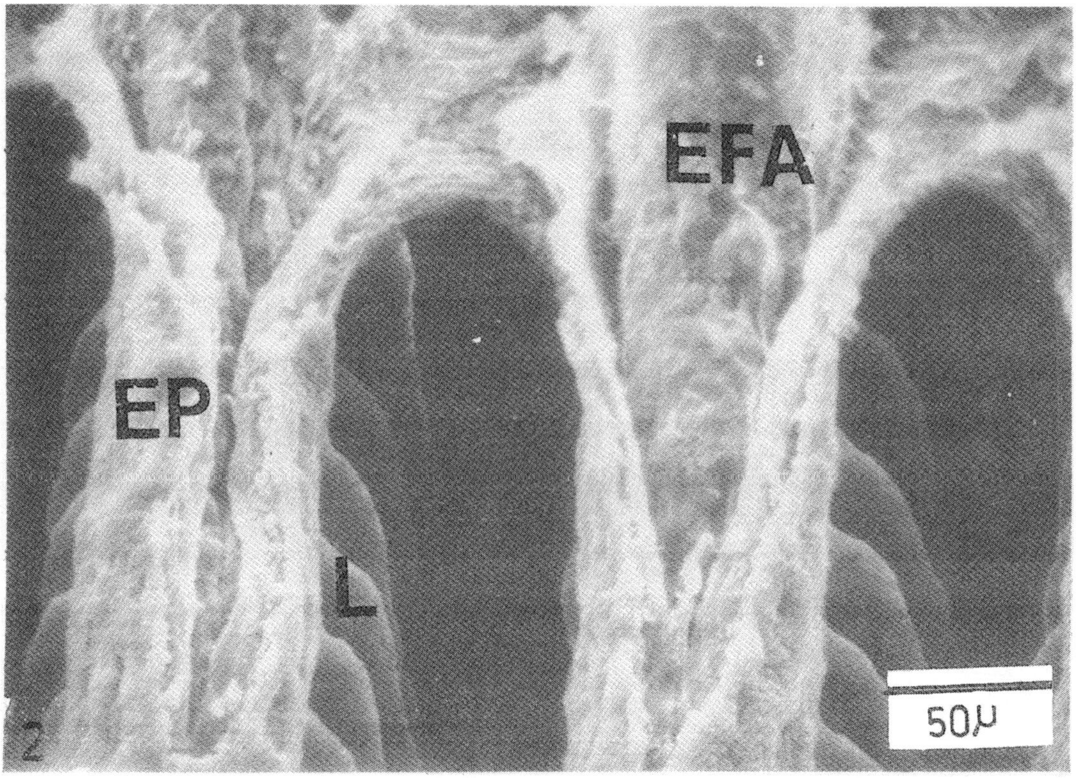

Fig. 2 : SEM photomicrograph showing rupture of filament épithelium exposina efferent filament artery after exposure of fish to pH 5.3 for 96 h. EFA - Efferent filament artery, EP - Epithelium, L - Secondary gill lamella.

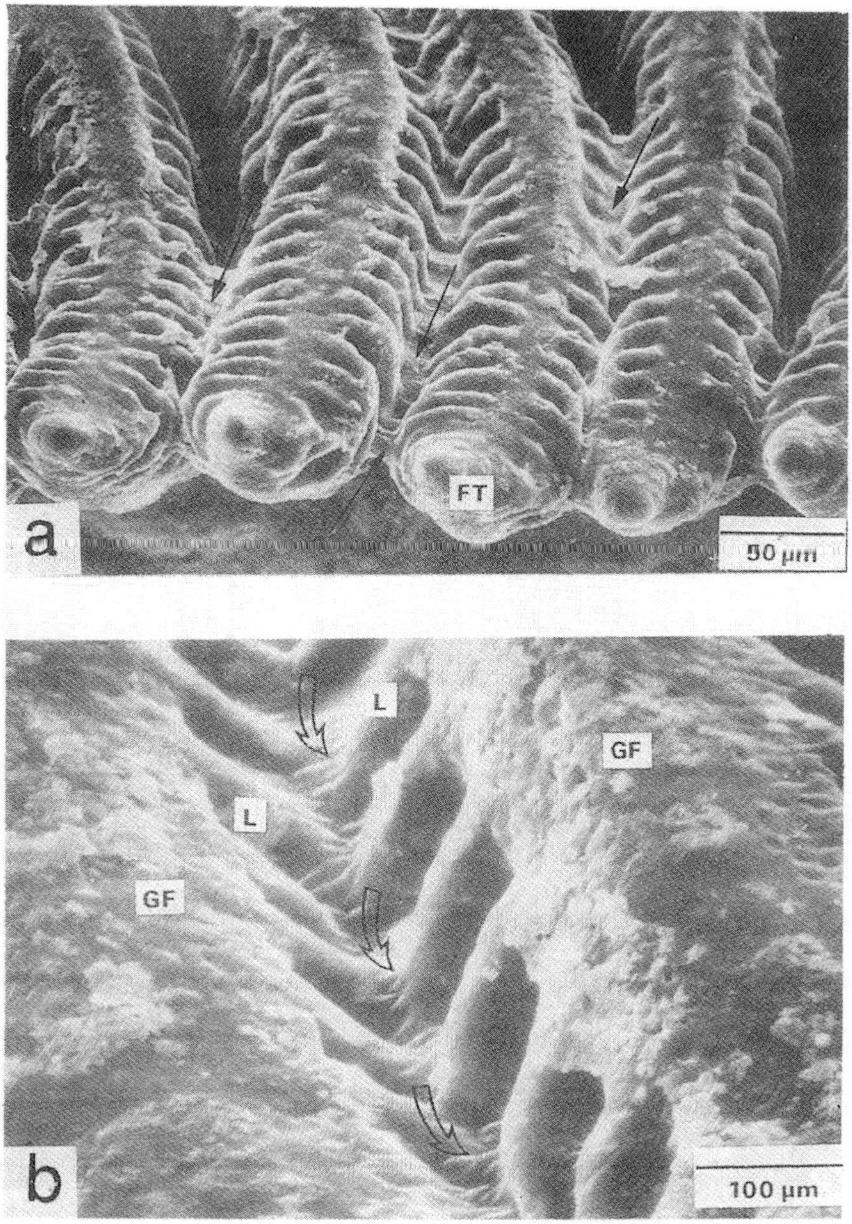

Fig. 3 : (a) Fusion of lamellae of filament tips of *Channa punctata* after 360 h exposure to pH 6.6. FT - Filament tip, Arrow - Lamellae of adjacent filaments fused. bar = 50 μm.
(b) Two adjacent gill filaments of *Channa punctata* showing series of fused lamellae after 15 days exposure to pH 6.0 GF - Gill filament, L - Lamellae, bar = 100 μm.
(After Munshi and Singh, 1992)

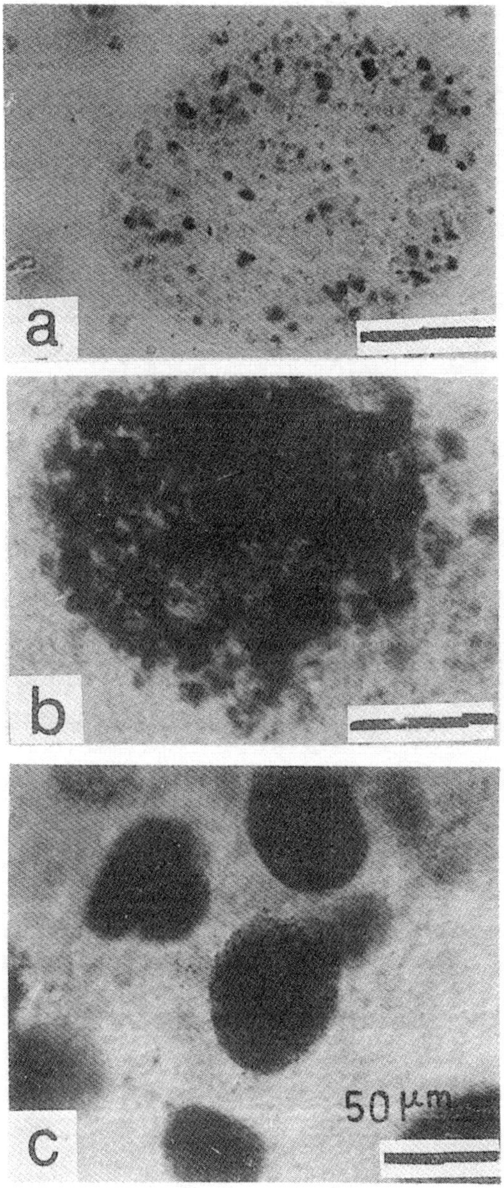

Fig. 4 : (a) Thin section of a part of spleen of Control fish showing the Prussian blue reaction in a large macro phase scale. bar = 30 μm.

(b) Smear of the spleen showing the Prussian blue reaction in the macrophages of fish exposed to 0.248 ppm of $HgCl_2$ for 7 days. bar = 30 μm.

(c) The Prussian blue granules have aggregated into large mass in the macrophages of Spleen of fish exposed to 0.248 ppm of $HgCl_2$ for 15 days. bar = 50 μm.

The gill area/g body wt, has decreased significantly ($p<0.001$) to 49.160 mm^2/g in case of copper and 53.703 mm^2/g in mercury respectively from the control value of 78.415 mm^2/g. (Singh and Munshi, 1990).

3.12 *Oxygen Consumption in Fishes under Sublethal Levels of Copper Sulphate Concentration in Water*

Oxygen Consumption and Ventilation rate in *Channa punctata* (Bloch) and *Mystus vittatus* were studied following exposure to Sublethal Levels of concentration of heavy metals by Singh and Singh (1979) and Singh (1996).

The effects of sublethal concentrations (0.5-2.0 ppm, LC_{50} : 2.13 ppm) of $CuSO_4$ on oxygen consumption and ventilation rate in *Channa punctata* were studied at various periods of short (hours) and long (days) term exposures. After 24 h acclimatization in different sublethal solutions of $CuSO_4$, VO_2 decreased significantly and the ventilation rate increased. On short-term exposure to different sublethal solution of $CuSO_4$, VO_2 showed no clear-cut trend and the ventilation rate increased significantly ($p<0.001$). After long-term (20 days) exposure of 0.5 ppm $CuSO_4$, VO_2 decreased gradually from 75.33 $mlO_2/kg/h$ (on 1st day) to 44.49 $mlO_2/kg/h$ (on 20th day) of exposure from the control value of 90.66 $mlO_2/kg/h$ (Table 5). Skidmore (1970) evaluated the effects of zinc sulphate on respiration and osmoregulation in rainbow trout, while Sellers *et. al.*, (1975) worked on the ventillatory activity, blood O_2 and pH of the same fish.

3.13 *Macrophages and Xenotoxicity*

Macrophages are mononuclear phagocytes present in spleen, kidney, heart and the mesentery of fish (Ellis *et. al.*, 1976) with high pinocyteic property (Ellis, 1977). Normally they are not found as a component population of circulating leucocytes.

Acid phosphatase and Perl histochemical tests (Pearse, 1968) were performed on smear squash and sections of the spleen and kidney to demostrate the density and distribution of macrophages and their intensity of reaction in the control and experimental fish (Munshi *et. al.*, 1990).

The Perl reaction showed the Prussian blue granules in the splenic and renal macrophages which indicate the presence of iron in ferric (Fe^+) form temporarily stored in the macrophages as ferritin and heamosiderin owing to phagocytosis of affected erythrocytes (Fig. 4a-c). There was a marked increase in these granules in all the exposed fish but they were more densely packed in the macrophages of the fish exposed for 15 days than in those exposed for 7 days. (4 b & c). Increased intensity of Purssian blue reaction in the splenic and renal macrophages after exposure to mercury suggests destruction of RBC in the blood due to mercury pollution. By 15 days the granuls aggregated to large blue coloured masses.

4. Water Management Strategy

We also must deal with the problem of diminishing waters in the Subernarekha river in unison with the control of industrial pollution load. Increasing the water volume in the rivers will help to dilute various pollutants. Water flow can be increased in interlinking all the small river channels in the area. In order to reduce pressures of water demand on the river, small reservoirs, tanks and bundhs should be suitably constructed at various places to entrap the rain water. This is specially important for the Chotanagpur plateuau, where rainfall is restricted only to a certain period of the year. The stored water can be utilized for irrigation and other purposes and can be released into the rivers during dry months. A very big dam is under the process of construction at Chandil, without proper ecological assessment. The recycling of waste in industries is most important.

5. Biologic Control of Heavy Metal Pollution

Levels of three heavy metals (Cu, Zn & Pb) were determined in sampled water, surface soil and some aquatic macrophytes (*Vallisnaria spiralis, Potamogeton pactinatus, Potamogeton pusillus* and *Hydrilla verticellata*) collected from three sites of river Subernarkekha at Ghatsila, a nutural sink for the effluents of Hindustan Copper Limited (HCL) and Uranium Corporation of India (UCI). The concentration of metals were maximum in water (Cu, 0.704; Zn, 0.788; Pb, 0.864 g/ml) and soil (Cu, 1112.10; Zn, 3228.00; Pb, 95.00 g/g) at Dahigora, the effluents discharge zone. The concentration of metals was also maximum in plants (2055.1, 661.0, 1194.0, 560.5 g/g of Cu; 1751.2, 1065.0, 176.1, 380.0 g/g of Zn and 82.1, 29.0, 40.4, 82.0 g/g of Pb in *V. spiralis, P. pectinatus, P. pusillus and H. verticellata* respectively) collected from Dahigora. The test plant (*Hydrilla verticellata*) absorbed Cu in laboratory scale model of batch system and therefore suggestive of a scavenger (Tables 6, 7) (Datta Munshi, Dutta and Singh, 1989).

6. Acknowledgement

This research has been possible because of committed work of several research personnels in the team - Dr. Narendra K. Singh. Dr. Manish C. Varma, Dr. Anita Pandey, Dr. Alaknanda Singh, Dr. G.R. Dutta, Dr. Mahesh Singh, Dr. Gopal K. Kumar, Dr. Prabhat K. Roy. The authors acknowledge the financial support received from the C.S.I.R., Govt. of India. This review paper was prepared with the financial assistance received from INSA, New Delhi.

References

APHA (1981). Standard Methods for Examination of Water And Waste Water, American Public Health Association (APHA)

Brooks, R.R. and Rumsey, D. (1974). Heavy metals in some New Zealand commercial sea fishes. New Zealand Journal of Marine and Freshwater Research 8, 155-166.

Brown, M.W., Thomas, D.G., Shurben, D., Solbe, J.F. De, L.G., Kay, J., and Cryer, A. (1986). A comparison of the differential accumulation of cadmium in the tissue of three species of freshwater fish *Salmo gairdeneri, Rutilus* and *Noemachilus barbatulus.* comp. Biochem. Physiol. 84C, 213-217.

Bryan, G.W. (1976). Heavy metal contamination in the seafish. In *Marine Pollution* (Ed. Johnson, R.) Academic Press, London, p. 185-291.

Chevalier, G., Gauther, L & Moveau, G. (1985). Histopathological and electron microscopic study of gills of brook trout *Salvelinus fontinalis* from acidified lakes. *Canadian Journal of Zoology.* 63. 2062-2070.

Datta Munshi, J., Dutta, G.R. and Singh, N.K. (1989). Biological control of heavy metal pollution by some aquatic macrophytges in river Subernarekha. J. Freshwater Biol. 1(1) ; 25-31.

Daye, P.G. and Gardside, E.T. (1976). Histopathologic changes in superficial tissues of brook trout, *Salvelinus fontinalis* (Mitchill) exposure to acute and chronic levels of pH. Can. J. Zool., 54. 2140-5155.

Dora, M.M. and Roy, N.N. (1987). Investigation of water quality of Subernarekha River for irrigation. Indian Health, 29: 292-298.

Ellis, A.E. ; Munro, A.L.S. and Roberts, R.J. (1976). Defence mechanism in fish. 1. A study of the phagocyte system and the fate of intraperitoneally injected particulate material in the plaice (*Pleuronectes platessa*), Journal of Fish Biology, 8, 67-78.

Ellis, A.E. (1977). The leucocytes of fish : A review. Journal of Fish Biology, 11, 451-491.

Eller, L.L. (1975). Gill lesions in freshwater teleosts. In *The Pathology of Fishes* (Ribelin, W.E. and Magaki, G. eds.) pp. 305-330, Madison : University of Wisconsin Press.

Establier, R. (1970). Copper, Iron, Manganese and Zinc levels in serveral organs of *Thunnus thynnus*. Inv. pesq. 34. 399-408.

Leino, R.L. Wilkinson, P. and Anderson, J.G. (1987). Histopathological changes in the gills of pearl Dale, *Somotilus margarita* and fathead minnows, *Pinephales promelas* from experimentally acidified Canadian lakes. Canadian Journal of Fisheries and Aquatic Sciences, 44, 126-134.

McDonald. D.G. (1983). The effects of H^+ upon the gills of freshwater fish. Canadian Journal of Zoology, 61, 691-703.

Mishra, A., Munshi J.S.D. and Singh Mahesh (1994a). Heavy Metals Pollution of River Subernarekha in Bihar Part I : Industrial Effluents. J. Freshwater Biol. (3) : 197-199.

Mishra, A., Munshi J.S.D. and Singh Mahesh (1994b). Heavy Metal Pollution of River Subernarekha in Bihar Part II : River Sediments. J. Freshwater Biol. 6(3) : 201-204.

Mishra, A., Munshi J.S.D. and Singh Mahesh (1994c). Heavy Metal Pollution of River Subernarekha in Bihar Part III : River Water. J. Freshwater Biol. 6(3) : 205-208.

Munshi, J.S.D. & Dube, S.C. (1973). Oxygen uptake capacity of gills in relation to body size of air-breathing fish, *Anabas testudineus*. Acta. Physiol. 44(2) : 113-123.

Munshi, J.S.D., Mishra, A., Datta Munshi, J., Singh, N.K., Dutta, G.R. and Singh, M. (1990). Estimation of heavy metal pollution and status of plankton community in the river ecosystem of Subernarekha. In *Impact of Environment on Animals and Aquaculture* (Manna, G.K. and Jana, B.B. eds), pp. 101-106 Saraswaty Press.

Munshi, J.S.D., Singh, N.K., Mishra, N., and Ojha, J. (1990). Cytology of macrophages in normal and mercury treated air-breathing fish, *Channa punctata* (Bloch.) J. Fish Biol. 37 : 651-653.

Munshi, J.S.D. and Singh, A. (1992). Scanning electron microscopic evaluation of effects of low pH on gills of *Channa punctata* (Bloch). Journal of Fish Biology, 41, 83-89.

Muir, B.S. and Hughee, G.M. (1969). Gill dimension for three species of tunny, *J. Exp. Biol.* 51 : 271-285.

NEERI, (1986). Manual on water and waste water analysis : National Environmental Engineering Research Intitute, Nehru Marg, Nagpur-20.

Olson, K.R., Bergman, H.L. and Fromm, P.G. (1973). Uptake of methy1 mercury chloride by trouts : a study of uptake pathways into the whole animal and uptake by erythrocytes in vitro : J. Fish. Res. Bd. Can. 30. 1293-1299.

Pandey, Anita, Kunwar, G.K., and Munshi, J.S.D. (1990). Integumentary chromatophores and Mucous glands of fish scale as indicator of heavy metal pollution : a simple scale method for evaluation of quality of freshwater. J. Freshwater Biol. 2(2) : 117-122.

Pearse, A.G.E. (1968). *Histochemistry, Theoretical and Applied* London. J. and A. Churchill.

Ruparelia, S.G., Verma, Y., Pandey, C.B., Sathawara, N.G., Shah, G.M., Parikh, D.J. and Chatterjee, B.B. (1987). Trace metal contents in water and the fish *Sarotherodon mossambica* of Kankaria. Environ. & Ecol. 5(2) : 294-296.

Roy P. K., Munshi, J.S.D. and Datta Munshi, J. (1986). Scanning electron microscopic evaluation of effects of Saponin on gills of climbing perch, *Anabas testudineus* (Bloch) (Anabastidae, Pisces), Indian Journal of Experimental Biology, 24 : 511-516.

Singh, A., (1996). Oxygen consumption and ventilation rate in *Channa puntuata* (Bloch.) following exposure to sublethal levels of concentration of copper sulphate. J. Freshwater Biol. 8(2) : 89-93.

Singh, A., and Munshi, J.S.D. (1990). The effects of heavy metals on gills of fishes : a morphometric study. J. Freshwater Biol. 2(1) : 69-72.

Singh, Alakhnanda (1989). Heavy metal induced pathomorphological alterations of mucous glands on the skin of a cat fish, *Heteropneustes fossilis* (Bloch.). J. Fresh water Biol. 1(2) : 97-102.

Singh, Alaknanda, and Munshi, J.S.D. (1992). Effects of sublethal mercuric chloride exposure on melanophores of a catfish, *Heterpneustes fossilis* (Bloch). J. Environ. Biol. 13(4) : 303-308.

Singh, D.K. and Singh, C.P. (1980). Pollution studies on river Subernarekha around industrial belt of Ranchi (Bihar) India J. Environ. Health. 33 : 26-33.

Singh, N.K., Dutta, G.R., Singh, M. and Kishore, K. (1989). Assessment of water quality of the River Subernarekha at Ghatsila with reference to heavy metal pollution. Journal of Freshwater Biology, 1, 113-120.

Singh, A. & Munshi, J.S.D. (1990). The effects of heavy metals on gills of fishes : a morphometric study. J. Freshwater Biol. 2(1) ; 69-72.

Singh, A. and Munshi, J.S.D. (1993). Melanophore diamensions of skin of *Hateropneustes fossilis* (Block) as indicator of pH of ambient fresh waters. Proc. Zool. Soc., Calcutta, 46(2) : 99-103.

Singh, A. & Munshi, J.S.D. (1996). Scanning electron microscopic evaluation of the effects of short and long term exposure of copper and mercury on *Channa punctata* (Bloch). *J. Freshwater* 8(1) : 51-55.

Singh, N.K., Varma, M.C., and Munshi, J.S.D. (1990). Accumulation of copper, Zinc, lead, iron and cadmium in certain freshwater of river Subernarekha, J. Freshwater Biol. 2(3) : 189-193.

Singh, S.R. & Singh, B.R. (1979). Changes in oxygen consumption of a siluroid fish (*Mystus vittatus*) put to different concs. of some heavy metal salts. Indian J. Exp. Biol. 17(3) : 274-276.

Sellers, C.M.J., Health, A.G. & Bass, M.L. (1975). The effect of sublethal concs. Of Cu and Zn ventilatory activity, blood O_2 and pH in rainbow trout, *Salmo gairdneri.* Water Res., 9: 401-408.

Skidmore, J.F. (1970). Respiration and osmoregulation in rainbow trout with gills damaged by Zinc sulphate. J. Exp. Biol. 52 : 481-494.

Varma, M.C., Singh, N.K., and Munshi, J.S.D. (1991). Algal species diversity as an index of water quality or river Subernrekha at Ghatsila, J. Freshwater Biol., 3(3) : 235-239.

Varma, M.C., Singh, N.K., and Munshi, J.S.D. (1992). Impact of industrial effluents on auatic biota of river Subernrekha at Ghatsila. Proc. Acad, Environ. Biol., 1(1) : 87-94.

Wharfe, J.R. and Brock, W.L.F. Vanden. (1977). Heavy metals in macro-invertabrates and fish from the low Medway esturay, Kent Mar. Poll. Bull. 8.31-34.

Proceedings of the National Seminar on Environmental Biology, April 03-05, 1998
***Edited By :* A.K. Aditya & P. Haldar**
DAYA PUBLISHING HOUSE, DELHI

Digestive Enzymes in *Porcellio laevis* (Isopoda : Crustacea) as Indicator of Heavy Metal Toxicity in Soil

S. Joy, S.K. Maity and V.C. Joy
Department of Zoology (DSA, UGC)
Visva-Bharati University, Santiniketan 731 235, West Bengal

Abstract

Terrestrial isopods are ideal bioindicators of environmental pollution because of their unique capacity to store heavy metals in hepatopancreas, the main site of production of digestive enzymes. This paper demonstrates chronic sublethal toxicity of cadmium sulphate, on the rate of production of digestive enzymes in *Porcellio laevis*, a cosmopolitan detritivore arthropod species reared on leaf litter under microcosm conditions. The rate of production of enzymes (µg/mg tissue/hour) in uncontaminated litter was in the order amylase>cellulase>protease. Exposure to cadmium treated litter produced strong physiological stress as evidenced from higher amylase and cellulase activity within 7 days. Thereafter the level of enzymes decreased when compared to the rate in untreated control set. The low profile of protease activity was not affected by the pollutant. The enzyme activity ratios with respect to total tissue proteins also confirmed higher amylase and cellulase activites when exposed to the toxicant. The observed toxicity manifestation through physiological response has potential scope for biomonitoring the stress of heavy metal pollution in soil subsystem.

Introduction

Heavy metals accumulate in the environment in polluting level due to technological activities like mining, industries, combustion of fossil fuels, garbage and waste water disposal. The presence of pollutants can be highly toxic to the normal functioning of terrestrial ecosystem. Soil is inhabited by a wide range of organisms comprising different trophic levels. Among them phylum Arthropoda outnumber others in density and diversity. Predominance of detritivore and fungivore groups namely Collembola, Acari, Ispoda and Diplopoda occur in most natural soil conditions. Available literature demonstrate that heavy metal contamination is detrimental to microbial and funal communities in soil affecting their population density, species diversity and biological functioning (Frank et al., 1993 ; Larink, 1997). protection of ecological functions is guarenteed only if the ecological structure and equilibrium are maintained. Heavy metal containing organic waste materials namely flyash, industrial effluents, sewage sludge and garbage materials are used in our country as fertilizer in agricultural fields and for soil management and land reclamation purposes. These are rich in inorganic nutrients but at the same time contain high percentage of heavy metals like Copper, Cadmium, Zinc, Mercury, etc. It has been demonstrated that soil animals are ideal bioindicators of chemical pollutants and that arthropods like isopods are suitable model for ecotoxicological experiments (Van Straalen and Van Gestel, 1993). Studies on heavy metal toxicity in terrestial arthropods probably started with the discovery by Wieser

(1961) that hepatopancreas of *Porcellio scaber* (Isopoda) collected from abandoned mine sites in England contained highest concentration of copper ever recorded from the soft tissues of any terrestrial animal. Hopkin and Martin (1984) found that accumulation and storage of heavy metals in the hepatopancreas of Isopoda is a detoxification mechanism and when concentrations are extremely high the organ becomes white or grey in colour instead of the normal yellow/ brown colour. Recent studies by Coenenstass (1997) and Alikhan (1997) have demonstrated that heavy metal toxicity can affect the metabolic rate and reproductive success in terrestrial isopods. In the present study an attempt has been made to compare the stress of chronic toxicity of a heavy metal on the rate of production of digestive enzymes in *Porcellio leavis,* a cosmopolitan isopod species, under microcosm conditions.

Materials and Methods

Adults of *P. laevis* collected from natural litter beds were acclimatized in the laboratory on decomposing leaf litter. They were used for the microcosm experiments in small polythene rearing vessels containing leaf litter of *Anthocephalus chinensis* pretreated with 0.01% (100 ppm) solution of cadmium sulphate in double distilled water. Both treated and untreated control sets were maintained under constant temperature (27±1°C) and moisture (approx.20%) conditions in a B.O.D incubator. Batches of specimens drawn at regular intervals from the microcosms were used for quantitative estimation of amylase, cellulase and protease following Bernfield (1955), Sadasivam and Manikam (1992) and Anson (1938) respectively with minor modifications wherever necessary. Whole body homogenate of known live wight of the animals centrifuged at 7,000 rpm at 4°C was used as the enzyme source. The catalysing property of amylase in the hydrolysis of starch into reducing sugar was estimated spectrophotometrically against standard curve of pure maltose. Similarly, cellulase activity was estimated by incubating the enzyme source with carboxymethyl cellulose and optical density compared against standard curve of D-glucose. Protease activity was measured from the amount of amino acids released from bovine serum albumin and a standard curve of pure glycine was used for conversion of value into µg/mg tissue/hour. The enzyme activities were expressed with respect to the amount of total tissue proteins estimated by Lowry et al., (1951) method. The observed differences between sets in the rate of production of enzymes and amount of tissue proteins were compared statistically using 't' test.

Results and Discussion

Findings are summarised in Table 1 and Figures 1 and 2. The direct and residual toxicity of cadmium sulphate at different concentration with respect to percentage mortality of *P. laevis* are given in Figure I. It was found that the heavy metal at 0.1 per cent did not produce any direct mortality of the animals with in 96 hours of exposure to the treated leaf litter. Similarly, the residual toxicity of the treated litter gradually decreased with respect to time and dilution of the toxicant. Figure I also gives an idea of the rate of faecal production by *P. laevis* at different intervals. The animals registered high rate of feacal production in the untreated control set. On the other hand, the surviving individuals of the treated set produced very little faecal pellets. However, a gradual increase in the amount of faecal pellets with respect to progress of litter decomposition and dilution of the toxicant could be noticed. This study showed that cadmium sulphate could produce some chronic toxicity in the animals through food source. Therefore, a sublethal concentration of cadmium sulphate at the rate of 0.01 per cent (100 ppm) was selected for the detailed microcosm experiment on the assumption that higher survivability of the animals

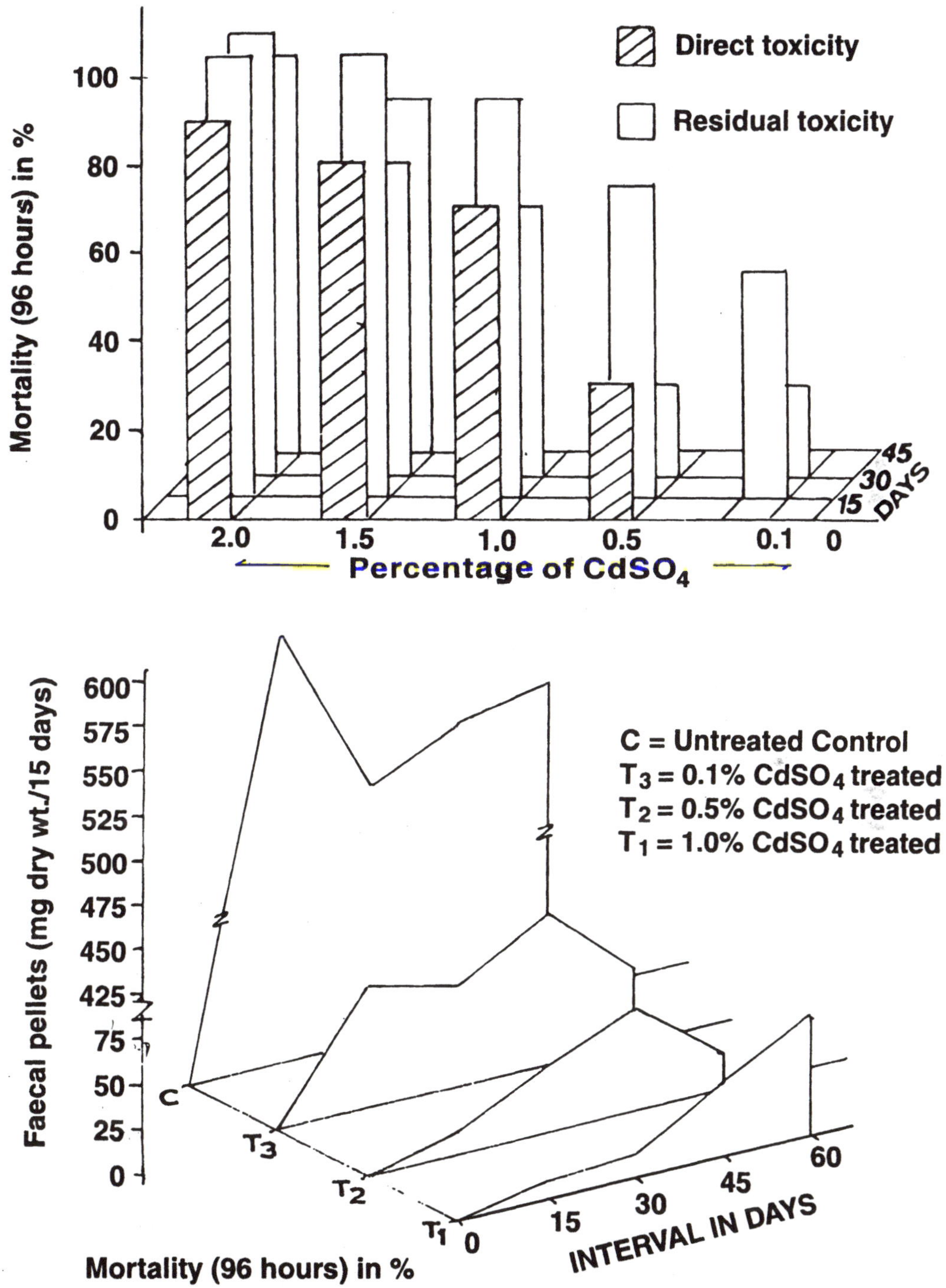

Fig. 1 : Toxicity of cadmium sulphate on *Porcellio laevis* with respect to mortality and faecal production rates.

on less toxic leaf litter would facilitate higher rate of feeding so as to produce some chronic effect on digestive enzyme activity. Coughtrey et.al. (1997) suggested that terrestrial isopods are good biological monitors of environmental contamination by heavy metals. The rate of production of digestive enzymes by decomposer arthropods can vary depending on the nature of food eaten. Joy et. al. (1998) demonstrated that decomposer arthropods like *Anoplodesmus saussurei* (Diplopoda) are able to modify their enzyme production rates depending upon the availability of nutrients in the substratum. Rushton and Hassall (1983) showed that isopods are very selective about the variety of food they ingest, the preference being expressed by more frequent visits if given a choice and by higher rates of ingestion. Joy and Joy (1991) while comparing the rates of ingestion, ejestion and assimilation of *P. laevis* on the decomposing leaf litter of 10 different tree species observed increased rates of faecal production with respect to the duration of decomposition in most litter types. The average rate of production of important digestive enzymes in the whole body homogenate of *P. leaveis* (µg/mg tissue/hour) under untreated microcosm conditions are presented in Figure 2. The animals showed comparatively higher amylase activity when compared to cellulase and protease. Joy and Joy (1997) while comparing the effect of controlled microbial decomposition of leaf litter on feedng and digestive enzyme production by *P. laevis* observed about 3 to 5 times higher amylase activity than cellulase. This is in full confirmity with the fact

Table 1 : Rate of production of digestive enzymes in *Porcellio laevis* under heavy metal toxicity

Digestive enzymes	Experimental sets	Interval in days		
		7 days	14 days	28 days
Amylase	Control	130.81 ± 38.48	163.71 ± 86.45	218.75 ± 19.74
(µg/mg	treated	318.73 ± 33.52	180.38 ± 7.09	176.25 ± 21.14
tissue /	X1–X2	187.92	16.67	42.50
hour)	t-value	21.02	0.547	4.15
	df	25	13	14
	p	< 0.001	n.s	0.< 0.001
Cellulase	Control	72.33 ± 13.44	51.00 ± 3.54	77.87 ± 21.34
(µg/mg	treated	92.37 ± 8.93	40.86 ± 5.52	50 25 ± 16.92
tissue /	X1–X2	20.04	10.14	27.62
hour)	t-value	4.92	4.29	2.87
	df	29	13	14
	p	< 0.001	< 0.001	0.< 0.025
Protease	Control	4.87 ± 1.15	8.44 ± 3.17	3.80 ± 0.79
(µg/mg	treated	5.00 ± 2.64	6.90 ± 1.73	4.70 ± 1.89
tissue /	X1–X2	0.13	1.54	0.90
hour)	t-value	0.18	1.33	1.39
	df	30	17	18
	p	n.s	n.s	n.s
Tissue	Control	57.11 ± 13.57	50.11 ± 7.75	44.20 ± 5.82
proteins	treated	56.08 ± 9.44	38.56 ± 6.00	41.56 ± 4.25
(µg/mg /	X1–X2	1.03	12.25	2.64
tissue	t-value	0.29	3.11	0.552
	df	73	50	48
	p	n.s	0.< 0.001	n.s

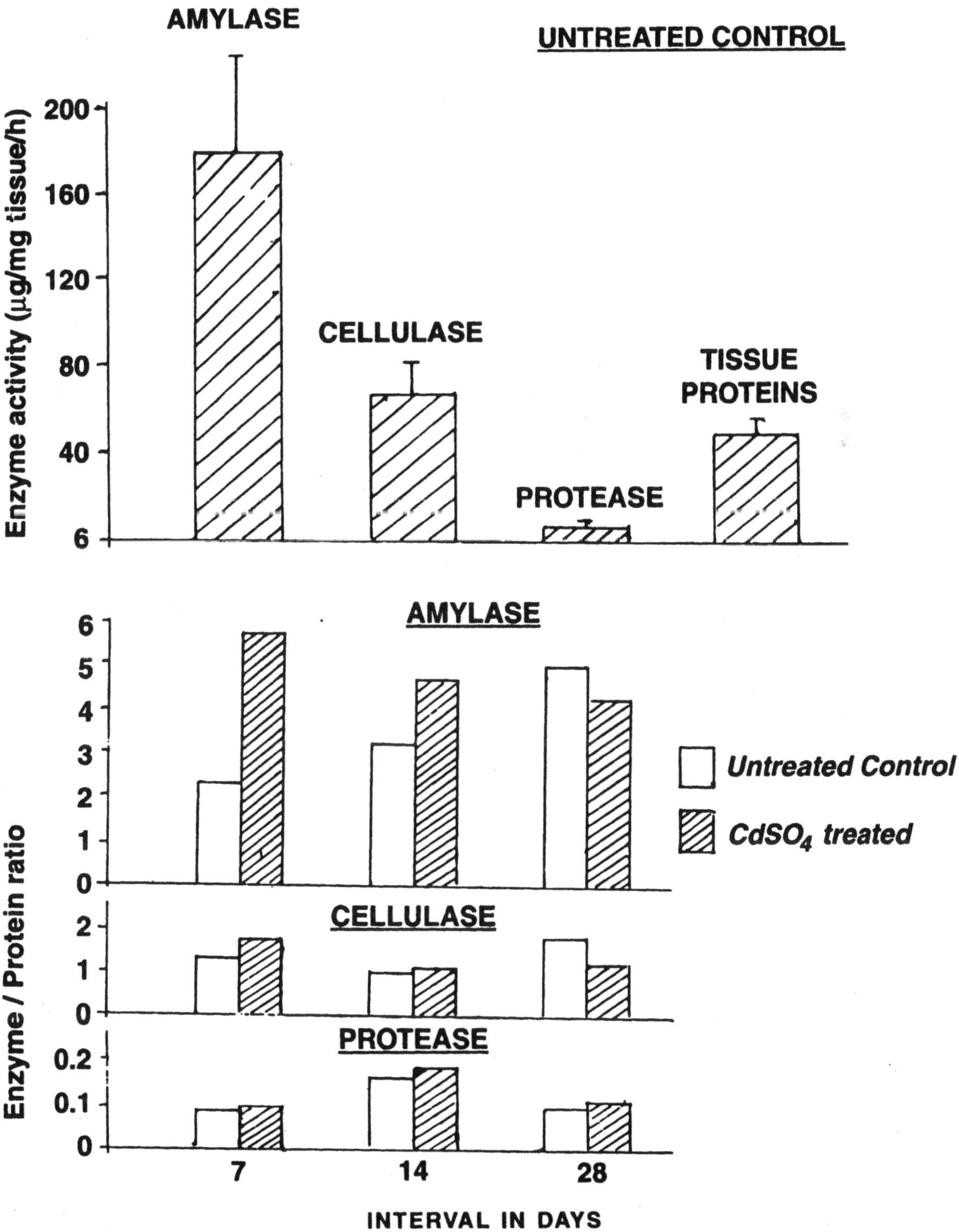

Fig. 2 : Digestive enzyme activity by *Porcellio laevis* in untreated control and cadmium sulphate treated microcosms.

that detritivore arthropods are able to utilize easily assimilable carbohydrate resources of decomposing organic matter for their growth and reproduction. It is known that these animals depend on symbiotic gut microflora for digestion of cellulose and other complex materials and that they always resort to coprophagy as a means to suppliment nutritional requirements. The high rates of faecal production and poor assimilation efficiencies of isopods and diplopods as demonstrated by several workers also indicate their inability to digest complex materials. In order to evaluate the effect of pollutants under near natural conditions, model ecosystems namely 'microcosms' are designed that are easy to manage and manipulate in the laboratory. Van Straalen and Van Gestal (1993) showed that microcosm studies are useful in validating the results of simple tests to complex systems. Bengtsson et al. (1988) conducted microcosm studies on the influence of soil animals and heavy metals on litter decomposition and demonstrated direct relationship between population abundance, heavy metal concentration and weight loss of litter.

A detailed consideration of the experimental data with respect to rate of production of digestive enzymes and the amount of tissue proteins in *P. laevis* under $CdSO_4$ treated and untreated microcosm conditions is given in Table - 1. The animals registered highly significant increase in the rate of production of amylase and cellulase wthin 7 days of exposure to the treated leaf litter. This may be due to strong physiological stress of the toxicant. It is known that soluble portion of food enter into the lumen of hepatopancreas for digestion and absorption and therefore direct illeffect of the toxicant at the site of production of digestive enzymes is possible. The rate of production of amylase in normal animals gradually increased with respect to the progress of litter decomposition indicating availability of more nutrients. However, in the treated animals the amylase activity decreased within 14 days and the absence of any further increasing trend indicated prolonged illeffect of $CdSO_4$. A more or less similar decrease in cellulase activity suggest that the toxicant may have some inhibitory effect on the symbiotic gut microflora also, because the isopods are poor producers of cellulase by themselves. Table - 1 also shows that protease activity was negligible when compared to the carbohydrate digesting enzymes. Similarly, the amount of tissue proteins also fluctuated without much difference except for a significant decline in the treated animals during 14th day of the experiment. The enzyme activity ratio (Figure-II) also confirmed the illeffect of $CdSO_4$ as evidenced by initial increase in enzyme activity followed by reversed effect due to chronic toxicity. According to Alikhan (1997) recognition of environmental toxicants depends upon the observation on biological damage by the background concentration of toxicants to which organisms are exposed in the contaminated ecosystem. Hopkin and Martin (1984) described 2 main routes for absorption of heavy metals in the digestive tract of terrestrial isopods, the metals in liquid fraction is directly absorbed by 'S' and 'B' cells of hehatopancreas, whereas enzymatic and microbial action release metals from solid food in the hind gut and passes into the haemolymph. Simkiss (1983) suggested that zinc, cadmium and copper may pass across cell membrances more than a million times faster than sodium and postassium. Inspite of such excellent capacity to concentrate heavy metals the isopods are yet to be extensively used in toxicity evaluations. The present quantitative comparison of digestive enzymes in *P. laevis* appears to be a potential biomonitoring test criteria for estimating the bioactivity of heavy metal pollution in soil subsystem. According to Lebrun (1980) this type of ecologically oriented direct approach gives more valuable and quick information on the toxicity and persistence of pollutants over the conventional and expensive chemical methods, since the results are based on actual bioactivity of the toxicants.

Conclusions

Microcosm experiments demonstrated sublethal toxicity of cadmium sulphate on the rate of production of digestive enzymes in *Porcellio laevis,* a cosmopolitan detritivore arthropod species. When reared on uncontaminated leaf litter the animals secreted higher amount of amylase than cellulase and protease. Short duration exposure to the heavy metal contaminated litter could induce significant increase in amylase and cellulase activities. However, chronic exposure to the treated litter decreased the enzyme activities to significantly low levels. The observed toxicity manifestation through physiological stress has potential scope for biomonitoring heavy metal pollution in soil.

Acknowledgements

The authors are grateful to the Head, Department of Zoology (DSA, UGC) Visva-Bharati University for laboratory and other facilities. S.J. is thankful to the UGC and Visva-Bharati for Research Associateship (Environmental Biology) under the DSA programme.

References

Alikhan, M.A. (1997). A comparative study of the fecundity and juvenile survival in *Porcellio spinicornis* (Porcellionidae : Isopoda) from a nickel-copper contaminated and uncontrolled site in Northeastern Ontario, Canada. Fourth symposium on the Biology of Terrestrial Isopods. Haifa-Israel, 1997. 15pp.

Anson, M.L. (1938). Gen. Physiol., 22:79.

Bengtsson, G., Berden, M. and Rundgren, S. (1988). Influence of soil animals and metals on decomposition processes : A microcosm experiment. J. Environ. Qual. 17, 113-119.

Bernifield, P. (1955). In: Methods on Enzymology Vol.-I. Eds : Colowick S.P. and Kaplon. N.O., Academic press, New York. p. 149.

Coenen-stass, D. (1997). Effects of heavy metals (Pb, Cd) on the metabolic rate of *Porcellio scaber* (Oniscidea, Isopoda, Crustacea), Fourth Symposium on the Biology of Terrestrial Isopods. Haifa-Israel, 1997 21pp.

Coughtrey, P.J., Martin, M.H. and Young, E.W. (1977). The woodlouse *Oniscus aselle* as a monitor of environment cadmium levels. Chemosphere. 6: 827-832.

Frank, V., Tamova, G., and Takacsova, (1993). Effects of cadmium and mercury on growth and differentiation of *Trichoderma viride.* Zentralbl. Microbiol. 148: 229-232.

Hopkin, S.P. and Martin, M.H., (1984). Heavy metal in woodlice. In : The biology of terrestrial isopods. Eds. Sulton, S.L. and Holdich, D.M., Calrendon Press, Oxford, 143-166.

Joy, S. and Joy, V.C. (1991). Food preference and growth rate of *Porcellio laevis* reared on decomposing leaf litter. Entomon 16(2) : 107-113.

Joy, S. and Joy, V.C. (1997). Feeding and digestive enzyme production by *Porcellio laevis* under inhibited microbial decomposition of leaf litter. J. Ecobiology, 9,: 177-183.

Joy,S., Sarkar, K., Pramanik, R. and Joy, V.C. (1998). Digestion of carbohydrate resources on decomposing leaf litter by *Anoplodesmus saussurei* (Humb.), (Polydesmida: Diplopoda). Ecol. Env. and cons. 4(3) : pp. (117-122).

Larink, O. (1997). "Springtails and mites : Important knots in the food web and soils" In : Fauna in soil ecosystems, recycling processes, nutreint fluxes and agricultural production. Marcel Dekker, Inc. 225-264.

Lebrun, P.H. (1980). A preliminary study of the use of some soil mites in bioassays for pesticide residue detection. In biology as related to land use practices. Ed. Dindal, D.L., Proc. VIIth Int. Soil.Zool.Colloq., Syracuse. : 42-55.

Lowry, O.H., Rosebrough, N.J., Farr, A.L. and Randall R.J. (1951). Protein measurements with folin phenol reagent. Biol. Chem. 193: 265-275.

Rushton, S.P. and Hassall, M. (1983). Food and feeding rates of the terrestrial isopod *Armadillidium vulgare* (Latreille). Oecologia, 57, 415-419.

Sadasivam, S. and Manickam, A. (1992). Biochemical methods for agricultural sciences. Wiley Eastern Ltd. New Delhi, 246 pp.

Simkiss, K. (1983). Lipid solubility of heavy metals in saline solutions. J. mar. Biol. Assn. U.K. 63: 1-7.

Van Straalen, N.M. and Van Gestel, C.A.M. (1993). Soil invertebrates and microorganisms. Handbook of ecotoxicology. Vol-I. Ed: Calow, P., Blackwell scientific publications, Oxford., : 251-277.

Wieser, W. (1961). Copper in isopods. Nature, Lond. 191:1020.

Proceedings of the National Seminar on Environmental Biology, April 03-05, 1998
***Edited By* : A.K. Aditya & P. Haldar**
DAYA PUBLISHING HOUSE, DELHI

Influence of Environmental Factors on the SGR of *Catla catla* Grown in Sewage Ecosystem Investigated with Multiple Regression Analysis

A.K. Roy, M. Rout, P.K. Saha*, A.K. Datta**
Computer Section, CIFA, Kausalyaganga, Bhubaneshwar-751002 (Orissa)
* Pond Monitoring and Environment Division, CIFA, Bhubaneswar-751002
** Wastewater Aquaculture Division of CIFA, Rahara - 743186, West Bengal

Abstract

In order to quantify the impact of various environmental factors on the Specific Growth Rate (SGR) of *Catla catla,* multiple regression analysis was performed on a data set that was collected from a sewage fed carp culture experiment. The dependent variable considered for the study was SGR and the independent varibales were DO, BOD, pH, P_2O_5,NH_4-N, alkalinity, chloride, free CO_2, temperature, plankton & benthos of pond water and initial weight of the test species. Following various methods of stepwise regression a model was identified which relates SGR of *Catla catla* to initial weight, temperature and benthos (R^2=0.4578; P<.01; n=39). Out of the above mentioned twelve variables, BOD, P_2O_5, CO_2, temperature, benthos & plankton showed positive impact and the rest negative impact on the SGR of *Catla catla.* To test the violations of regression assumption and validation of the selected variables examined the fitted model through basic regression diagnostics. Unexplained causes of variations are also discussed. It appeared that the temperature followed by initial weight are the most important variable to influence the SGR of Catla in sewage ecosystem.

Introduction

It is believed that in a pond environment a host of factors interact dynamically and influence the fish growth and production. Some of the environmental factors are uncontrollable while other management variables like stocking density, manure, fertiliser, feed application, aeration, pond design etc. may be controlled as per requirement. But interaction of various factors and their resulting effects on fish growth are seldom clearly understood. In order to make the behaviour of these systems more predictable which themselves undergo internal changes over the culture cycle, mathematical models capable of describing the fish pond ecosystem precisely are necessary. Some modelling applications to aquaculture relating to the growth of some fishes of commercial importance are known (Allen, et. al., 1984; Bernard, 1986; Balchen, 1987; Cuenco, 1989). Multivariate analysis in aquaculture data have also been attempted (Prein, 1987, 1990, 1993; Milstein et. al., 1993; Mair & Pauly 1993; Costa-Pierce et. al., 1993). In India, multivariate approach of aquaculture data analysis are meagre (Moitra, 1977; Jana and De, 1983; Eknath and Doyale, 1985 and Bhowmik et. al., 1989). In the present study an attempt has been made to analyse a set of data of some aquaculture experiments and to identify the influence of various environmental factors on the growth rate of *Catla catla*, with the help of multiple regression analysis and development of suitable models for predicton of growth.

Materials and Methods

Dataset for the present multiple regression analysis was taken from the culture experiments conducted at Wastewater Aquaculture Division of Central Institute of Freshwater Aquaculture, Rahara W.B. The original experiments were conducted to test the efficacy of sewage in carp culture compared to other treatments. The experiments were conducted in the ponds (.06 to .33 ha) during August, 1982 to March, 1983 treated with sewage effluent only. All the ponds were stocked at a constant stocking density of 8000 fingerlings/ha and at a constant stocking ratio of catla 2 : rohu 2.5 : mrigal 2 : silver carp 2 : grass carp 0.5 : common carp 1. No supplementary feed was done in the ponds. Altogether thirtynine sets of monthly growth sampling data on both length and weight, as well as, corresponding representative values of biotic and abiotic factors were collected. In the present study dependent varibale considered was specific growth rate (SGR) which was calculated as follows.

$$SGR = \frac{\log W_2 - \log W_1}{t} \times 100$$

where W_1 = Initial weight of the fish (g)

W_2 = weight of the fish (g) at the end of the growth experiment

t = Duration of growth experiment (days)

The independent variables considered were DO, BOD, pH, P_2O_5, NH_4-N, alkalinity, chloride, free CO_2, temperature, plankton and initial weight of the stocked speices. To quantify the effect of biotic and abiotic factors on the SGR of catla a hypothised multiple linear regression model of the form.

$$Y = a+b_1X_1+b_2X_2+........+b_{12}X_{12}$$

was assumed where Y = Specific growth rate (% per day) of *Catla catla* and b_1, b_2.....and b_{12} are the regression coefficients of the independent variables X_1, X_2,...........X_{12} respectively,
again where

X_1 = DO in ppm

X_2 = BOD in ppm

X_3 = pH

X_4 = NH_4-N in ppm

X_5 = P_2O_5 in ppm

X_6 = Alkalinity in ppm

X_7 = Chloride in ppm

X_8 = Free CO_2 in ppm

X_9 = Temperature in °C

X_{10} = Plankton in No./l (log transformed value)

X_{11} = Benthos in No./m^2 (log transformed value)

X_{12} = Initial weight of Catla

Collected data were entered in spread sheets. Multiple regression analysis using procedures like forward selection, backward elimination and stepwise regression were performed with the help SPSS (MS-Windows 6.0) software package. Statistical criteria for entry into and removal from an equation were those of default values of criteria subcommand of regression procedure denoted by key words PIN (0.05), POUT (0.10) and tolerance (0.01). Related parital regression values, coefficient of determination values (R^2), and other assoicated test statistic were used as per the requirement of the selection procedures. Examination of residuals were done following the methods of Draper and Smith (1981).

Results and Discussions

Descriptive Statistics

Summary of descriptive statistics of all variables used in the present analysis are presented in Table 1. It can be seen from the table that average values of dissolved inorganic phosphorus and nitrogen content of the experimental pond water are congenial for optimum fish growth. But a lot of fluctuations in the values of temperature, plankton and benthos (log transformed) are observed.

Table 1 : Descriptive statistics of different variables used in the present analysis (n = 39)

Variables	Unit	Mean	Std. Dev.	Minimum	Maximum
Av. P	ppm	0.34	0.45	0.01	2.10
SGR	% Per day	0.43	0.55	0.02	2.20
NH_4N	ppm	0.94	1.22	0.00	6.00
FREE CO_2	ppm	3.45	2.37	1.00	10.00
lnPL	No./l	5.32	1.95	2.30	9.79
lnB	No./l	7.31	1.60	3.91	10.79
DO	ppm	7.33	2.59	1.20	14.40
pH	–	7.97	0.37	7.00	8.50
BOD	ppm	10.01	4.35	3.60	28.00
TEMP	°C	24.10	4.36	17.80	30.20
CHLORIDE	ppm	110.38	12.05	100.00	150.00
INILWT	g	152.66	104.76	17.00	409.10
ALKALINITY	ppm	228.72	65.52	126.00	390.00

SGR : Specific growth rate
lnB, lnPL : Log transformed value of Benthos and Plankton.
INILWT : Initial weight of catla.
TEMP : Temperature

Correlation Matrix

Correlation martix of all the thirteen variables used in the present analysis for derivation of model for growth rate of catla are presented in Table 2. From the correlation matrix it can be seen that SGR is having highly significant positive relationship with both the DO and temperature and also significant positive relationship with P_2O_5 which are realities. The significant positive relationship of SGR with benthos may be a chance event. Among independent variables significant positive relationship of P_2O_5 with each of BOD and pH are worth mentioning but the strong

Tab e 2 : Correlation matrix of the variables used for model of growth rate of *Catla catla* (n = 39)

	BOD	DO	pH	P_2O_5	NH_4N	ALK	CHLO	INILWT	lnB	lnPl	CO_2	TEMP	SGR
BOD	1.000												
DO	−0.284	1.000											
pH	0.248	−0.095	1.000										
P_2O_5	0.353*	0.188	0.348*	1.000									
NH_4N	0.202	0.208	0.057	0.668*	1.000								
ALK	0.528*	−0.223	0.456**	0.536**	0.266	1.000							
CHLO	0.034	−0.239	0.056	−0.282	−0.201	−0.139	1.000						
INILWT	0.369*	-0.499**	−0.117	−0.029	0.117	0.269	0.182	1.000					
lnB	0.008	[illegible]	0.089	0.490**	0.481**	0.163	−0.204	0.051	1.000				
lnPL	0.089	[illegible]	0.033	0.288	0.236	0.148	0.180	0.223	0.500*	1.000			
CO_2	0.118	−0.326*	−0.065	−0.147	0.063	−0.106	0.164	0.395*	−0.186	−0.191	1.000		
TEMP	−0.297	0.242	−0.052	0.006	−0.216	0.001	−0.038	−0.005	0.053	0.061	−0.060	1.000	
SGR	−0.100	0.410**	0.093	0.362*	0.117	0.059	−0.258	−0.261	0.426**	0.221	−0.078	0.493**	1.000

* Significant at 5% level

** Significant at 1[illegible] level

relationship between NH_4-N and P_2O_5 may be spurious. Significant positive relationship between benthos with each of DO, P_2O_5 and NH_4-N are justified. The positive relationship of alkalinity with pH is reasonable but the same with each of BOD and P_2O_5 are of no consequences. Significant positive relationship between initial weight and BOD and negative relationship with DO may be acceptable. Significant negative relationship of free CO_2 with DO is logical and positive relationship with initial weight may be acceptable. Significant positive relationship between plankton and benthos may exist. It is observed from multiple regression analysis with all the twelve variables that BOD, P_2O_5 , temperature, benthos and plankton have positive impact and the rest are having negative impact on the SGR of *Catla catla.*

Selection of Predictor Variables and Model Building

Summary statistics of multiple regression analysis using backward elimination procedure performed for the purpose of extracting information from fish culture data and for identifying and quantifying the role of predictor variables are presented in Table 3.

Table 3 : Summary of backward elimination of variables at different steps with F, Sign. F, R^2 and SE values

Step No.	Variables removed	F	Sign. F	R^2	SE
1-12	Entered all variables	2.9295	0.0106	57.48	0.4301
13	Removed X_3 (pH)	3.3179	0.0054	57.47	0.4221
14	Removed X_1 (DO)	3.7793	0.0026	57.44	0.4146
15	Removed X_6 (Alkalinity)	4.3198	0.0012	57.28	0.4082
16	Removed X_4 (NH_4–N)	4.9431	0.0006	56.86	0.4033
17	Removed X_5 (P_2O_5)	5.7197	0.0003	56.36	0.3990
18	Removed X_{10} (PLANKTON)	6.4017	0.0002	54.55	0.4008
19	Removed X_7 (CHLORIDE)	7.5559	0.0001	53.38	0.3928
20	Removed X_8 (CO_2)	8.7371	0.0001	50.69	0.4050
21	Removed X_2 (BOD)	10.8397	0.0000	48.16	0.4093

Variables in the equation : *INILWT, TEMP, BENTHOS*

From the table it is clear that at steps 1 to 12 all the twelve variables were entered for regression analysis which together explains 57.48% of the total variation in the specific growth rate of *Catla catla.* Subsequently, the backward elimination procedure proceeded upto step 21 based on POUT criterion and removed the varibales X_3, X_1, X_6, X_4, X_5, X_{10}, X_7, X_8 and X_2 at step nos. 13, 14, 15, 16, 17, 18, 19, 20 and 21 respectively. It can be seen from the Table 3 that removal of variables like pH, DO, alkalinity, NH_4N, P_2O_5 from step 13 to 17, there is a reduction of only 1.12% in R^2 value indicating their negligible influence on SGR of catla. But removal of variables like plankton, chloride, free CO_2 and BOD at steps from 18 to 21 has reduced R^2 value to the extent of 8.20%, rightly justifying the relative importance of these variables in sewage-fed fish culture system. Lastly, after elimination, the procedure selected the remaining variables X_{12} (Initial Wt), X_9 (Temp) X_{11} (Benthos) as the predictor variables in the regression equations at setp no. 21. Therefore, the fitted multipe linear regression equation takes the form:

$$\hat{Y} = -1.7949+0.0587\ X_9 + 0.1418\ X_{11} - 0.0015\ X_{12}$$

Interpretation and Validation of the Model Using Regression Diagnostics

The aforementioned equation is capable of explaining 48.16% of the total variation in the growth rate of *Catla catla*. It is noteworthy that selection of temperature (24.27%) in the prediction equation as significantly important variable is fully justified due to the important role of this parameter in sewage fed pond culture dynamics. INITWT (7.84%) is also an important variable selected for the regression equation which is negatively related to the growth of the test species. Benthos (16.05%) may be having some indirect role on the growth of catla. Similar study on the same species with eleven independent variables could account for only 44.84% of the total variability of SGR (Roy et al., 1997). Incidentally, both the forward and stepwise regression procedures also identified the same variables along with same coefficients and constants as done by backward elimination procedure. It is interesting to note that biotic factors (plankton, benthos and initial weight) contributed 25.7% of the total variability and the rest 31.78% are due to the environmental factors considered in the study.

The present model could explain only 48.12% of the total variability of SGR of catla. The presence of multicollinearity, as is evidenced from Table 2, is taken into account during selection of variables through use of tolerance values, the most frequently used indicator of interdependency. Casewise plot of standardised residuals do not indicate any major violation of assumption nor the presence of outliers. It is seen that 22 out of 39 cases have yielded negative residuals which means that the observed values for those cases are less than the expected. The model can be termed as a good fit because a good number of residuals were observed close to zero. Further, distribution of standardised residuals and normal partial plot of standardised dependent variables do not show any violation of normality. Again, plotting of standardised residuals against standardised predicted values do not show any discernable pattern indicating the violation of assumptions of homogeneity of variance.

Unexplained Variance

One of the reasons of unexplained variance may be caused by cases in which important variables like solar radiation, wind velocity and photoperiod are not taken into consideration. Another reason may be the imprecision and inaccuracy of measuring average values per interval. In the present case some of the measured variables which are highly fluctuating such as DO and pH, their average effect on fish growth during the growth interval is expressed in the form of a single mean value, based on available individual measurements in the interval. These sample values can not be termed representative in true sense, therefore, it is likely that they will not adequately reflect the true effects on the SGR.

Conclusion

In the present study, identification and quantification of the prediction variables like temperature, initial weight and benthos through multiple regression analysis fitting of model will be helpful for prediction of growth of *Catla catla.*

Acknowledgement

The authors are thankful to the Dr. S. Ayyappan, Director, CIFA, Kausalyaganga for constant encouragement and for extending all facilities.

References

Allen, P.G., Botsford, L.W., Schuur A.M. and Johnston, W.E. (1984). Bio-economics of aquaculture. *Dev Aquacult. Fish. Sci.* **13,** 351 p. Elsevier, Amsterdam.

Bernard, D.R., (1986). A survey of the mathematical models pertinent to fish production and tropical pond aquaculture, pp. 231-243. *In* J.E. Lannan, R.O. Smitherman and G. Tchobanoglous (Eds.) *Principles and practices of pond aquaculture.* Oregon State University Press, Corvalis, Oregon 252 P.

Bhowmick, M.L., Johnston, Rai, S.P., Roy A.K. and Tripathi, S.D. (1989). Role of some morpho-edaphic factors in fish production in rural ponds. *Proc. National Seminar on Forty Years of Freshwater Aquaculture.* held at CIFA, Kausalyaganga, Bhubaneswar. pp. 94-97.

Costa-Pierce, B.A., Van Dam, A. A. and Kapeleta M. V. (1993). Multiple regression analysis of growth of Tilapia rendalli in polyculture with *Oreochromis shiranus* as affected by water quality and pond inputs, p. 88-96. *In* M. Prein, G. Hulata and D. Pauly (eds.) *Multivariate methods in aquaculture research: Case Studies of Tilapias in Experimental and Commercial Systems.* ICLARM *Stnd. Rev.* **20,** 221 P.

Draper, N.R. and Smith, H. (1981). *Applied Regression Analysis,* 2nd Edition, New York, John-Wiley & Sons., Inc.

Cuenco, M.L. (1989). *Aquaculture systems modelling: an introduction with emphasis on wormwater aquaculture. ICLARM Stnd. Rev.* **19,** 46 P.

Eknath, A.E. and Doyle, R.W, (1985). Maximum likelihood estimation of "unobservable" growth and development rates using scale data: application to carp aquaculture in India. *Aquaculture* **49**: 55-71.

Jana, B.B. and De, U.K. (1983). Primary productivity of phytoplankton, environmental covariates and fish growth in West Bengal ponds under polyculture system. *Int. Rev. Gesamt. Hydrobiol.* **68**: 45-58.

Mair, G. and Pauly, D. Multivariate analysis of growth of Tilapia, *Oreochromis aureus* and *O. niloticus,* Cichlidae, reared in recirculating systems (1993). P. 97-104. *In* M. Prein, G. Hulata and D. Pauly (eds.). *Multivariate Methods in Aquaculture Research Case Studies of Tilapias in Experimental and Commercial Systems. ICLARM Stnd. Rev.* **20,** 221 P.

Milstein, A., A. Goldman and G. Hulata. (1993). Multiple regression analysis of fish seed and profit in commercial fish farms in Israel, pp. 161-177. *Ibid.*

Moitra, S.K. (1977). An econometric model of prawn production. *Proc. World Mariculture Soc.* **8**: 741-146.

Prein, M. (1987). Multivariate models of Tilapia and carp growth as an application of microcomputers in aquaculture research. *Naga ICLARM Q.* **10**(2): 7-8.

Prein, M. (1990). Multivariate analysis of Tilapia growth experiments in ponds: case studies from Phillipines, Israel, Zambia and Peru. Christian-Albrechts-Universitat. Kiel, 125, Doctoral dissertation.

Prein. M. and Pauly, D. (1993). Two new approaches for examining multivariate aquaculture data: the "Extended Bayley Plot" and Path analysis, pp. 32-49. *In* M.

Prein, G. Hulata and D. Pauly (eds.). Multivariate Methods in Aquaculture Research: Case Studies in Tilapias in Experimental and Commercial Systems. *ICLARM Stnd. Rev.* **20,** 221 P.

Prein, M. and Milsteinn A. (1993). Multiple regression and Rath analysis of Tilapia growth in commercial fish farms in Israel, P. 178-188.ICLARM Stnd. Rev. **20,** 22 P.

Roy, A.K., Rout, M., Saha, P.K. and Datta, A.K. (1997). A multiple regression model for prediction of growth of Catla catla reared in sewage-fed ponds (Abs.). *In : All Orissa Conference on "Decission making models and their application"* organised by Deptt. of Statistics, Utkal University, 20-21 December, 1997.

Proceedings of the National Seminar on Environmental Biology, April 03-05, 1998
Edited By : A.K. Aditya & P. Haldar
DAYA PUBLISHING HOUSE, DELHI

Impact of Anthropogenic Activities on Hugli Estuary with Special Reference to the Distribution Pattern of Intertidal Macrozoobenthos

Sobhana Paul*, Abhijit Mitra & N.C. Nandi***
* Zoological Survey of India, M-Block
New Alipore, Calcutta-700053
** Department of Marine Science, Calcutta University,
35, Ballygunge, Circular Road, Calcutta - 700 019

Abstract

The Hugli river system has been under continuous anthropogenic stress due to discharge of domestic and industrial waste waters as well as various other human activities. An attempt has been made herein to show these activities happening in and around the river and to highlight the general physico-chemcial and biological (macrozoobenthos) conditions of the three different zones viz., upper, middle an lower zones. Data pertaining to faunal composition of intertidal macrozoobenthos, a parameter widely used as bioindicator in pollution studies, show differential distribution in the three zones of the estuary and especially at the outfall regions. In general, macrozoobenthos are represented mainly by molluscs, crustaceans and polychaetes in the upper and middle zones and by actinarians, nemertines, molluscs, polychaetes, crustaceans, echiurids, sipunculids, asteroids and ophiuroids in the lower zone. Maximum number of benthic species (142) was recorded from the marine zone i.e. zone III while the existence of poor biological quality in zone I (38) and Zone II (64) was observed. The impact of pollution was found to be localized around outfalls At the sewage outfall region of Tolly's nullah, the occurrence of tubificid worms *Limnodrilus hoffmeisteri* in large numbers was noticed along with a low value of Shannon Wiener species diversity (0.18-0.24)

Introduction

The Hugli, a tidal estuary, is a major source for fisheries and is intensively used for industries, transport and various other purposes. It is also used for disposal of huge amount of waste waters arising from industries and municipalities. These human activities coupled with reduction in freshwater flow over the years, all impinge upon the water quality and benthic habitat. Pollution studies on river Hugli are made mostly at the industrial outfalls (Seth and Bhaskaran, 1950; Ghosh and Basu 1968; Basu *et al.*, 1973; Ray *et al.*, 1977; Ghosh *et al.*, 1977; 1980; Ray *et al.*, 1979) but pollution due to non point sources and domestic outfalls has received very little attention (Basu, 1965). The objective of the present study is to summarise all these anthropogenic activities occurring in the tidal stretches of Hugli river along with the general physico-chemical conditions, as well as, distribution pattern of benthic organisms occurring in the estuary.

Study Area and Methodology

The tidal stretch of Hugli estuary which is divided into three zones on the basis of salinity gradient (Dutta *et al.*, 1973) viz., Zone - I (upper zone) extending from Nabadwip to Baranagar (Calcutta), Zone - II (middle zone) from Baranagar to Diamond Harbour and Zone - III (lower zone) comprising entire Sunderbans and a tract below Diamond Harbour, is considered for the present study. Both benthic faunal analysis and physico - chemical analysis were done collecting samples from each of the upper, middle and lower zones of the estuary. Benthic samples were collected using a box type sampler covering an area of 15 cm x 15 cm. Water samples collected in clean glass bottles were analyzed as per standard methods (APHA, 1985) and/or using a photometer SQ 118 (Merck, Germany). A sewage outfall region (Tolly's nullah outfall region) was thoroughly investigated to show the pollution problem affecting the benthic biota.

Utilization Scenario of River Hugli

Throughout its entire stretch, both tidal and non tidal, the river is used for bathing and washing clothes, as well as, for drinking water supply, religious activities and industrial uses viz., distillery, textile, viscose rayon, pulp and paper, tannery, integrated steel, etc. (Ghosh *et al.*, 1973). Besides all these domestic and industrial uses, the river is exploited for various waste water discharge purposes. There are numerous open drainages from the different towns located on its banks opening into the river directly as non point sources of pollution. Sewage and domestic wastes are discharged directly to the estuary. Both treated and untreated industrial waste water from pulp and paper, distillery and yeast, matches, petroleum oil refinery, textiles, metal and steel, paints and varnish industries etc. discharge into the estuary.

Results

The data on nine physicochemical characteristics of water recorded from upper, middle and lower zones are represented in Table -1. The water temperature follows nearly the same trend in all the three zones, minimum during the winter season and maximum in summer. The pH value remains alkaline throughout the year the three zones ranging from 7.2 to 8.9. Salinity was found to increase gradually from the upper to lower zone viz., 0.02-0.04, 0.02-1 and 1.36 to 17.89 ppt respectively. The dissolved oxygen value in the three zones remained more than 4 mg/l except in the effluent outfall regions where dissolved oxygen was present only in traces, ranging between 0.9 mg/l adn 4.5 mg/l at Tolly's nullah outfall. The alkalinity value varied between 60 mg/l (in zone I and II) and 280 mg/l (zone II). The phosphate content was higher in the upper and middle zone (0.1 to 0.8 mg/l and 0.03 to 0.9 mg/l respectively) compared to the lower zone (1.85 to 3.65 µg/l). The nitrate content of the estuary also followed the same trend as phosphate. BOD and COD values were relatively high in the upper and middle zones compared to the lower zone of the estuary.

In Tolly's nullah outfall region the dissolved oxygen content was seen to fall as low as 0.9 mg/l also. In most part of the year, the dissolved oxygen value remained below 3 mg/l. The BOD and COD values were recorded as high as 50 mg/l and 179 mg/l respectively at this station.

Qualitative analysis of benthos presented a composite structure of the fauna in the three zones (Table 2). The species composition markedly varied according to the zones. Altogether 38 species in the upper zone, 64 species in the middle zone and 142 species in the lower zone, representing different taxonomic groups have been encountered. The benthic macrofauna inhabiting the upper and middle zone are though dominated by molluscs, polychaetes and

Table 1 : Physico-chemical Characteristics of the Estuary

	Parameters	Upper Zone (I)	Middle Zone (II)	Lower Zone (III)
(1)	Water temperature (°C)	19-33	20.7 – 35.5	22.8 – 33.00
(2)	pH	7.3 – 8.9	7.2 – 8.8	7.99 – 8.33
(3)	Salinity (ppt)	0.02 – 0.04	0.02 – 1	1.36 – 17.89
(4)	Dissolved oxygen (mg/l)	4.0 – 8.5	4.0 – 7.0	4.87 – 6.82
(5)	Alkalinity (mg/l)	60 – 180	60 – 280	180 – 265
(6)	Phosphate	0.1 – 0.8 mg/l	0.03 – 0.9 mg/l	1.85 – 3.65 μg/l
(7)	Nitrate	1.2 – 6.4 mg/l	2 – 10.7 mg/l	12.8 – 30.08 μg/l
(8)	BOD (mg/l)	1 – 14	1 – 48	2 – 4
(9)	COD (mg/l)	5 – 32	11 – 39	—

Table 2 : Benthic Community of the Estuary

	Class	Upper Zone No. of species	Middle Zone No. of species	Lower Zone No. of species
(1)	Gastropoda	16	31	32
(2)	Bivalvia	6	6	20
(3)	Scaphopoda	–	–	1
(4)	Polychaeta	6	10	45
(5)	Oligochaeta	2	3	1
(6)	Crustacea (crabs)	3	3	11
(7)	Insect larvae	3	3	3
(8)	Actinaria	–	–	3
(9)	Nemertinea	–	–	2
(10)	Echiura	–	–	2
(11)	Sipuncula	–	–	1
(12)	Asteroidea	–	–	2
(13)	Ophiuroidea	–	–	1
(14)	Intertidal fishes	2	8	18
	Total	**38**	**64**	**142**

crustaceans, their composition and density varied from station to station. In the lower zone the benthic macrofauna are represented by molluscs, polychaetes, actinarians, nemertines, crustaceans, echiurids, sipunculids, asteriods and ophiuroids. The gobiid fishes are also invariably represented in the benthic community of the three zones.

The macrobenthic molluscs are not evenly distributed in the estuary. Species like *Littorina scabra, Onchidium typhae, Stenothyra deltae, Thiara (Thiara) scabra, Assiminea francesiae* have wider distribution. The first three speices are found in the middle, as well as, lower zones while *Thiara* is found in upper and middle zones. *Assiminea* has extended its distribution in all the three zones of the estuary. Species like *Bellamya bengalensis, Digoniostoma cerameopoma, Thiara (Melanoides) tuberculata, Thiara (Tarebia) lineatus, Thiara (Tarebia) granifera, Brotia costula, Lymnaea (Pseudosuccinea) accuminata, L. (P) luteola* and *Indoplanorbis exustus* are mainly found in the upper zones with a little extension to the middle zone while species like *Telescopium (T.) telescopium, Cerithidea (C.) cingulata, Natica tigrina, Littorina (Littorinopsis) scabra scabra,* are found mainly on zone III with little extension to zone II or middle zone. Except few species such as *Parreysia favidens, Corbicula straitella, Modiolus straitulus,* bivalves are mainly found in the lower zones, the common forms being represented by *Meretrix meretrix, Pelecyora trigona., Tellina iridescens, Solen brevis etc. Dentalium octangulatum* is the only representative of Scaphopoda found in the lower zone.

Among the polychaetes, majority of species are found in the lower zone and the number of species gradually decreases towards the upper zone. The most commonly occurring species of polychaetes are *Namalycastis fauveli, Namalycastis indica, Dendronereides heterpoda, Naeanthes meggitti, Nephtys oligobranchia, Nephtys polybranchia, Dendronereis aestuarina, Neanthes chingrighattensis, Telehsapia annandalei, Lumbrineris pollydesma, Mastobranchus sp., (cf. indicus)* of which the first six speices are mostly found in zone I and zone II. The intertidal oligochaete fauna is comparatively scanty and are represented mainly by *Branchiura* and *Limnodrilus.*

The crustacean fauna in the estuary is contributed by very few isopods but largely by decapods (crabs) which are the permanent members of the benthic community. The crabs of common occurrence in zone I and zone II are of genus *Sesarma.* Most of the species of crabs are recorded from zone III which include mainly *Sesarma bidens, Metaxplax dentipes, M. distincta. Ocypoda ceratophthalma, Dotilla blanfordi, Macrophthalmus brevis, M. erato* and *Metapograpsus messor.*

Species of actinaria; echiura, sipuncula, asteriodea and ophiuroidea are found in lower zone only which is entirely the marine zone. Actinarians are quite abundant and *Edwardsia zonesi* is a dominant actinarian in the lower zone. Nemertineans, Echiurids, Sipunculids, Asteriods and Ophiuroids are represented by the species of genus *Nemertea, Anelassorhynchus, Phascolosoma, Astropecten* and *Ophiactis* respectively.

The intertidal fishes, viz., mudskipper, gobiids, bleniids and eels are mostly recorded from the lower zones. The commonly occurring gobiid species in the upper zone is *Odontamblyopus rubicundus,* in the middle zone are *Apocryptodon madurensis, Parapocryptes macrolepis, O. rubicundus, Pisdonophis boro, Apocryptes bato, Scartelaos histophorus, Boleophthalmus boddarti, Periophthalmus chrysopilos* and in the lower zone are *P. boro, Periophthalmodon tredecemaradiatas, Pseudapocryptes laceolatus, Taenioides cirratus, Trypanchenichthys typus, Parapocryptes serperaster.* Gobiids and mud eels are mostly seasonal residents of the benthic community.

Pollution Problems Affecting Water Quality and Biota

Physico-chemical and biological studies done around the discharge point of industries and sewage outfalls give a clean picture of the effect of pollution on estuarine biota. It has been found that, of the average daily discharge of effluents from various industries, pulp and paper industry alone contributes 49.86% whereas miscellaneous industries only 22.5% (Ray, 1981). The pollution effect can be illustrated by following examples from industries (Ghosh *et al.*, 1973 Table 3) and from sewage outfall region (Tolly's nullah outfall region, present study).

Table 3 : Pollution Load (in terms of B.O.D.) of Hugli Estuary from Various Sources

	Sources	Vol. of Wastewater Million Gallons Per Day (MGD)	B.O.D.s / Day T / Day
(1)	Pulp and paper	39.29	28.59
(2)	Textile	4.48	7.26
(3)	Distillery	0.145	10.24
(4)	Yeast	0.0035	0.05
(5)	Tannery	2.50	4.54
(6)	Rubber	2.40	0.43
(7)	Paints & Varnish	2.54	1.49
(8)	Match	0.24	0.04
(9)	Shellac	0.09	0.39
(10)	Metal & Steel	2.54	0.20
(11)	Combined wastes of Bagh Canal, Rishra (8 industrial units & municipal wastes)	20.0	12.24
(12)	Hydrogenerated vegetable	3.12	0.56
(13)	Domestic & Municipal	175	40.08
	Total	**252.34**	**106.11**

Source : Ghosh *et al.* (1973)

The daily average pollution load with respect to B.O.D was estimated to be 106.11 of which industrial waste as a whole contribute to 66.03 of daily average pollution load as B.O.D the major contributor being pulp and paper industry.

The physico-chemcial and biological parameters recorded at various sampling stations with respect to sewage outfall (Tolly's nullah) are given in Table 4. It is evident from the Table 4 that the dissolved oxygen values at station 2 remained below the critical limit (1.47 - 3.7 mg/l) coupled with high concentration of BOD (25 - 30.5 mg/l) and COD (38.11-78.29 mg/l) created a stressed condition. The maximum number of benthos (11, 474) was found at station 2 of which tubificid worms *Limnodrilus hoffmeisteri*, a pollution indicator organism contributed more than 80% of total benthos. This prevailing condition at station 2 might be due to receiving of large quantity of sewage through Tolly's nullah, which has also lowered down the species diversity value considerably (0.18 - 0.24).

Table 4 : Hydrobiological Conditions around the Sewage Outfall Region

Station	Season	D.O. mg/l	B.O.D. mg/l	C.O.D. mg/l	Macrobenthos No./m^2	Sp. Diversity
Above	PRM	5.52	7.6	25.19	2,948	2.01
Outfall	M	5.1	5.0	28.09	304.22	1.60
Station 1	PTM	6.67	5.25	38.11	170.19	2.07
At	PRM	1.47	30.5	78.25	5,518.75	0.18
Outfall	M	1.95	25.0	47.78	11,474.92	0.23
Station 2	PTM	3.7	28	38.11	7755.19	0.24
Below	PRM	3.4	8.0	22.56	202.26	2.36
Outfall	M	3.95	7.5	31.77	55.49	2.03
Station 3	PTM	6.15	10.5	38.06	80.16	2.19

An indication of recovering tendency of pollution was observed at station 3 about 100 m downstream of station 2 as apparent from the rise in dissolved oxygen (3.4 - 6.15 mg/l), lowering of BOD (7.5 - 10.5 mg/l) and increase in species diversity value (2.03-2.36). The COD values at station 1 and 3 were low in premonsoon, as well as, monsoon compared to station 2 but in postmonsoon they showed a high value of 38.11 mg/l at station 1 and 38.06 mg/l station 3. In respect of COD, station 3 failed to show much improvement but the pollution indicator organism- *L. hoffmeisteri* was rare and recorded only occasionally in very few number. Variety of benthic fauna such as *Indoplanorbis exustus, Bellamya bengalensis, Neritina violacea, Septaria lineata* etc., found at station 1 completely disapperaed in station 2 and again made their appearance at station 3.

Discussion

Ellis (1937) and Klein (1973) have pointed out that the pH values between 6.7 and 8.4 are suitable while pH values below 5.0 and above 8.8 are detrimental. The pH limit during the present investigation was within this range in middle and lower zone while in upper zone slightly higher pH (8.9) was recorded.

The level of oxygen in any aquatic system is of great importance to the inhabiting organisms. It has been recommended that a minimum of 4 mg DO per litre should be maintained in estuarine and coastal waters for healthy growth of fish population (FWQA, 1968).

In estuarine environment DO concentrations varying from 3 to 6.9 mg/1 have been reported by Zingde *et al.*, (1979). These reports suggest that the estuarine system under present investigation was not under stress with respect to the dissolved oxygen level. However low concentration at the outfall regions does recall attention.

On the basis of alkalinity values, Philipose (1960) has classified the water of India in 3 categories, viz. (a) 4-50 ppm alkalinity as 'low', (b) 50-100 ppm alkalinity as moderate and (c) 100-600 ppm alkalinity as 'high'. Accordingly, the present river with its alkalinity ranging from 60-280 mg/l could be placed in high category of alkalinity types.

With regards to the phosphate content it was recommended that 0.01 mg inorganic phosphate per litre is the maximum possible concentration if the hazards of supporting undersirable growth of aquatic organisms is to be avoided (Pitcarin and Hawkes, 1973). According to this, the upper and middle zone have crossed the limit. The nitrate content during present investigations was within the acceptable range (50 mg/l - IS 1974) in all the three zones.

The BOD and COD levels of the estuarine water were though lower in comparison to the effluent discharge regions, the values were still significant. Devi *et al.*, (1979) reported COD value as high as 3797 mg/l for a polluted estuarine water. Martin (1970) considered water body with BOD level exceeding 8 mg/l to be moderately polluted.

From the results of the present study it is clear that most of the benthic species were reported from the marine zone i.e. zone III or lower zone. Kinne (1966) reported that salinity was the ecological master factor affecting the distributional patterns of estuarine organisms. This is a normal situation in the estuary but increasing industrialization and domestic sewage from agglomerated human population contribute to the fast deteriorating environment of this large riverine system. Inspite of the fact that the impact of pollution was localized around outfalls, the existence of poor biological quality in zone I and zone II was observed compared to the other regions in lower zone which corroborates the study of Khan (1995). In the present study, large numbers of oligochaete worms - *Limnodrilus hoffmeisteri* dominated in the sewage outfall area of Tolly's nullah (Station 2, Table 4) is in conformity with Hynes (1960). This finding finds further support from the works of Brinkhurst (1969) and Das (1978), who reported oligochaeta as the indicator of pollution because of their presence in large numbers in water bodies grossly polluted with organic matter and inert suspended solids.

Ray *et al.*, (1979) reported that the conditions in case of soda process pulp and paper mill improved within 100 m below the outfall. Ghosh et al. (1976) recorded that the recovery distance was 50-100 m in case of rayon factory. The present study around a sewage outfall area in Calcutta also indicates that the recovery distance is located at about 100 m distance which suggests the autopurification potential of Hugli river.

Acknowledgement

The authors are thankful to Director, Zoological Survey of India for providing laboratory facilities and to Ministry of Environment and Forests, New Delhi for Senior Research Fellowship to one of us (S.P.).

References

APHA. (1985). *Standard Methods for the Examination of Water and Waste Water.* 16th Edn. American Public Health Association. New York. pp. 1268.

Basu, A.K. (1965). Observation on the probable effects of pollution on primary productivity of Hugli Matla estuaries *Hydorbiologia.* 25: 302-316.

Basu, A.K., Rao, C.S.G., Dhaneshwar, R.S., Mukherjee, G., Kutty, M.K.A., Ray, D. and Trivedy, R.C. (1973). Studies on Hugli estuarine pollution in the vicinity of a tissue mill. *Proc. Symp. Environmental Pollution.* NEERI, Nagpur. pp. 26 - 40.

Brinkhurst, R.O. (1969). Changes in the benthos of lakes Ericonfario, April 16 - 17, 1968. *Bull. Buffalo Soc. Nat. Sci.* 25: 45 - 65.

Das, S.M. (1978). High pollution in lake Naintial, U.P. as evidenced by biological indicator. *Sci & Cult.* **44**: 236-237.

Devi, K.S., Venugopal, P., Ramani, K.N., Lalitha, S. And Unnithan, R.V. (1979). Hydrographic features and water quality of Cochin backwaters in relation to industrial pollution. *Ind. J. Mar. Sci.* 8 : 141 - 145.

Dutta, P., Laha, G.C. Mitra, P.M., De, Chaudhury, A., Pandit, P.K., Chaudhury, A.R., De, R.N. Saha, B.K. Mazumdar, H.S. Sarkar, N. D., Mandal, N.C., Bhattacharya, G.P., Namasudra, A.K., Ghosh, S. P. and Paul, A.R. (1973). Fishery resources of the Hooghly - Matlah esturarine system. Bull No. 19, Dec. Central Inland Fisheries Research Institute (ICAR) Barrackpore, West Bengal.

Ellis, M.M. (1937). Detection and Measurement of stream pollution. In L.E. Knep *et al.*, (Eds.) *Biology of Water Pollution.* pp. 129 - 198. USDI. FWPCA. Cincinnati.

FWQA (Federal Water Pollution Control Administration), 1968. Water quality criteria. Report of the National Technical Advisory Committee to the Screatary Interior. 66 - 92. Washington D.C., April 1, 1968.

Ghosh, B.B. and Basu, A.K. (1968). Observation on estuarine pollution of the Hugli by the effluents from a chemical factory complex at Rishra, West Bengal. *Environ Health.* 10 : 204-218.

Ghosh, B.B., Ray, P. and Gopalakrishan, V. (1973). Survey and characterisation of Waste waters discharged into the Hugli estuary. *J. Inland Fish. Soc. India.* 4 : 82 - 101.

Ghosh, B.B., Ray, P., Haldar, D.D. and Bagchi, M.M. (1976). Some observations on the pollution of Hugli estuary in the vicinity of a rayon industry with a note on the toxicity of the effluents. ACMRR/ IASO Expert Consultation on bioassay with aquatic organisms in relation to pollution problem. Dubronik, Yugoslavia, 16 - 19 Nov. FAO publ. F1 : ACMRR / BE / 7 - 18.

Ghosh, B.B., Ray, P., Bagchi, M.M. and Haldar, D.D. (1977). Effect of biotic conditions on the Hugli estuary due to disposal of effluents from sulphite pulp and paper industry. *Proc. Seminar on Industrial wastes.* Dec. 8 - 9, Calcutta, CMDA / NEERI : 167 - 201.

Ghosh, B.B., Bagchi, M.M. and Saha, S.B. (1980). Effect of industrial pollution by soda process pulp and paper mill wastes on primary productivity of Hugli estuary near Tribeni. *Proc. Symposium on Coastal Aquaculture.* Cochin, India. Jan 12 - 18.

Hynes, H.B.N. (1960). *The Biology of Polluted Waters.* Liverpol Univ. Press Liverpol, U.K.

Indian Standard (1974). *Tolerance limits for Inland Surface Water subject to Pollution* (first revision). Indian Standards Institution, New Delhi.

Khan, R.A. (1995). The Pollution problem of Hugli estuarine system. In : *Hugli Matlah estuaries.* Estuarine Ecosystem Series, Part 2. Zoological Survey of India. pp. 497 - 542.

Kinne, O. (1966). Physiological aspects of animals life in estuaries with special references to salinity. *Neth J. Sea Res.* 3 : 222 - 244.

Klein, L. (1973). *River Pollution II* - causes and effects (5th imp.). Butterworth and Co. Ltd.

Martin, D.M. (1970). *Marine chemistry.* Vol. 1, Marcel Dekker Inc., N.Y.

Philipose, M.T. (1960). Fresh water phytoplankton of inland fisheries. *Proc. Symp. Algol. I.C.A.R.* New Delhi. 1960. Pp. 272 - 291.

Pitcairn, C.E.R. and Hawkes, H.A. (1973). The role of phosphorus in the growth of Caldophora. *Wat Res.* 7 : 159 - 171.

Ray, P., Ghosh, B.B. and Bagchi, M.M. (1977). The effects of variations in physico-chemical parameters on plankton biomass around distillery outfall in the Hugli estuary. *Proc. Seminar on Industrial Waters.* Calcutta. Dec. 8 - 9, CMDA/NEER 1 : 141 - 153.

Ray. P., Ghosh, B.B. and Bagchi, M.M. (1979). Effects of pulp and paper mill wastes (soda process) around the outfall in the Hugli estuary. *Proc. Symposium Environmental Biology, Acad of Environmental Biology.* India : 453 - 464.

Ray, P. (1981). Impact of Man on Hugli estuary. In : C.K. Varshney (Ed.) *Water Pollution Management Review.* New Delhi : 79 - 119.

Seth, G.K. and Bhaskaran T.R. (1950). Effects of Industrial waste water disposal on the sanitary conditions of Hugli river in and around Calcutta. *Indian J. Med. Res.* 38 ; 341.

Zingde, M.D., Trivedi S.K. and Desai, B.N. (1979). Physico - chemical studies of coastal pollution of Bombay. *Ind. J. Mar. Sci.* 8 : 271 - 277.

Proceedings of the National Seminar on Environmental Biology, April 03-05, 1998
***Edited By :* A.K. Aditya & P. Haldar**
DAYA PUBLISHING HOUSE, DELHI

Cumulative Effect of Bactericide and Insecticides on Nutrient Release from Leaf Litter by Detritivore Soil Arthropods

R. Pramanik, K. Sarkar & V.C. Joy*
* Corresponding author
Soil Zoology Laboratory
Department of Zoology (DSA, UGC)
Visva-Bharati University
Santiniketan - 731 235, West Bengal

Abstract

The rate of mobilization of important nutrients from decomposing leaf litter as influenced by the feeding effect of *Anoplodesmus saussurei* (Polydesmida : Diplopoda) and *Porcellio laevis* (Isopoda : Crustacea) under the impact of a broad spectrum bactericide (Ampicillin sulphate) and two cyclodiene insecticides (endosulfan 35EC and heptachlor 20EC), were compared under laboratory microcosm conditions. The rate of relase of nitrate nitrogen, calcium and organic carbon were negligible in the control sets without detritivore fauna. However, microbial action on leaf litter was sufficient to release potassium into the sand medium Application of Ampicillin and insecticides in the control sets did not produce any uniform and statistically significant impact on the release of nutrients. Comminution by *A. saussurei* and *P. laevis* resulted in very high and rapid rate of nutrient release due to their voracious feeding habit. The bactericide could affect the positive influence of decomposer arthropods in enhancing the nitrification process. Both insecticides had some inhibitory effect - the ill effect of heptachlor 20EC being prominent on the release of calcium and that of endosulfan 35EC prominent on nitrates. On the other hand, the combined effects of Ampicillin with the insecticides were highly inhibitory to the release of all the nutrients. The study demonstrated that the cumulative impact of environmental toxicants could adversely affect the synergistic functional significance of decomposer organisms in soil.

Introduction

Anthropogenic interferences like agricultural practices can produce drastic changes in the faunal and floral characteristics of soil by reducing population abundance and species diversity. It is now evident that soil animals are very much involved in nutrient cycling as primary and secondary decomposers of organic matter through synergistic interactions with microflora (Lee, 1991). From the ecological point, the most deleterious human interference in agroecosystem is the application of pesticides. Several workers have demonstrated that chlorinated hydrocarbons, cyclodienes, organophosphates etc. not only reduce the soil animal population but also disturb the prey predator balance. According to Ramade (1987) there is an increasing urgency for a synthesis to be made on the problems of environmental contamination within the purview of an emerging specialist science namely Ecotoxicology. It aims to study the different forms of environmental contamination with opportunities not only for evaluating the importance of the effects of contamination on ecosystem but also for predicting to a certain extent the future consequences of a particular pollutant. Assessment of the ecotoxicological impact of chemical substances on biological processes above the level of population requires model ecosystems under controlled laboratory conditions. Calow (1993) has recommended the use of microcosm to monitor the side effects of toxicants on mineralization process. Among detritivore soil fauna arthropods constitute a functionally significant group because of taxonomic diversity and direct feeding habit on decomposing organic matter. The present study deals with the impact of two broad spectrum cyclodiene insecticides and a bactericide on the functional significance of two species of detritivore arthropods in releasing important nutrients from decomposing leaf litter under laboratory microcosm conditions.

Materials and Methods

Anoplodesmus saussurei (Polydesmida : Diplopoda) and *Porcellio laevis* (Isopoda : Crustacea) were selected as model detritivore soil arthropods because of their feeding habit on decomposing leaf litter. Adult specimens were collected in large numbers from natural litter beds and maintained in the laboratory on decomposing leaf litter. The microcosms were prepared in polythene rearing vessels (250 cm^3) containing 30g sterile river sand as substratum and 2g leaf litter cuttings as food source for the animals. The moisture content of the sand bed and leaf litter was adjusted to 20% (approximate) by adding distilled water. The vessels were grouped into feeding and control sets and each feeding vessel received 5 specimens of *A. saussurei* and 10 specimens of *P. leavis*, both comprising animals of same age and size group. The leaf litter in 50% of the vessels were also treated with a standardized volume of 0.34 ml of the pesticides in sublethal concentration (0.001625% of heptachlor 20EC, 0.00005% of endosulfan 35EC) and 2ml of the bactericide (1% solution of ampicillin sulphate) in different combination. The untreated vessels received a similar amount of distilled water. Earlier studies in this laboratory have shown that administration of 1% ampicillin solution on leaf litter does not affect the normal activity of decomposer arthropods. All the feeding and control vessels of both treated and untreated sets were maintained at constant temperature (27±1°C) condition in a B.O.D incubator. From each set 3 vessels were randomly sampled at fortnightly intervals for quantitative estimation of nutrients incorporated into the sand bed. All the litter fragments and specimens were discarded before processing the sand samples for analysis. Available nitrate nitrogen was estimated spectrophotometrically by phenol-di-sulphonic acid method and the optical density measured at 420 nm was compared against standered curve of pure potassium nitrate (Jackson, 1962). Micronutrients namely calcium and potassium were extracted from the samples in 0.5 (N) HCI following the standard USDA method and estimated using a digital flame photometer calibrated against pure calcium carbonate and potassium chloride respectively. Organic carbon content was estimated following the rapid titration method of Walkley and Black (Piper, 1942). The observed differences in the amount of nutrients between different sets were statistically analysed using 'tw' test, a modified version of 't' test (Snedecor and Cochran, 1967).

Results

Important findings are incorporated in Tables 1 and 2 and Figure I. The effect of insecticides and bactericide on the rate of mobilization of nutrients in the sand medium of laboratory microcosm sets are summarized in Table 1. A comparison between SL and SLB sets showed that Ampicillin may affect the release of potassium and organic carbon into the sand medium.

Table 1 : Effect of bactericide and insecticides on the rate of mobilization of nutrients in different microcosm sets

Sets	Interval in weeks	Calcium (μg/g)	Potassium (μg/g)	Nitrate Nitrogen (μg/g)	Organic carbon (%)
SL	2	17.5 ± 1.2	77.8 ± 6.9	1.2 ± 0.2	0.02 ± 0.01
	4	28.0 ± 4.7	101.5 ± 6.2	1.1 ± 0.0	0.02 ± 0.01
	6	21.8 ± 1.3	111.8 ± 11.2	2.6 ± 0.5	0.07 ± 0.01
SLB	2	24.7 ± 4.0	81.7 ± 19.6	2.5 ± 0.4	0.03 ± 0.01
	4	29.3 ± 0.6	74.7 ± 16.5	2.7 ± 0.7	0.02 ± 0.0
	6	30.1 ± 0.5	79.0 ± 5.4	3.0 ± 0.7	0.02 ± 0.0
SLE	2	21.0 ± 1.2	98.5 ± 9.0	1.0 ± 0.1	0.02 ± 0.0
	4	32.3 ± 0.7	126.1 ± 18.4	1.2 ± 0.2	0.01 ± 0.0
	6	26.1 ± 2.9	131.4 ± 13.9	1.8 ± 0.3	0.05 ± 0.02
SLH	2	17.5 ± 1.2	85.5 ± 4.4	1.1 ± 0.0	0.02 ± 0.0
	4	31.1 ± 3.3	109.7 ± 1.4	1.1 ± 0.0	0.02 ± 0.01
	6	23.7 ± 1.8	116.7 ± 14.9	2.3 ± 0.5	0.04 ± 0.02
SLEB	2	23.7 ± 1.2	51.3 ± 2.5	1.1 ± 0.0	0.04 ± 0.0
	4	31.7 ± 2.1	138.0 ± 14.0	1.1 ± 0.0	0.04 ± 0.01
	6	40.2 ± 3.1	200.0 ± 12.0	1.1 ± 0.0	0.04 ± 0.02
SLHB	2	17.7 ± 1.2	59.3 ± 1.2	1.0 ± 0.3	0.04 ± 0.0
	4	28.0 ± 3.0	173.7 ± 1.2	1.4 ± 0.2	0.06 ± 0.01
	6	36.7 ± 3.2	273.5 ± 1.5	1.8 ± 0.2	0.08 ± 0.01
SLAP	2	38.9 ± 7.9	150.4 ± 9.3	3.9 ± 0.1	0.30 ± 0.07
	4	719.4 ± 266.7	199.6 ± 37.2	18.3 ± 4.4	0.59 ± 0.0
	6	980.0 ± 121.3	233.3 ± 14.9	92.9 ± 13.8	0.76 ± 0.6
SLAPB	2	81.7 ± 14.0	175.7 ± 58.9	2.8 ± 0.1	0.29 ± 0.05
	4	361.3 ± 233.0	193.7 ± 26.6	4.6 ± 1.0	0.59 ± 0.0
	6	602.7 ± 52.3	215.3 ± 7.3	6.9 ± 0.7	0.70 ± 0.0
SLAPE	2	51.7 ± 22.8	150.4 ± 9.0	2.0 ± 0.2	0.27 ± 0.03
	4	350.0 ± 58.3	237.5 ± 9.0	9.0 ± 1.6	0.56 ± 0.01
	6	688.3 ± 242.8	228.4 ± 26.6	43.1 ± 4.7	0.65 ± 0.13
SLAPH	2	74.6 ± 1.9	147.8 ± 2.6	2.0 ± 0.2	0.34 ± 0.05
	4	171.1 ± 58.7	180.1 ± 25.1	8.3 ± 2.5	0.48 ± 0.03
	6	513.3 ± 157.8	198.9 ± 19.5	74.2 ± 5.7	0.66 ± 0.05
SLAPEB	2	28.0 ± 3.0	79.3 ± 2.5	0.6 ± 0.0	0.07 ± 0.01
	4	109.7 ± 19.5	159.0 ± 0.0	3.0 ± 0.5	0.11 ± 0.02
	6	150.3 ± 20.2	210.0 ± 3.4	5.6 ± 0.5	0.12 ± 0.02
SLAPHB	2	31.0 ± 0.0	61.3 ± 1.5	1.0 ± 0.3	0.06 ± 0.0
	4	126.3 ± 71.2	144.0 ± 15.5	7.0 ± 0.14	0.13 ± 0.01
	6	120.4 ± 50.5	198.5 ± 17.2	11.0 ± 1.0	0.14 ± 0.01

S = Sand; L = Leaf litter; E = Endosulfan 35 EC

H = Heptachlor 20 EC; B = Bactericide (ampicillin sulphate)

P = *Porcellio laevis*; A = *Anoplodesmus saussurei*

However, the insecticides did not produce any notable change in the mobilization rate of nutrients in the control sets. The feeding activity by *A. saussurei* and *P. laevis* on the leaf litter resulted in very high and rapid rate of nutrient release in SLAP set probably due to their voracious feeding habit. A comparison between SLAP and SLAPB sets revealed that the bactericide could adversely affect the positive influence of decomposer arthropods in enhancing the nitrification process and mobilization of calcium. However, in the case of potassium and organic carbon no ill effect of Ampicillin could be observed suggesting that the release of water soluble components is not regulated by bacteria only. Application of the insecticides were also harmful to the fauna mediated nutrient mobilization from decomposing leaf litter. The effect of endosulfan was more prominent on the nitrification process whereas heptachlor could produce substantial decrease in the rate of release of calcium into the sand medium. On the other hand, the insecticides did not affect potassium and organic carbon. Table 1 also includes information on the combined effect of insecticides and the bactericide on the release of nutrients from leaf litter. The cumulative effect of Ampicillin with the insecticides (SLAPEB and SLAPHB sets) affected potassium leaching at least in the initial stage of the experiment. On the contrary, the combined effect of insecticides with the bactericide could produce synergistic deleterious effects in the mobilization of calcium, nitrates and organic carbon.

The average amount of nutrients estimated from the sand medium of different experimental sets are included in Figure 1. Except in the case of potassium all other nutrients maintained very low levels in the sets without decomposer arthropod fauna and the effect of insecticides and the bactericide was also insignificant. However, the combined feeding effect of *A. saussurei* and *P. laevis* was remarkable leading to enhanced availability of calcium, nitrates and organic carbon but the faunal effect could not produce any substantial increase in the availability of potassium. The effect of Ampicillin on the fauna mediated nutrient release was very prominent in the case of nitrates whereas individual application of the bactericide and the insecticides could not produce any appreciable difference in the amount of organic carbon. However, both insecticides had some ill effect on the mobilization of calcium and nitrate nitrogen, the effect of heptachlor being prominent inhibiting the release of calcium and the effect of endosulfan prominent in affecting the release of nitrates. The cumulative effect of the insecticides and the bactericide, on the contrary, produced drastic decline in the release of all nutrients and the effect was discernable in the case of potassium also.

The statistical validity of the observed differences in the amount of nutrients between different microcosm sets is summarized in Table 2. The deviations represent increase or decrease in the average amount of nutrients between sets as illustrated in Figure 1. As observed earlier the effect of insecticides and the bactericide in the sets without decomposer arthropods on the rate of release of different nutrients were not uniform and therefore mostly insignificant (combination i to Q). Comparison between SL and SLAP sets indicated highly significant increase in the amount of all nutrients released as a result of feeding activity. The faunal effects were even evident between the control and the feeding sets treated with Ampicillin, endosulfan and heptachlor (combinations 11 to 13). On the other hand, comparison between treated and untreated sets of decomposer arthropods (combinations 14 to 16) showed inhibitory effects on the mobilization of nutrients except potassium. Comparison between the effect of bactericide with insecticides on the faunal activity (combinations 17 to 19) indicated insignificant effects except in the case of nitrate nitrogen where the bactericide could inhibit release of the nutrient to highly significant low level. However, comparison between the effect of bactericide with the combined effect of bactericide and insecticide recorded significant inhibitory effects (combinations 22 & 23).

Table 2 : Summary of 'tw' test showing statistical significance of the differences in the amount of nutrients between different microcosm sets

Combinations	Calcium			Potassium			Nitrate Nitrogen			Organic Carbon		
	Dev.	tw	P	Dev.	tw	P	Dev.	tw	P	Dev.	tw	P
1. SL vs SLB	+ 5.6	0.42	< 0.02	−18.3	0.63	< 0.01	+ 1.1	0.69	< 0.01	−0.02	0.44	< 0.02
2. SL vs SLE	+ 4.1	0.28	n.s.	+ 21.7	0.40	< 0.05	− 0.3	0.19	n.s.	− 0.01	0.17	n.s.
3. SL vs SLH	+ 1.7	0.10	n.s.	+ 7.0	0.14	n.s.	− 0.2	0.11	n.s.	− 0.01	0.17	n.s.
4. SLB vs SLE	− 1.5	0.13	n.s.	+ 40.0	0.81	< 0.01	− 1.4	1.0	< 0.01	+ 0.01	0.22	n.s.
5. SLB vs SLH	− 3.9	0.27	n.s.	+ 25.3	0.60	< 0.01	− 1.3	0.81	< 0.01	+ 0.01	0.22	n.s.
6. SLE vs SLH	− 2.4	0.15	n.s.	− 14.7	0.28	n.s.	+ 0.1	0.06	n.s.	0.0	0.0	n.s.
7. SL vs SLEB	+ 9.4	0.52	< 0.01	+ 32.8	0.30	n.s.	− 0.6	0.67	< 0.01	0.0	0.0	n.s.
8. SL vs SLHB	+ 4.9	0.25	n.s.	+ 71.8	0.54	< 0.01	− 0.3	0.18	n.s.	+ 0.02	0.34	< 0.05
9. SLEB vs SLHB	− 4.5	0.21	n.s.	+ 39.0	0.20	n.s.	+ 0.30	0.38	< 0.05	+ 0.02	0.40	< 0.05
10. SL vs SLAP	+ 557.0	1.08	< 0.01	+ 97.4	1.24	< 0.01	+ 36.7	0.70	< 0.01	+ 0.51	0.87	< 0.01
11. SLB vs SLAPB	+ 320.4	1.11	< 0.01	+ 116.2	2.13	< 0.01	+ 1.8	0.56	< 0.01	+ 0.50	2.0	< 0.01
12. SLE vs SLAPE	+ 336.8	0.77	< 0.01	+ 86.7	0.84	< 0.01	+ 16.8	0.71	< 0.01	+ 0.46	1.56	0.01
13. SLH vs SLAPH	228.9	0.75	< 0.01	+ 71.6	1.26	< 0.01	+ 26.7	0.67	< 0.01	+ 0.46	1.92	< 0.01
14. SLAP vs SLAPB	− 231.0	0.29	n.s.	+ 0.5	0.01	n.s.	33.8	0.62	< 0.01	− 0.03	0.06	n.s.
15. SLAP vs SLAPE	− 216.1	0.23	n.s.	+ 11.0	0.09	n.s.	− 20.2	0.27	n.s.	− 0.06	0.11	n.s.
16. SLAP vs SLAPH	− 326.4	0.41	< 0.02	− 18.8	0.21	n.s	− 10.2	0.11	n.s.	− 0.06	0.12	n.s.
17. SLAPB vs SLAPE	+ 14.9	0.02	n.s.	+ 10.5	0.10	n.s.	+ 13.6	0.52	< 0.01	− 0.03	0.06	n.s.
18. SLAPB vs SLAPH	− 95.4	0.16	n.s.	− 19.3	0.28	n.s.	+ 23.6	0.56	< 0.01	− 0.03	0.07	n.s.
19. SLAPE vs SLAPH	− 110.3	0.15	n.s.	− 29.8	0.28	n.s.	+ 10.0	0.16	n.s.	0.0	0.0	n.s.
20. SLEB vs SLAPEB	+ 64.2	0.78	< 0.01	+ 19.8	0.13	n.s.	+ 2.0	0.73	< 0.01	+ 0.06	0.13	n.s.
21. SLHB vs SLAPHB	+ 198.6	1.00	< 0.01	− 34.2	0.18	n.s.	+ 4.9	0.75	< 0.01	+ 0.05	0.67	0.01
22. SLAPB vs SLAPEB	− 252.4	0.71	< 0.01	− 45.3	0.44	< 0.02	− 1.5	0.26	n.s.	−0.42	0.61	< 0.01
23. SLAPB vs SLAPHB	− 122.5	0.26	n.s.	− 60.3	0.54	< 0.01	+ 1.7	0.20	n.s.	− 0.41	1.46	0.01
24. SLAPEB vs SLAPHB	+ 129.9	0.50	< 0.01	− 15.0	0.10	n.s.	+ 3.2	0.38	< 0.05	+ 0.01	0.02	n.s.

df = 9 in all cases

n.s. = not significant

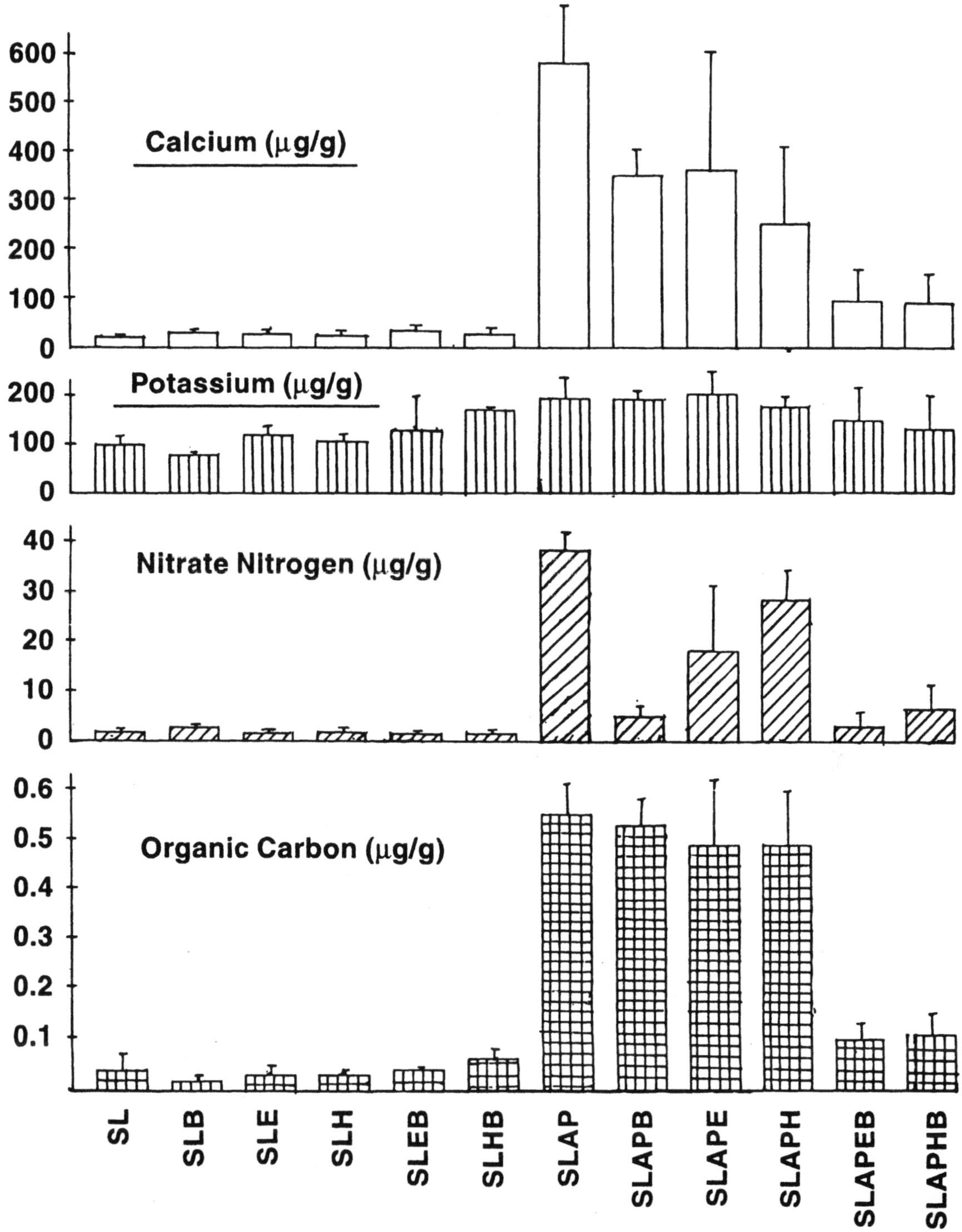

Fig. 1 : Average amount of imporant nutrients incorporated into the sand medium of different microcosm sets.

The study showed that the application of bactericide was inhibitory to the fauna mediated mobilization of nitrate nitrogen, whereas the combined effect of bactericide with endosulfan/haptachlor were inhibitory to the release of all the nutrients.

Discussion

Assessment of ecotoxicological impact of chemical substances on biological processes above the level of population requires model ecosystems under controlled laboratory conditions. Calow (1993) has emphasised on the possibility of monitoring the side effect of toxicants on mineralization processes using such microcosms. Standardization of ecotoxicological methods requires the use of substrates of known physicochemical composition and therefore, the present study was conducted using sterile pure river sand in the mocrocoms which served as a good medium for adequate moisture and as a natural substratum for decomposer organisms. Similarly, leaf litter was used as the common food source for both microflora and the decomposer arthropod species so that the effect of environmental toxicants on their interactions could be easily compared. Detrtivore arthropods namely *P. laevis* and *A. saussurei* were selected on the basis of their direct litter feeding habit. Bano and Kishnamoorthy (1987) conducted litter decomposition and humification studies using *Jonespeltis splendidus* (Diplopoda) and Vink (1995) used *Procellionides pruinosus* (Isopoda) as representative of soil fauna responsible for comminution of litter. Microflora have an important role in the ecosystem through mineralization and conversion of nutrients. Levin et al. (1989) studied the effect of pollutants of species interaction and decompostion processes such as nutrient mineralization and found that essential inorganic elements like nitrogen could be inhibited by toxic substances. In the present study the feeding activity by *A. saussurei* and *P. laevis* on leaf litter could enhance the release of nitrates, calcium and organic carbon. Seastedt (1984) also showed that among different nutrients the movement of nitrates and calcium are greatly influenced by soil organisms whereas potassium in plant litter is completely water soluble and can leach out without the activity of microbes and soil fauna. Verhoef and Meintser (1991) noticed significant increase in the rate of leaching of ammonium, nitrates, potassium and calcium by the addition of soil arthropods in leaf litter. As observed in the present study, several workers have demonstrated that application of insecticides is not inhibitory to the microbial activity in soils. However, application of insecticides even at very low doses could produce harmful effects on decomposer arthropod fauna as evidenced from significantly low levels of nutrients released in the treated microcosms. Pradhan and Mishra (1986) demonstrated that carbaryl in normal agricultural dose (1 kg/ha) though does not affect the density and biomass of earthworm significantly, produced some physiological stress in the animals resulting into increased rate of respiration. Attempts were made in the present study to clarify the role of soil arthropod fauna in nutrient release by partially inhibiting the role of microbes on the decomposing leaf litter. The results indicated that the cumulative effect of Ampicillin (broad spectrum bactericide) and the 2 cyclodiene insecticides was highly detrimental as seen from the very low level of nutrients released. Anderson et.al. (1983) found that litter grazed by millipedes released 10 to 20 times more ammonia than ungrazed litter, whereas sodium and potassium were leached easily and rapidly from litter and faunal impact on their cycling was minimum. On the contrary, the present study established that comminution of plant litter by detritivore fauna is important in exposing the biologically bound calcium in plant cells for microbial degradation. In the present study inhibition of bacterial activity on litter could adversely affect the mobilization of calcium and nitrates during feeding by the arthropod species. The application of insecticides along with bactericide could produce a cumulative harmful effect on the fauna mediated nutrient release and was evident in the case of all the 4 nutrients considered for the study.

Acknowledgements

The authors are grateful to the Head, Department of Zoology (DSA, UGC), Visva-Bharati University for providing laboratory and other facilities. They are also grateful to the Ministry of Environment and Forests, New Delhi, for financial assistance in the form of a major research project.

References

Anderson. J.M., Ineson, P. and Huish. S.A. (1983). Nitrogen and cation mobilization by soil fauna feeding on leaf litter and soil organic matter from deciduous woodland. Soil Biol. Biochem., 15 : 463-467.

Bano. K. and Krishnamoorthy. R.V. (1981). Relative rates of litter decomposition and humification of various soils by the millipedes, *Jonespeltis splendidus.* Progress in Soil Biology and Ecology in India. Ed. Veeresh, G.K., UAS, Tech. Series, 37 : 87-92.

Calow, P. (1993). Handbook of ecotoxicology. Ed. Calow, P. Vol. 1, Blackwell Scientific Publication, Oxford.

Jackson, M.L. (1962). Soil chemical analysis. Asia Publishing House, New Delhi.

Lee, K.E. (1991). The role of soil fauna in nutrient cycling. In : Advances in management and conservation of soil fauna. Eds. Veeresh, G.K., Rajagopal, D. and Viraktamath, C.A. Oxford and IBH Publishing Co. Pvt. Ltd. New Delhi, : 462-472.

Levin, S.A., Harwell, M.A., Kelly, J.R. and Kimball, K.D. (1989). Ecotoxicology : Problems and approaches. Springer Verlag. New York.

Piper, C.S. (1942). Soil and plant analysis. Hans Publ. Bombay.

Pradhan, S.C. and Mishra, P.C. (1986). Changes in population and metabolism of earthworms following application of carbaryl insecticide to a grassland site. Proc. Nat. Sem. of G. Waste utiliz. Vermicomp. Part B. Verms and Vermicomposting. Eds. Dash, M.C., Senapati, B.K. and Mishra, P.C., : 111-121.

Ramade, F. (1987). Ecotoxicology, John Wiley and sons, Great Britain.

Seastedt. T.R. (1984). The role of microarthropods in decompostion and mineralization process. Ann. Rev. Entomol. 29 : 25-46.

Snedecor, G.W. and Cochran, W.G. (1967). Statistical methods. 6th edition. Oxford & IBH Co. Pvt. Ltd. Calcutta.

Verhoef, H.A. and Meintser, S. (1991). The role of soil arthropods in nutrient flow and the impact of atmospheric deposition. In : Advances in management and conservation of soil fauna. Eds. Veeresh, G.K., Rajagopal, D. and Viraktamath, C.A. Oxford & IBH Publishing Co. Pvt. Ltd. New Delhi ; 497-506.

Vink, K. (1995). A laboratory method to test side effects of pesticides on tropical litter decomposition. In : Tropical Isopods; Ecological aspects and ecotoxicological application. Ed. Vink. K., Virje University Publication. Amsterdam : 93-106.

Proceedings of the National Seminar on Environmental Biology, April 03-05, 1998
***Edited By :* A.K. Aditya & P. Haldar**
DAYA PUBLISHING HOUSE, DELHI

Importance of Biodiversity

J.R.B. Alfred
Director, Zoological Survey of India
New Alipur, Calcutta - 700 053

Biodiversity is variety of life and its many processes. It includes all life forms from bacteria, fungi and protozoa to higher plants, insects, fishes, birds and mammals. Biodiversity also includes countless millions of races, subspecies, and local variants of species and the ecological processes and cycles that link organisms into populations, communities, ecosystems, and ultimately the entire biosphere.

The foundation of biodiversity is genetic conservation. Genetic variation affects a species physical characteristics, productivity, resilience to stress and long-term evolutionary potential. Genetic diversity means that species contain characteristic levels of genetic variation within and among their populations, including that of extreme populations.

A more easily recognised element of biological diversity is the distinct species. Associations in an area are another recognizable element of biological diversity. We refer to these associations as biological communities and usually recognize them as distinct stands, patches or sites such as forests, meadows, fencerows, ponds, areas and wetlands. Communities form the biotic parts of ecosystems. The variety of species in an ecosystem, usually referred to as its species composition is intertwined with its structural and functional and hence with the diversity of its ecological processes. Biologically diverse contain sufficient compositional, structural and functional variety that they are assured a high prospect of continued presence and ecological influence in an area.

Finally, at the geographic scale of regional landscapes diversity includes variety in the kinds of biological communities and a biogeography (that is patterns, sizes, shapes, juxta positions and inter connectedness) that provides for free, natural interchange of individuals throughout the area. Many species, especially those with specialized habitat affinities or those that a migratory or wide-ranging, can only be sustained in variable numbers and distribution in very large wideland areas.

Humans have a great influence on the variety and health or life around them. For many this means a responsibility to care for the richness of life on Earth. This has been stated in many ways across the centuries, though rarely better than by Aldo Leopold (1953:147) : "If the land mechanism as a whole is good, then every part is good, whether we understand it or not. If the biota, in the course of eons has built something we like but do not understand, then who but a gool would discard seemingly useless facts? To keep every cog and wheel is the first precaution of intelligent tinkering."

This history of this planet goes back about 5 billion years. The planet is finkite, and while there have been major changes in the chemical composition of the atmosphere, until very recently, the puritine state of air, water, and soil was maintained by the interaction of the global network of organisms that exploited them. One organism's wastage can be another organism's food in the intricate web of interconnections.

Human beings arose as a species out of web of life a mere 200,000 years ago or so. We are made up of the same basic atoms in the same proportion, the same macromolecules, and the same cellular plan as all other living things because we share common ancestors. The genetic blueprint written in nucleic acids is a universal genetic code.

Life evidently first appeared on Earth more than 3(4) billion years ago, and for the next 1.5 billion years, only the single bacteria and blue-green algae, or centotrophic prokaryotes existed.

The first fossils organisms with cells, a nucleus and chromosomes (eukaryotes) are from rocks about 1.4 million. These are thought to have arisen from prokaryotes through symbiotic associations (Marylis, 1970).

About 640 million years ago, there was a sudden appearance of a diverse assemblage of multicellular eukaryotic organisms.

Since the Cambrian, there has been continuous evolution and extinction of life. Over a half billion species may have existed over the past 600 million years. Current estimate are that there are between 5 and 30 million species.

Calculation show that for organisms inhabiting shallow marine waters, the longivity of species is usually less than 10 million years (6 million in echinoids, 1.9 million in graptolites, 1.2 to 2 million in ammonites).

If the average survival of a species is about 5 million years and eukaryotic life has been around for over 600 million years, then there could have been at least 120 complete changes of species during the time of metazoan existence.

Since the Cambrian began, the average species extinctions rate has been 9% per million years. The earliest known mass extinction occurred about 650 million years ago, when a worldwide event eliminated as much as 70% of the plankton diversity on earth.

The distinctive survival strategy evolved by one species - the *Homo sapiens* - was a highly developed brain. From the brain emerged self awareness, memory, curiosity, imagination, and creativity. The human brain had an unprecedented capacity for abstraction, to create ideas that existed only in a mind that became as real to that mind as a rock or tree.

It was this capacity for abstraction that enabled us to invent notions of tribes, loyalty, nations justice, peace, war, love, equality, god, heaven and hell. And it was also responsible for creating economies an idea the value in human activity and natural "resources" that we can exploit and which symbolize that value in currency. When biological diversity is depleted, all the money in the world won't substitute it.

This is the challenge we face. In discussions of environmental issues, we each bring all of the baggage of our personal values and belief systems reinforced by an lifestyles and what we do for a living. Earlier even efforts were not motivated by the desire to maintain an ecological balance between nature and man.

Fortunately for us in India – our so called illiterate villager's perspective was always evident that their village habitat should be in an ecological balance. They understood that ecosystems have their own linkage, all other species have as much right their habitat as the human and the human communities living there should also be self-contained as much as possible.

Maintaining and managing forests to provide goods and services (biodiversity) to the society was one of the original goals for setting aside forest lands, certainly much before the advent of scientific forestry in our country during the 19th century. One of the earliest records of setting aside forests for production of timber, other forest produce and conservation of wildlife dates back to between 250-230 B.C. and is attributed to Kautilya (Ranagarajan, 1992). Likewise Emperor Ashoka (252 B.C.) laid down in stone his edicts for conservation of wildlife, including setting aside reserves for this purpose (Mackminon *et. al.* 1986; Forest Research Institute, 1961, Saharia, 1981).

The seventeenth century A.D. marked a watershed in revolutionary thought that began to dismantle the long standing conviction that the natural world had been created solely for human use (Mabey, 1995). Ethical concern received strong support from the developing discipline of natural history, that transcended the intitial interest in plant and animal taxonomy during the seventeenth and eighteenth century to later explore the frontiers of the functioning of the natural world marked by Darwin's theory of evolution by natural selection.

Probably worldwide, an intellectual debate relating to nature and its conservation was initiated with the establishment of the Yellowstone Natural Park (8983 sq. km.) established in 1872.

This concept has been passed on to generation after generation. A number of instances where men and women have sacrified their lives have been recorded in Indian history. The case of the "Bishnois" sacrificing their lives to prevent the destruction of trees by loggers drew worldwise sympathy. Another well known movement is the : "Chipko" (*i.e.* "hug the trees") in Garhwal district of Uttar Pradesh. This movement was in protest of logging and the local women expressed this by logging the trees that was earmarked for felling.

Moreover, in our country, traditional farming practices are identitifed, domesticated and preserved for a wide range of plants and animals which gave rise to the rich heritage of agrobiodiversity. Women exchanging farm conserved seeds at the time of important ceremonies is a wide spread practice throughout rural India. Seasonal dietary practices/restrictions, closed periods of animal harvests, taboos or hunting pregnant/gravid animals are still prevalent to a large measure in the country.

India's concern and commitment towards environmental conservation was reinforced in 1976 by the 42nd Amendment of the Constitution of India. Article 48a under the Directive Principles of State Policy and Article 51A(g) of the Fundamental Duties in the Constitution, states that "the State shall endeavour to protect and improve the environment and to safeguard forests and wildlife in the country" and "to protect and improve the natural environment including forests, lakes and rivers and wildlife and to have compassion for the living creatures."

What is all this for? Biodiversity, the provident of millions of years of biological evolution, has always provided mankind with food, fibre, shelter, medicines and socio-cultural environment. It is quite clear tha sustainable economic use of these genetic assets and other natural resources will depend on (or it may indeed demand) fundamental changes in the way humans chose to interact with each other and other species co-habiting planet Earth. The rate of genetic erosion is quite incompatible with not only the desired human dignity and material welfare but, more importantly, with the integrity of the biosphere and planet Earth. It, therefore, calls for charting a new path, with a new vision, vigour and renewed mandate.

A particular case of utility of biodiversity is seen among the Ladakhis. Ladakh is a Tibetan, Buddhist region in the north Indian states of Jammu & Kashmir with rugged elevations of 3,500 meters and more, poor soil conditions, less than 10 centimeters of rainafall per annum, temperatures of – 40°C and very brief growing periods. And yet the Ladakhis have developed a diversity of crops, animals, agricultural and social practices that allow them to survive in these harsh conditions.

A key feather of their success is that they have selected and planted suitable crops in particular niches. Barley is grown everywhere except in the highest villages while apricots, apples and potatoes, are grown in the lower valleys. Peas are harvested by hand with nitrogen-rich nodules left in the soil. Fields are irrigated with glacier run off and enriched with human night soil. In this diversified setting, many barley plants have more grains pre stalk than mot Europen varieties and cereal yields average about 10 metric tons per hectare compared to yeilds of 1 ton per hectare in India and Africa, 2.2 tons per hectare in North America and 1.5 tons in the former Soviet Union. The Ladakhis also raise a variety of animals for dairy products, meat, wool and transport. Stray social cooperation also enables the Ladakhis to use these methods successfuly. (Natiers-Hedge *et. al.* 1991).

India's immense biological diversity encompasses ecosystems, populations, species and their genetic make-up. This diversity can be attributed to the vast variety in physiography and climatic situations resulting in a diversity of ecological habitats ranging from tropical, sub-tropical, temperature, alpine to desert. As a result according to world biogeographic classification, India represents two of the major realms (the Palaearctic and Indo-Malayan) and three biomass (Tropical Humid Forests, Tropical Dry/Deciduous Forests and Warm Deserts/Semi-Deserts). These include 12 biogeographic regions. However, the Wildlife Institute of India has proposed a modified classification – which divides the country into 10 biogeographic regions : Trans-Himalayan, Himalayan, Indian Desert, Semi-Arid, Western Ghats, Deccan Peninsula, Gangetic Plain, North East India, Island and Coasts.

The Culture of Conservation

India has a rich tradition of respect for and conservation of diversity of life. These traditions are responsible for the presence, through the length and breadth of the country, of hundreds of thousands of trees such as banyan and peepal, now recognized by ecologists as a significant keystone resource. They have permitted the persistence of macaque and langur monkeys and peafowl in the midst of villages and towns.

Such traditions have also protected entire biological communities in sacred groves and sacred ponds. For instance, a few years ago, scientists of the Botanical Survey of India discovered a new species of a leguminous climber, *Kunstrelia keralinsis* in a sacred grove in the densely inhabited plains of Kerala.

What is further encouraging is that these practices are today being adopted to the modern secular context. Thus many shifting cultivator communities from the North-eastern states of Manipur and Mizoram have revived protection sacred groves as safety forest serving as large belts of firebreaks exclusively than villages (Gadgil, 1997).

I hope the fact that we have assembled reflects a desire in all of us to find common ground, to establish a shared understanding of language, assumptions and goals. Biological diversity keeps the planet going and offers the wonders that evoke our great love of nature. The writing is on the wall for many species, for many indigenous cultures and for many ways of making a living. We should be working to preserve them all.

Reasons for Retaining Biodiversity

Perhaps it is redundant to ask why we should retain biodiversity. While there are many reasons, I shall briefly list seven major ones. *First,* protecting biodiversity keeps options open for future generations. *Second,* it has economic values in everything from medicine to pest control. *Third,* it has aesthetic values – enjoying life for its own sake. *Fourth,* a species is gone forever once it has become extinct. We must save the blueprints of the pieces of the ecosystem. *Fifth,* we need "indicator species", meanig those species that are more sensitive than ourselves to environmental changes that could warn us of environmental degradation. *Sixth,* there are moral values: everything has a right to live. And seventh, people want it. Biodiversity is a hot topic today as confirmed by recent task forces, symposia and polls.

Section III

Invertebrate Biodiversity : Aquatic Environment

Proceedings of the National Seminar on Environmental Biology, April 03-05, 1998
Edited By : A.K. Aditya & P. Haldar
DAYA PUBLISHING HOUSE, DELHI

The Problem of Hydra Photoresponse

C. Taddei-Ferretti, C. Musio, S. Santillo & A. Cotugno
Istito di Cibernetica, CNR, Via Toiano 6,
80072 Arco Felice, Italy

Introduction

The Cnidarian Hydra, known since Trembley (1744), is a fresh water animal, living in clear brooks, it exists only in the polyp form. The usual way of reproduction through budding may change to sexual phases, driven by environmental conditions such as ionic components of the medium (Loomis, 1961). Abnormal budding patterns may be induced by chemicals (Aditya, 1981).

Such animals represent the first tissue grade construction, after the cellular grade of sponges. The gastrovascular coelenteron derives from the gastrula archenteric cavity, while the mouth corresponds to the blastopore, it is a dipblastic animal, that consists of two cell layers (ectoderm and endoderm) separated by a thin non-cellular mesoglea and by a network of neurons.

The phylogenetic separation of functions into different cells and the specialization of cells into different types are not complete. In fact, the ephithelial cells have also the function of active transport of ions (linded to the osmoregulation processes related to the fresh water environment), and they bear also myofibrils (thus, they are called epitherliomuscular cells).

The myofibrils are arranged in an orthogonal way in one epithelial layer with respect to the other. They are arranged longitudinally with respect to the long axis of animal in the ectoderm and circularly in the endoderm so that the ectoderm and endoderm constitute an agonist-antagonist system, and their alternate activity leads to the shortening and elongation of the animal body respectively.

The cells of the neural net are bipolar and multipolar neurons, that are slightly more concentrated at the two animal ends, and especially at the proximal one. However, no ganglia are present, so that the neural net is almost isotropic from the structural point of view. Only a very thin sub-oral neural ring lies at the base of the tentacles in some species. Several neurotransmitters have been identified (Grimmelikhuijzen & Westfall, 1995).

Ciliated sensory neurons potrude towards the external environment, having mechano-or chemo-receptive function. No photoreceptive cells have been identified up to now, neither with anatomical technique, nor with microspectrophotometric ones.

The cnidoblasts containing nematocysts, that are among the most complex organoids of the animal kingdom, are receptive-effective independent units utilized by the animal to adhere to the substrate or paralize the pray (Lenz, 1966, pp. 106-13; see review in Cotugno, 1994).

The behaviour of the animal consists mainly of continuous periodic alternations of a phase of body shortening and a phase of body elongtion, usually lasting respectively half a minute and few minutes. Such movements are linked to osmoregulation, food ingestion and debris ejection through the mouth. Occasionally, locomation movements by sumersaulting are observed. Living preys, usually little Arthropods, are captured by tentacles, while reciprocal exploring behaviour of two individuals through tentacle movements have been observed (Taddei-Ferretti *et. al.*, 1999). Different bioelectric activities are linked to the body shortening, body elongation and tentacle movements (Passano, 1962; 1963; 1964; 1965; Rushorth & Burke, 1971).

In spite of the fact that no photoreceptive cells have been identified the animal presents a clear photoresponse. Phototactic movement have been observed since the end of the last century (Wilson 1891) light pulses may either stop a body shortening activity in progress, or vice versa evoke an event of shortening (Passano, 1962; 1963; 1964; 1965).

At this point, the research of our group since more than two decades of year has been devoted to deeply investigate the influence of light on the behaviour of the animal and to afford the problem of the identification of the cellular elements responsible of its photoresponse. Our results are summarized below.

Materials and Methods

The main experiments have been performed with *Hydra vulgaris*, while data also from *Chlorohydra viridissima* and *H. magnapapillata* have been obtained in some cases.

The photic stimuli that we have used are light pulses of defined intensity level, duration, polarity (i.e. light or darkness pulses) and chromatic content (i.e. light wavelength ranging from 350 to 700 nm), while also the intensity level and chromatic content of the back round illumination was controlled.

Thermic or chemical stimuli have been added during some experiments, in order to investigate the interactions due to multimodal stimulation. The effect of electric stimuli has been tested.

The bioelectric activity has been recorded by electrophysiological techniques from different types of preparations : **(a)** in order to precisely monitor the response time along the natural period of body shortening-elongation, the bioelectric activities linked to body shortening and body elongation of the animal has been picked up through extracellular electrodes; **(b)** membrane potentials and membrane currents of enzymatically dissociated single cells have been picked up respectively by classical techniques (intracellular recordings) and advances ones (patch clamp recordings).

Results

Behaviour Expression

We have ascertained that, in addition to the two already known compound biopotentials linked one to the ectodermal myofibril activity and the other to the endodermal one, a big slow wave could be recorded. It has the same frequency of the shortening-elongation behaviour and a variation trend opposed at the two ends of the animal, being electrically always more negative at the distal end with respect to the proximal one (Taddei-Ferretti *et al.*, 1976).

There are differences between the ontogenetic development of the body shortening bioelectric activity and that of the big slow wave (Taddei-Ferretti, 1981).

The behavioural period is influenced also by the steady temperature level (Taddei-Ferretti *et al.*, 1977).

By appylying $MnCl_2$, the polarity of the bioelectric activity linked to the body shortening is reversed and later it is totally inhibited (Taddei-Ferretti & Cordella, 1976b).

Electric pulses of different polarities applied at different ends of the animal elicit different phases of the behavioural period (Taddei-Ferretti *et al.*, 1976; 1987a). A long body shortening phases is elicited by a positive pulse applied at the distal end or a negative pulse at the proximal end. Viceversa, a long body elongation phase is elicited by a negative pulse applied at the distal end or a positive pulse at the proximal end. Various after-effects are also observed.

Photoresponse

The reponse to a light pulse may be expressed as the direction (i.e. advance or delay) and amount of the dephasing of the subsequent behavioural event (e.g. the beginning of a body shortening phase). Such response depends strictly on the light pulse intensity, duration, polarity, and wavelength. In addition, after a light pulse of unchanged intensity, duration, polarity and wavelength, unexpectedly the response time is not always the same, but it depends also on the phase of application of the light pulse along the period of the behavioural shortening-elongation events (Taddei-Ferretti & Cordella, 1975; 1976a; Taddei-Ferretti *et al.*, 1978, 1987a; 1988b). Phase response curves characteristic for each stimulus type may be constructed taking into account all the phases of stimulation.

With repetitive periodic light pulse stimulation of suitable frequency, the shortening-elongation activity may be either entrained to the stimulation period (Taddei-Ferretti & Cordella, 1976a), or supressed (Taddei-Ferretti *et al.*, 1994).

The duration of the shortening-elongation period varies inversely with the wavelength of the steady background illumination level (Taddei-Ferretti & Cotugno, 1988). We have ascertained this fact not only under laboratory conditions, but also in natural environmental ones: in fact, the behavioural shortening-elongation period is different at dawn, noon, dusk and night.

If then, always expressing the response in terms of the dephasing of the subsequent behavioural event, one has normalized to one the behavioural period in all background illumination conditions, the phase response curve characteristic of the application of one type of pulse stimulus should not change when the background illumination wavelength is different. However, contrary to the logical expectation, this fact does not hold, (Taddei-Ferretti *et al.*, 1992b).

Furthermore, non-linear types of interactions have been observed between two light pulses that differ for what it concerns their wavelength and duration, if they are applied either simultaneously, or in close succession (Taddei-Ferretti & Cotugno, 1991; Taddei-Ferretti *et al.*, 1991).

We tested the effect of various light stimulations on different naturally white strains of Hydra, on green strains (having symbiotic algae) and on aposymbiotic white strains (artificially obtained from green strains). Qualitative and quantitative differences of the responses were observed in the three strain types. The presence of algae has a higher value than the genetic factor for what it concerns the response differentiation (Taddei-Ferretti *et al.*, 1987b).

We have ascertained that the combination of the effects of the variation of the temperature and of the background illumination wavelength simultaneously applied is such that the differences among the values of the behavioural period at different wavelengths are higher at lower temperatures (Taddei-Ferretti *et al.*, 1997).

A light stimulus influences the shape of the big slow wave, but after the application of

$MnCl_2$ the polarity of the big slow wave variation in consequence of a light stimulus is reversed (Teddei-Ferretti & Cordella, 1987b).

We have begun a systematical analysis of the various cell types responsive for the phototransduction, by using both intracellular recording and patch clamp recording. Up to now we have recorded membrane currents in the whole cell configuration with steady light conditions from enzymatically dissociated ectodermal epithelial cells (Santillo *et al.*, 1996).

Discussion

Behaviour Expression

The results on electric stimulation have to be related to the different potential level of the above big slow wave of bioelectric activity when recorded at the two ends of animal. It has to be related also to the different sites of pacemaking of the two waves of myofibril activity beginning at the two ends of the animal, i.e. to the origin site of ectodermal and endodermal miofibril activity located respectively at the distal and proximal ends.

The complex of our results and of other researchers' has been interpreted by our group in terms of the role of the nervous net (Taddei-Ferretti & Chillemi, 1987; Taddei-Ferretti & Musio 1999; Taddei-Ferretti *et al.*, 1988a; 1992a). The slight difference of its neuron concentration at the two ends of animal, especially at the proximal one, may constitute a first step of specialization into two pacemaking centres that inhibit each other and each one exerts its activity at the ends of the received inhibition, eliciting the alternate activity of the body shortening and body elongation miofibrils, arising respectively at the distal and proximal ends of the animal.

Photoresponse

The results of light pulse stimulation combined with the variation of background wavelength is not exerted isotropically along all the phases of the behavioural period.

The results of light pulse stimulations application of variation of temperature and background wavelength indicate that the interaction between the effects of temperature and light wavelength is not linear due to the overall change of ion channel dynamics at high temperature levels.

The effects of the application of $MnCl_2$ indicate that Ca^{2+} plays a role in Hydra motility and probably in its light response.

The bulk of the results of our group and of other researchers on Hydra photoresponse speak in favour of the obvious existence of phototransductive moleculues that mediate the light changes of the environment, modulating the animal behaviour in order to drive it at its best. Carotenolkds, taken with food, are stored in the epithelial cells of Hydra and their level decreases with an increase of the illumination intensity level (Frigg, 1970). In spite of these facts, we were unable, collaborating with colleagues of another institute, to identify the photoreceptive cells by means of microspectrophtometric investigations.

The failure of this part of investigation lead us to the hypothesis that in Hydra the phototransductive molucules are to segregated at limited specialized zones of the photoreceptive cells, but may be uniformly distributed on the membrane of scattered photoreceptive cells. Such cells, that are still unknown, could be either cells of the nervous system, or epithelial cells. In fact, in other animal species these types of cells may bear also photoreceptive molecules, being clear examples of extraocular photosensitivity (Musio, 1966).

For this reason, we have begun to analyse the bioelectric activity of single cells, in order to characterize them in darkness condition and later compare their bioelectric activity under various light stimulations. Our preliminary results constitute the first step to identify the photoreceptive cells and determine the dynamic properties of their light-sensitive ion channels.

References

Atiya A.K., (1981). Induced abnormal budding in Hydra. Zool. Anz., Jena 206, 5/6, S. 334-340.

Cotugno A., (1994). L' Hydra : un preparato sperimentale mulipurpose nella ricerca biologica. In : F, Allocati, C. Musio & C. Taddei Ferretti (eds.), Cibemetica Biologica, Rosenberg & Sellier, Torino, pp. 131-138.

Frigg M., (1970). Vorkommen and Bedeutung der Ksarotinoide bei Hydra. Z. Physiol., 69 : 186-224.

Grimmelimhuijzen C.J.P., Westfall J.A., 1995 – The nervous systems of Cnidarians. In : O, Breidbach & W. Kutsch (eds.), The Nervours Systems of invertebrates; An Evolutionary and Comparative Approach, Brikhauser, Basel, pp. 7-24.

Lentz T.M., (1961). The Cell Biology of Hydra, Noth-Halland, Amsterdam.

Loomis W.F., (1961). Feedback factors affecting sexual differentiation in *Hydra litoralis*, In : H.M.

Lenhoff & W.F. Loomis (eds.), The Biology of Hydra and of Some other Coelenterates, University of Miami Press, Coral Gables, Fl, USA, pp. 337-362.

Musio C., (1996). Extraoccular photosensitivity in invertebrates : a look into biophysical processes and functional mechanisms, In : C. Taddei-Ferretti (ed.), Biophysics of Photoreception : Molecular and Phototransductive Events, World Scientific, Singapore, New Jersey, London, Hong Kong, pp. 245-262.

Passano L.M., Mcullough C.B., (1963). Pacemaker hierarchies controlling he behaviour of hydras nature, 1991 : 1174-1175.

Passano L.M., McCullough C.B., (1964). Co-ordinating systems and behaviour in Hydra. I. Pacemaker system of the periodic comtractions, J. Exp. Biol., 42 : 205-231.

Rushforth N.B., Burke D.S. (1971). Behavioral and electrophysiological studies of Hydra. II. Pacemaker activity of isolated tentacles. Biol. Bull., 140 : 502-519.

Santillo S., Taddei-Ferretti C., Nobile M., (1996). Resting potentials recorded in the whole cell configuration from epithelial cells of Hydra. Italian Journal of Zoology, 64 : 7 11.

Taddei-Ferretti C., (1981). *Hydra attenuata* rhythmic activity. Ontogeny, mutual intraction and modulation by light of the different bioelectric activities. VI Natl. Congr. of GNCB, Perugia, Italy, pp. 153-154.

Taddei-Ferreti C., Chillemi S., (1987). Modulation of *Hydra attenuata* rhythmic activity. V.A. revised interpretation. Biol., Cybern, 65 : 225-235.

Taddei-Ferretti C., Cordelia L., (1975). Modulation of *Hydra attenuata* rhythmic activity. Photic stimulation. Arch. Ital. Biol. 113 : 107-121.

Taddei-Ferretti C., Cordella L., (1976a). Modulation of *Hydra attenuata* rhythmic activity II. Phase response curve. J. Exp. Biol., 65 : 737-751.

Taddei-Ferretti C., Cordella L., (1976b). Inhibition of Hydra Motility, VI J. Natl. Congr. Cybern and Biophys, Siena, Italy, pp. 89-94.

Taddei-Ferretti C., Cotugno A., (1988). Wavelength effect on background illumination upon pulse stimulation effect in Hydra. IX Natl. Congr. of GNCB, Trento, Italy, p. 102.

Taddei-Ferretti C., Cotugno A., (1991). Interaction between the effects of light pulses of different chromatic content in *Hydra attenuata*. Hydrobiologia, 216/217 : 589-593.

Taddei-Ferretti C., Musio C., (1999). Neural net of Hydra and the modulation of its periodic activity. In : J. Mora-Mira & R. Moreno-Diaz (eds.), WANN'99 Springer, Berlin, In press.

Taddei-Ferretti C., Chillemi S., Cotugno A., (1987a). Modulation of *Hydra attenuata* rhythmic activity. IV. The mechanism responsible for rhythmic activity. exp. Biol., 46 : 133-140.

Taddei-Ferretti C., Chillemis., Di Maio V., (1988a). The functional structure of the mechanism responsible for the contraction-relaxation activity of *Hydra attenuata*. In : R. Trappl (ed.) Cybernetics and Systems '88, Kluwer, Dordrecht, Boston, London, pp. 413-420.

Taddei-Ferretti C., Cordella L., Chillemi S., (1976). Analyusis of Hydra contraction behaviour. In : G.O. Mackie (ed.) Coelenterata Ecology and Behaviour, Plenum, New York, London, pp. 685-694.

Taddei-Ferretti C., Cordella L., Chillemi S., (1978). Hydra simple nervous system and behaviour. In : R. Trappl, J. Klir & L. Ricciardi (eds). Progress in Cybernetics and Systems Research Vol. III, Hemisphere, Washington, Dc. pp. 643-646.

Taddei-Ferretti C., Musio C., Chillemi S., (1992a). A descriptive model of the interactions in the nervous network of Hydra In : M. Marinaro & G. Scarpetta (eds.), A Structure : from Physics to General Systems Vol. 2 World Scientific, Singapore, New Jersey, London, Hong Kong, pp. 113-130.

Taddei-Ferretti C., Musio C., Cotugno A., (1991). Different weights of the influence of different light pulse position inside a stimulation in Hydra. In : C. Frediani (ed.), Cybermetics and Biophysics Italian Conference, Compositori, Bologna, pp. 123-128.

Taddei-Ferretti C., Chillermi S., Di Maio V., Musio C., (1994). Suppression of Hydra periodic shortening activity by repetitive phase delay. Plogr. Biochem. Biotechnol, 2(1/2) : 147-154.

Taddei-Ferretti C., Di Maio V., Cotugno A., Durante M., (1987b). Photoresponse of symbiotic and aposymbiotic Hydra, VI Natl. Congr. SIBPA, Viareggio, Italy, p. 62.

Taddei-Ferretti C., Di Maio V., Ferraro S., Cotugno A., (1988b). Hydra photoresponses to different wavelengths. In : R.H. Douglas, J. Moan & F. Dall Acqua (eds), Light in Biology and Medicine, Vol. 1, Plenum, New York, London, pp. 411-416.

Taddei-Ferretti C., Di Maio V., Musio C., Cotugno A., (1992b). Modulation of *Hydra attenuata* rhythmic activity. VI. Combined effects of background and pulse light wavelength. J. Photochem. Photobiol. B : Biol. 15 : 307-315.

Taddei-Ferretti C., Musio C., Santillo S., Cotugno A., (1997). Combined light and temperature effects on Hydra periodic behaviour. In : J.C. den hartog (ed.), Coelenterate biology, National Naturhistorisch Museum, Leiden, pp. 473-482.

Taddei-Ferretti C., Musio C., Santillo S., Cotugno A., (1999). Neuronal coding of nonhuman conscious-lide and human conscious activities, in : W. Backhaus (ed.), Neuronal Coding of Perceptual Systems, World Scientific, Singapore, New Jersey, London, Hong Kong, in press.

Trembley A., (1744). Memories pour servir a l'histoire d'un genre de polypes d'eau douce a bras en forme de cornes, J. & H. Verbeck, Leyden, The Netherlands.

Wilson E.B., (1891). The heliotropism of Hydra, Amer. Naturalist, XXV : 413-433.

Proceedings of the National Seminar on Environmental Biology, April 03-05, 1998
Edited By : A.K. Aditya & P. Haldar
DAYA PUBLISHING HOUSE, DELHI

Glimpses of the Biodiversity of *Culicoides* Insects (Diptera : Ceratopogonidae) in India

Debjani Gangopadhyay & Sujit K. Dasgupta
P.G. Zoology Deptt., Presidency College
86/1 College Street, Calcutta - 700073

Abstract

The paper brings to light the illustrated account of 7 new species of *Culicoides* (Diptera : Ceratopogonidae) of India after a critical reappraisal of those manuscript species – *annandalei, causeyi, distinctipalpis, magnithecalis, mukerjii, peculiaris* and *swaminathi*, of Majumdar and Dasgupta (Majumdar 1972). It also furnishes an updated check-list of other 54 valid *Culicoides* species of India citing their subgenus status and brief distribution data thereby showing the extent of species diversity of the genus in India. A total of 61 *Culicoides* species of India thus known validly to the date is found to belong to 10 subgenera of the genus *Culicoides* and as : *Avaritia* (18), *Beltranmyia* (1), *Diphaomyia* (7), *Haemophoructus* (1), *Hoffmania* (12), *Meijerehelea* (3), *Monoculicoides* (1), *Oecacta* (9), *Pontoculicoides* (1) and *Trithecoides* (7), with one species (*C. pelilioueniss*) having its subgenus status undetermined. To indicate their biological diversity, a summarisation of the breeding and feeding habit of the Indian *Culicoides* is furnished in the paper.

Introduction

In their most updated and critical check-list of world species of biting midges forming the nematoceran Diptera family of Ceratopogonidae, Borkent and Wirth (1997) comment that an updated knowledge of species is always the essential base for any future research decisions as our planet continues to experience a biodiversity crisis and frightening level of species extinction. They are able to show that till then the family Ceratopogonidae was known to be composed of 5155 extant and 205 extinct species, accommodated in 5 subfamilies and 6 tribes and 125 valid genera, on global basis.

The genus *Culicoides* Latreille, of tribe Culicoidini of subfamily Ceratopogoninae, is by far the largest genus of 1210 extant and 37 extinct species globally and those species can be accommodated in more than 40 species groups of 29 subgenera. The universal habit of avid haematophagy of females of *Culicoides* drawn attention of researchers who study these tiny insects to make accurate assessment of their diversity, biology and the loss they cause to our economy so that effective programme of management may be undertaken as necessary. Through their painful bites that may cause serious type of allergic dermatitis, *Culicoides* may retard the importance of tourist spots or residential areas or reduce the output of poultry and farm animals. Selected species of biting midges are found to be actual or potential vectors of disease pathogens in man and other animals including poultry and wild birds of which transmission of bluetongue virus (BTV) in domestic and wild ruminants by them is of prime importance even in India. Prasad and Srivastava (1995) observe that India has several outstanding breeds of cattle and buffaloes, but the presence of BTV in India puts trade barrier on commercial movements of those breeds, their germplasm and products.

The work-status on the *Culicoides* insects of Indian subcontinent is far from satisfactory. Through disparate, stray works since Kieffer (1910), a first hand account of the diversity of Indian *Culicoides* may be summarised and it is only during past three decades a few good works on the same are available that offer much-needed base in the Indian context of the subject (Majumdar 1972, Majumdar *et al* 1997, Chaudhuri 1985, Wirth and Hubert 1989, Dasgupta 1995, Paimpali 1996). But all these relate to imaginal biodiversity of Indian *Culicoides* from several States only, and with vast areas of our country still lying unexplored or weakly explored for *Culicoides* biodiversity, it is clear that our knowledge on the subject is far from sufficient. The present paper furnishes taxonomic descriptions of 7 new species of *Culicoides,* collected from Sikkim, Siliguri-Darjeeling area, North 24 Paraganas to Ranigunge area of West Bengal and Dehri-on-Sone in Bihar mostly in late sixties and left unpublished so long though diagnosis reported by Majumdar (1972). The check-list of other known *Culicoides* species of India and some more *Culicoides* species which are identified as new to science and await due reporting show the impressive biodiversity of *Culicoides* midges of India. Information on their biology known so far are also reviewed in brief. It should be clarified that adequate comparison of 7 new species described now in the paper was made by us again in light of our latest knowledge on the *Culicoides* of Southeast Asia (Wirth and Hubert, 1989).

Materials and Methods

The imago stage of insects of the genus *Culicoides* (family Ceratopogonidae, Diptera) extant in India forms the study-material of the present paper. The samples collected/obtained by employing standard techniques are processed for critical assessment of structural features through light microscopy and in those exercises, the procedures of Wirth and Hubert (1989) are followed liberally. The types of the species described now are all light-trapped samples in permanent slide-mounts which will be duly deposited to the National Collection at Zoological Survey of India, Calcutta, with representative additional material submitted colaterally to other depositories from the present custody of the whole material with the second author at Presidency College, Calcutta.

The text of the paper is mainly divided into two parts – Case study of *Culicoides* speciation in India through illustrated taxonomic account in essential format of 7 selected new species of *Culicoides* ascertained by us after critical scrutiny of our unpublished data, Enlistment of known species of Indian *Culicoides* and their distribution range in brief, and brief summarisation of various biological data reported so far about those species.

Results and Discussions

(I) Specific Descriptions

(A) Subgenus *Meijerehelea* Wirth & Hubert, Hegneri Group

1. *Culicoides magnithecalis* Majumdar and Dasgupta, n.sp. (figs. 1-4, Pl. 1; Lit. Ref.- Majumdar 1972 : 147)

Female wing-size 0.94 × 0.46 mm. Deep brown frontovertex with 10 setae. Eyes narrowly separated, bare. Antenna with sensoria on III, XI-XV segments, antennal ratio (AR) 1.23. Maxillary palp with Length/Width Ratio (L/W Ratio) of III segment 2.1, its sensory pit shallow though briaad, and Proboscis/Head Length Ratio (P/H Ratio) 0.58. Mandible of 14-15 teeth, Maxillarblade of 15-16 denticles. Brown pallor in scutum (thorax) dorsad variously intensify to form a distinct pattern; scutellum with 4 large and 6-8 small bristles. Femora-tibial part pale brown without any distinct, clear colour band; Hind Tibial Comb (HTC) of 4 unequal spines; Hind Leg Tarsal Ratio (HLTR) 2.2. Wings greyish with pale spots, setation and venation as figured. Haltere knob pale. Spermatheca one, ovoidal and enormously large though lightly sclerotized, measuring 0.166 × 0.112 mm. Male genitalia (external) well-formed, as figured, the inflated stem-tip with truncated end of aedeagus and mesal fusion of parameres at base outstanding.

Type Data, Comments : Holotype male, from Siliguri (College Campus), West Bengal, during June 1968; Allotype female and Paratypes of 1 male, 3 females – data same as type.

The present species resembles *C. hegneri* Causey, described extensively from six SE Asian countries including Sri Lanka, but it is taken as a new species in view of its outstanding genitalic features, and it is named after its enormously large single spermatheca.

(B) Subgenus *Avaritia* Fox, Orientalis Group

2. *Culicoides annandalei* Majumdar and Dasgupta, n.sp. (figs. 5-8, Pl. 1; Ref. Lit.– Majumdar 1972 : 83)

Female wing-size 0.95 × 0.46 mm. Grayish brown frontovertex with more than 20 setae scattered throughout. Eyes just touching, bare. Antenna with sensoira on III, XI-XV; AR 1.33. Maxillary palp with L/W Ratio of III palpomere 2.87, its sensory pit small, rounded and very shallow; P/H Ratio 0.64. Mandible of 12 teeth, apical few larger. Scutum brown, without a colour pattern dorsad, scutellum with 3 larger bristles. Legs brown to brownish, with bases of all femora and apex of hind tibia vaguely pale while subbasal pale bands occur in all tibiae; hind tibial comb of 5 unequal spines, the first from spurend longest; tarsal ratio 2.25. Wings light brown with fairly distinct pale spots at wing bases (engulfing arculus), at r-m (2 subconfluent spots), at cell R_5, cell M_1 and cell M_2 (3-4 spots in each); two radial cells unequal – I smaller than II; macrotrichia scanty, restricted to apical half; costa short, extends to 0.59 of wing length. Haltere knob pale. Abdomen brown; two developed spermathecae unequal, subspherical and moderately sclerotized (0.048 × 0.038 mm and 0.04 × 0.03 mm), with short neck plus a rudimentary spermatheca and a ring. Male unknown.

Type Data, Comments : Holotype female, by light trap (Coll. T. Sharma), from Renock, Sikkim, at an altitude of 3500 m above sea level in September 1968. It may be placed in the Orientalis Group of s.g. *Avaritia*. Though based on single female, we find its nearness to *C. hui* Wirth and Hubert of Taiwan and we contrast it clearly from its smaller sized pale spots in wings, a different colouring of femora and tibia and larger spermatheca rather broad than long in shape. We name it after Dr. N. Annandale, Superintending Zoologist of Zoological Survey of India, Calcutta, in early part of this century, for his valuable contributions in the development of Zoology and Entomology in India.

(C) Subgenus *Diphaomyia* Vargas, Clavipalpis Group

3. *Culicoides mukerjii* Majumdar and Dasgupta, n.sp. (figs. 9-11, Pl. 1; Ref. Lit.– Majumdar and Dasgupta 1972 : 151).

Female wing-size 0.78 × 0.37 mm. Frontovertex with 30-35 setae. Eyes very narrowly separated, bare. Antenna with sensoria on III, VII-X segments, AR 1.14. Maxillary palp with L/W Ratio of III segment 1.6, its sensory pit wide-mouthed, deep and P/H Ratio 0.53. Mandible of 8-9 teeth, Maxillar blade of 14 denticles. Scutum dorsad having a fixed pattern of brown pallor,

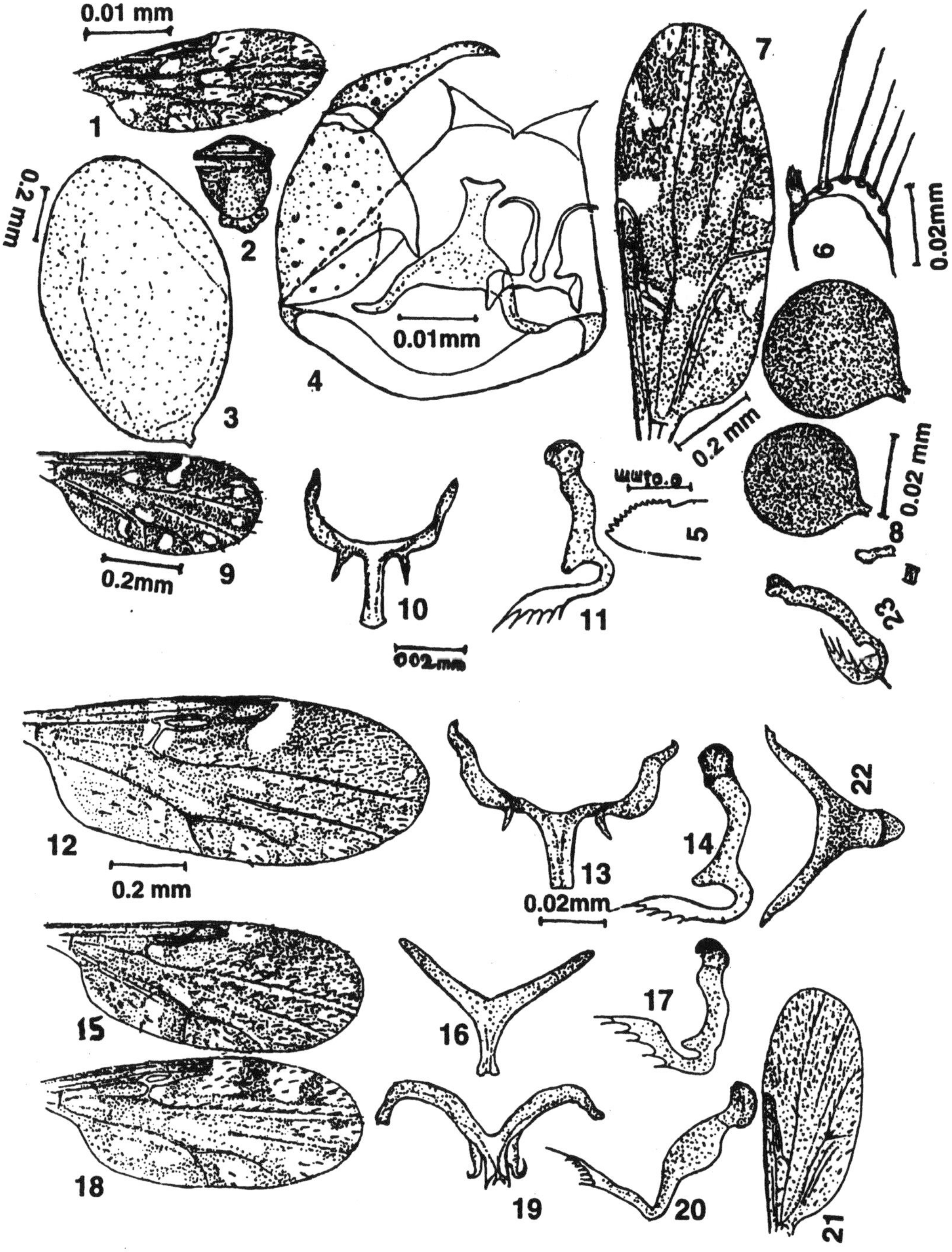

Plate 1, Figs. 1-23 : Camera lucida diagrams (except fig. 2 which is a freehand drawing of dorsal aspect of scutum showing colour pattern) of structural parts of *Culicoides* spp. – *magnithecalis* (figs. 1-4), *annandalei* (figs. 5-8), *mukerjii* (figs. 9-11), *peculiaris* (figs. 12-14), *causeyi* (figs. 15-17), *distinctipalpis* (figs. 18-20), *swaminathi* (figs. 21-23)

scutellum with 4 large and 12-16 small bristles. Hind legs with TR of 1.9, HTC of 4 unequal spines, first from spurend longest wings darksih brown with distinct, small pale spots with setation and venation as figured. Haltere pale. Two subequal, subspherical spermathecae moderately sclerotized with long necks and measuring 0.05 × 0.044 mm and 0.044 × 0.033 mm; a rudimentary spermatheca and a ring present. Male genitalic features of importance as figured.

Type Data, Comments : Holotype male, from Habra (College Campus), 24 Pargs (N), West Bengal, during May 1968; Allotype female, and Paratypes of 10 females and 10 males (+ Additional Specimens of 23 females, 16 males) – from Habra, Darjeeling, Kalimpong (West Bengal) and Renock (Sikkim) during 1968-69.

The present species resembles *C. distinctus* Sen and Dasgupta (of same species group and widespread in India and Thailand) but distinguished by smaller, rounded nature of pale spots in wings and conical processes on aedeagal shoulders of male genitalia being rather slim and long. The species is named in honour of the pioneer Indian worker on *Culicoides* midges, Late Dr. S. Mukerji.

(D) Subgenus *Diphaomyia* Vargas, Haematopotus Group

4. *Culicoides peculiaris* Majumdar and Dasgupta, n.sp. (Figs. 12-14, Pl. 1; Ref. Lit.– Majumdar 1972 : 172)

Female wing-size 1.15 × 0.58 mm. Frontovertex with 15 setae. Eyes narrowly separated, bare. Antenna with sensoria on III, VII-X, XII, XIV segments, AR 1.36. Maxillary palp with L/W Ratio of III segment 1.72, its sensory pit deep, sac-like, and P/H Ratio 0.56. Mandible of 14-15 teeth, maxillar blade of 17 denticles. Scutum dorsal with a fixed pattern of brown pallor, scutellum with 3 large and 12 small bristles. Hind legs with TR of 2.18, HTC of 5 unequal soines. Wings light brown with broad darkish pallor in upper half of cell R_5 and several spacious but feeble pale spots, setation and venation as figured. Haltere pale. Two grossly unequal, oval spermathecae moderately sclerotized with long necks and measuring 0.052 × 0.039 mm and 0.04 × 0.029 mm; a rudimentary spermatheca and ring present. Male genitalic features of importance as figured.

Type Data, Comments : Holotype female, from Darjeeling (Govt. College Campus) during July 1968; Allotype male and 2 female Paratype, with same data as type.

The present species is named *peculiaris* as it is a new species of a few outstanding features in having the distalmost pale spot in cell R_5 circular, very small and removed from wing margin and the aedeagus of male genitalia bearing rather short, conical processes on its shoulder while its caudal stem is simple and flt at tip.

(E) Subgenus *Oecacta* Poey, Similis Group

5. *Culicoides causeyi* Majumdar and Dasgupta, n.sp. (figs. 15-17, Pl. 1; Lit. Ref.- Majumdar 1972 : 101)

Female wing-size 0.87 × 0.42 mm. Frontovertex with 15 setae. Eyes almost touching, bare: Antenna with sensoria on III, V, VII-X segments, AR 1.46. Maxillary palp with L/W Ratio of III segment 1.9, its sensory pit very shallow though large, and P/H Ratio 0.5. Mandible of 10-11 teeth, Maxillar blade of 10-11 denticles. Scutum dorsal with a fixed pattern of brown pallor, scutellum with 3 large and 6-8 small bristles. Hind legs with TR of 2.4, HTC of 4 unequal spines. Wings shaded brown with very small to rather large pale spots, setation and venation as figured. Two ovoidal, moderately sclerotized spermathecae measuring 0.048 × 0.036 mm and

0.042 × 0.032 mm, a rudimentary spermatheca and a ring visible. Male genitalia with smooth-shouldered aedeagus devoid of any projecting processes, and as figured.

Type Data, Comments : Holotype male, from Habra, 24 Pargs (N), West Bengal, during May 1967; Allotype female, data as type; 4 males, 10 females as Paratypes (+21 females as Additional Specimens), collected during May 1968-69, from West Bengal Localities as – Darjeeling, Gobardanga, Habra, Itachuna, Siliguri.

The present species seems quite close to *huffi* Causey, described from India and several other SE Asian Countries, but it is distinct as a new species in several male genitalic features and the distalmost pale spot of cell R_5 quite removed from wing margin. It is named in honour of Late O.R. Causey who discovered several new *Culicoides* species of Thailand in late thirties.

(F) Subgenus *Oecacta* Poey, Segnis Group

6. *Culicoides distinctipalpis* Majumdar and Dasgupta, n.sp. (figs. 18-20, Pl. 1; Lit. Ref.- Majumdar 1972 : 116)

Female wing-size 0.74 × 0.35 mm. Frontovertex with 15 setae. Eyes very narrowly separated, bare. Antenna with sensoria on III - X segments, AR 1.21. Maxillary palp with L/W Ratio of III segment (very large) 1.85, its sensory pit broad-mouthed, shallow, and P/H Ratio 0.58. Mandible of 9 teeth, Maxillar blade of 13 denticles. Scutum dorsal with a fixed pattern of brown pallor, scutellum with 4 large and 4 small bristles. Hind legs with TR of 2.0, HTC of 4 unequal spines. Wings shaded pale brown with no conspicuous markings save for the small post-costal pale spot and this, the setation and venation as figured. Two subeqal, pyriform, moderately sclerotized spermathecae measuring 0.047 × 0.034 mm and 0.042 × 0.029 mm; a rudimenatry spermatheca and ring present. Male genitalia as figured.

Type Data, Comments : Holotype female, from Raniganj, West Bengal, during October 1967; Allotype male, data as type; 10 males and 10 females as Paratypes (+ Additional Specimens 31 males, 101 females) collected during 1967-72 from West Bengal localities as – Burdwan, Chinsura, Habra, Itachuna, Raniganj, Siliguri and also from Dehri-on-Sone (Bihar).

The present species is named *distinctipalpis* in view of its maxillary palpi having the third segment disproportionately large, feeble wing blotching and very distinctive male aedeagus.

(G) Subgenus *Oecacta* Poey, Shortti Group

7. *Culicoides swaminathi* Majumdar and Dasgupta, n.sp. (figs. 21-23, Pl. 1; Ref. Lit.- Majumdar 1972 : 232)

Female wing-size 0.88 × 0.46 mm. Frontovertex with 15 setae. Eyes broadly separated, bare. Antenna with sensoria on III, VIII - X segments, AR 0.95. Maxillary palp with L/W Ratio of III segment 2.2, its sensory pit broad-mouthed, shallow, and P/H Ratio 0.67. Mandible of 13 teeth, Maxillar blade of 13 denticles. Scutum dorsal with a fixed pattern of dark grey pallor, scutellum with 4 large and 4 small bristles. Hind legs with TR of 2.0, HTC of 5 unequal spines. Wings characteristically blotched with large pale spots and dark shade with setation and venation as figured. Two subequal, oval to subspherical, well sclerotized spermathecae with prominent necks and measuring 0.041 × 0.031 mm and 0.036 × 0.025 mm; a rudimentary spermatheca and ring present. Male genitalia as figured.

Type Data, Comments : Holotype female from Habra, 24 Pargs (N), West Bengal, during March 1967; Allotype male, data same as type, 9 females and 2 male as Paratypes collected during 1967 from Gobardanga and Habra, 24 Fargs (N), W.B.

The present species is a new species for its prominent wing blotching combined with aedeagus and parameres in male genitalia of highly characteristic features, and it is named in memory of Late C.W. Swaminath who co-authored Dr. R.O.A. Smith (Late) of School of Tropical Medicine, Calcutta, in describing Assam *Culicoides.*

(II) Biodiversity Data of Indian Culicoides

A check-list of valid species of *Culicoides* known so far from India is summarised here together with distribution range of those species in brief. Their feeding and breeding data are also furnished subsequently as available in literature (Abbreviations : ArP – Arunachal Pradesh, AP – Andhra Pradesh, WB – West Bengal).

(A) Subgenus *Avaritia* Fox

1. *actoni* Smith – 11 SE Asia countries to New Guinea, Japan, Australia includ. Sri Lanka, India (ArP, Assam, Bihar, Kerala, WB).
2. *autumnalis* Sen & Dasgupta – India (WB).
3. *boophagus* Macfie – 4 SE Asia countries includ. Sri Lanka, India (Kerala).
4. *brevitarsis* Kieffer – 5 SE Asia countries, Ryukyu Is, New Guinea, Sri Lanka, India (WB).
5. *certus* Dasgupta – India (WB).
6. *definitus* Sen & Dasgupta – India (WB).
7. *dumdumi* Sen & Dasgupta – Sabah, India (WB)
8. *fulvus* Sen & Dasgupta – widespread in SE Asia to Australia includ. India (Kerala, Tamil Nadu, WB).
9. *himalayae* Kieffer – India Sikkim (WB).
10. *imicola* Kieffer – African tropics, Israel with Sinai Peninsula, Sri Lanka, 3 SE Asia countries Includ. India (Kerala, WB).
11. *inexploratus* Sen & Dasgupta – India (WB).
12. *jacobsoni* Macfie – widespread in SE Asia to Japan, Ryukyu Is., India (Kerala).
13. *molestior* Kieffer – India (WB).
14. *odiosus* Kieffer – India (WB).
15. *orientalis* Macfie – widespread in SE Asia to New Guinea, India (Kerala, Sikkim, West Bengal).
16. *paraliui* Dasgupta – India (WB).
17. *sikkimensis* Dasgupta – India (Sikkim).

(B) Subgenus *beltranmyia* Vargas

18. *circumscriptus* Kieffer – Laos, Thailand to China, Japan, Eurasia, North Africa, India (Bihar, Orissa, WB).

(C) Subgenus *Diphaomyia* Vargas

19. *candidus* Sen & Dasgupta – India (WB)
20. *clavipalpis* Mukerjii – 6 SE Asia countries includ. India (Bihar, Kerala, WB).
21. *distinctus* Sen & Dasgupta – Thailand, India (Kerala, WB).
22. *huffi* Causey – African Tropics, Israel with Sinai Peninsula and 3 SE Asian countries to Sri Lanka, India (Kerala, Orissa, Tamil Nadu, WB).

23. *similis* Carter, Ingram & Macfie – Africa, Arab countries, Russia includ. Crimea, North Caucasus and Transcaucasus, Laos, Malaysia, India (Bihar, Kaharashtra, Andhra, WB).

(D) Subgenus *Haemophoructus* Macfie

24. *rariradialis* Dasgupta – India (Sikkim).

(E) Subgenus *Hoffmania* Fox

25. *indianus* Macfie – 4 SE Asia countries, India (Kerala, Maharashtra).
26. *innoxius* Sen & Dasgupta – 7 SE Asia countries to Sri Lanka, India (Bihar, Kerala, WB).
27. *isoregalis* Majumdar & Dasgupta – India (WB).
28. *neoregalis* Majumdar & Dasgupta – India (WB).
29. *pararegalis* Majumdar & Dasgupta – India (WB).
30. *peregrinus* Kieffer – Sri Lanka to Australia, New Guinea covering many SE Asia countries includ. India (Andhra, Assam, Kerala, Orissa, WB).
31. *pseudoregalis* Majumdar & Dasgupta – India (WB).
32. *quasiregalis* Majumdar & Dasgupta – India (WB).
33. *recurvus* Delfinado – 5 SE Asia countries includ. India (Assam).
34. *regalis* Majumdar & Dasgupta – India (WB).
35. *subregalis* Majumdar & Dasgupta – India (WB).
36. *sumatrae* Macfie – 10 SE Asia countries to Manchuria, New Guinea, India (Kerala).

(F) Subgenus *Meijerehelea* Wirth and Hubert

37. *arakawai* (Arakawa) – SE Asia countries from Manchuria, Japan to Siberia, India (Assam, Kerala, Maharashtra, WB).
38. *hegneri* Causey – 7 SE Asia countries includ. India (Burdwan, Habra & Itachuna fringe around Calcutta, Siliguri – all in WB).

(G) Subgenus *Monoculicoides* Khalaf

39. *homotomus* Kieffer – 4 SE Asia countries to China, Japan, India (WB).

(H) Subgenus *Oecacta* Poey

40. *drydaeus* Wirth & Hubert – 4 SE Asia countries, India (WB).
41. *fortis* Sen & Dasgupta – India (Assam, WB).
42. *majorinus* Chu – Sabah, Tibet, India (WB).
43. *oxystoma* Kieffer – very widespread in SE Asia to Australia, Pakistan and Middle East, Sri Lanka, India (AP, Assam, Bihar, Kerala, Delhi to other NW States, Tamilnadu, WB).
44. *rarus* Dasgupta – India (Sikkim).
45. *shortti* Smith & Swaminath – 5 SE Asia countries, India (Assam, WB).

(I) Subgenus *Pontoculicoides* Remm

46. *kamrupi* Sen & Dasgupta – Laos, Thailand, India (Assam).

(J) Subgenus *Trithecoides* Wirth & Hubert

47. *anophelis* Edwards – 9 SE Asia countries includ. Bangladesh China, Sri Lanka, India (Assam, Kerala, Bihar, Tamil Nadu, WB, AP).
48. *flaviscutatus* Wirth & Hubert – 9 SE Asia countries, Sri Lanka, India (Assam, Kerala, Maharashtra, WB).
49. *inornatithorax* Dasgupta – India (Sikkim).
50. *macfiei* Causey – 4 SE Asia countries, India (Kerala, WB).
51. *palpifer* Dasgupta & Ghosh – widespread in SE Asia, India (Assam, WB).
52. *parararipalpis* Dasgupta – India (Sikkim).
53. *raripalpis* Smith – 3 SE Asia countries to China, India (Assam, Kerala, Sikkim, WB).

(K) Unplaced subgenuswise

54. *pelilioueensis* Tokunaga – widespread in SE Asia, India (WB-Sagar Is).

To the above-listed 54 species are to be added 7 new species described in this paper in order to have the complete list of 61 *Culicoides* species of India to the date. A total of other 37 species, worked out by Majumdar (1972) and Chaudhuri (1985) as new species, await inclusion to reveal the species diversity of *Culicoides* midges in India.

The breeding habitat of Indian *Culicoides* in nature are definitely known for some 17 species only. The most common breeding substrate in nature is the muddy substrate of varying consistence at the edge of locally available drains, ditches, water pools of assorted descriptions starting from hoof-dented stray ones. *Culicoides* species – *arakawai, oxystoms, peregrinus, shortti* and *similis* breed in such substrates whose efficacy rises with higher organic content, or as for *C. huffi* that prefers grassy swamp dumped with trash. The dominant species *C. oxystoma* prefers muddy silts and outlet media off animal sheds and aviaries while *C. brevitarsis* likes cow pats, *C. imicola* the horse paddocks and *C. shortti* and *C. similis* the buffalo wallows. A variation is seen in *C. orientalis* that prefers old piles of manures (surface dry, hard but the inside moist) as breeding substrate.

The choice of vegetational material to breed into is not uncommon with the Indian *Culicoides*. The arboricolous species like *C. drydaeus* breeds freely inside moist rot-holes of wayside trees while the immatures of *C. anophelis* are obtained from water pools in leaf axils of *Colocasia* plants. The rotting stumps of banana plants act as patent breeding substrate for *Culicoides* like – *candidus, clavipalpis, distinctus, innoxius* and *palpifer* (also reared from decaying wild fruits). *C. circumscriptus* breeds both in muddy sites rich in organic content as well as in saline or coastal habitats, while *C. dumdumi* is unique by breeding in rotting sea weeds.

In respect of haematophagic diversity of females of Indian *Culicoides* species, following Jamnback (1965) that those with 4-6 sensillar antennomeres (per antenna) are mammalophilic, those with 8-13 sensillar antennomeres (per antenna) are ornithophilic and those bearing the intermediate number of 7 sensillar segments (per antenna) are of geneal blood-sucking habit, we may ascribe the anthropophilic habit to 7 such *Culicoides* species of India as – *definitus, distinotus, distinctipalpis* each with 8 sensillar antennomeres and 4 others such as *arakawai, certus, circumscriptus, pelilioueensis* each with 12 sensillar antennomeres/antenna. A mammalphilic habit may thus be ascribed to 4 other species as : 9 species each with 4 sensillar antennomeres/ antenna – *candidus, clavipalpis, fortis, homotomus, kamrupi, oxystoma, shortti, swaminathi, rarus*; 8 species each with 5 sensillar antennomeres/antenna – *actoni, brevitarsis, causeyi, distinctus, hegneri, imicola, mukerjii, raripalpis*; 24 species each with 6 sensillar antennomeres/antenna – *anophelis, annandalei, drydaeus, dumdumi, flaviscutatus, huffi, indianus, innoxius, inornatithorax, isoregalis, macfiei, majorinus, magnithecalis, neoregalis, orientalis, palpifer, pararegalis, peculiaris, peregrinus, pseudoregalis, quasiregalis, rariradialis, regalis* and *subregalis*. With seven sensillar antennomeres/antenna, the two *Culicoides* species, *insolens* and *similis*, belong to the category of general blood-sucking forms. A good corroboration of these contentions is noted in most of the cases of observed feeding habits of *Culicoides* species of India (Dasgupta 1995, Dasgupta *et al.* 1969, Wirth and Hubert 1989). In some cases, the per antenna sensillar antennomere number may vary within the range of a species which may naturally exhibit deviations in its haematophagic habit as well.

Acknowledgement

We thank our institutional authorities for permission and privileges for our exercises on Indian Ceratopogonidae over years. We are also grateful to our colleagues and friends in the department who extended various courtesies and help to us in completing our work. Our thanks are also due to the ceratopogonid experts of other places specially Dr. W.W. Wirth (Late), the topmost authority retired from the U.S. National Museum and Dr. A. Borkent, of Canada, for consultative help and donating us technical literature. Correspondence with Mr. T.M. Paimpali, of St. Stephens College, Uzhavoor (Kerala) who is also engaged in the study of Indian ceratopogonids is also found useful for which we thank him duly.

References

Borkent, A. and Wirth, W.W. (1997). World species of biting midges (Diptera : Ceratopogonidae). *Bull. American Mus. Natl. Hist., N.Y. No. 233, pp. 257.*

Chaudhuri, P.P. (1985). A taxonomic study of the *Culicoides* biting midges (Ceratopogonidae : Diptera : Insecta) of West Bengal. *Ph.D. thesis (Guide Dr. S.K. Dasgupta), North Bengal Univ., 242 pp.*

Dasgupta, S.K. (1995). Morphotaxonomic features and species of Indian *Culicoides* (Diptera : Ceratopogonidae). *As Chapter 7 between pp. 115-188 in the technical treatise, "Bluetongue : Indian Perspective" – a CCS Haryana Agric. Univ. Publn., editorship of G. Prasad and R.N. Srivastava.*

Dasgupta, S.K., Sharma, T., Mitra, A. and Dasgupta, B (1969). On the incidence and blood-meal of biting insects (Diptera, Ceratopogonidae, Simuliidae) of Darjeeling, India, *Rev. d'Entomol. de l'USSR, 48 : 770-773.*

Jamnback, H.A. (1965). The *Culicoides* of New York State. *New York State Mus. Bull., No. 399 : 1-154.*

Majumdar, B.C. (1972). Studies on systematics and distribution of *Culicoides* midges (Ceratopogonidae : Diptera) of eastern India. *Ph.D. thesis (Guide Dr. S.K. Dasgupta), Calcutta Univ.*, 265 pp.

Majumdar B.C.,. Dasgupta, S.K. and Gangopadhyay, L. (1997). Some new species of *Hoffmania* Fox subgenus of *Culicodes* biting midges (Diptera : Ceratopogonidae) from Darjeeling. *Journ. Bengal Natl. Hst. Soc. N.S., 16(2) : 27-34.*

Paimpali, T.M. (1996). Biting midges of Kerala – a preliminary report. *Hexapoda, 8(2) : 55-66.*

Prasad, G. and Srivastava, R.N. (1996). Bluetongue : Indian Perspective. *A CCS Haryana Agric. Univ. Publn. Hisar, 125004, India, 188 pp.*

Wirth, W.W. and Hubert, A.A. (1989). The *Culicoides* of Southeast Asia (Diptera : Ceratopogonidae). *Mem. American Entomol. Inst., No. 44, 508 pp.*

Proceedings of the National Seminar on Environmental Biology, April 03-05, 1998
***Edited By :* A.K. Aditya & P. Haldar**
DAYA PUBLISHING HOUSE, DELHI

Induced Abnormalities and Neoplasia in Planarians, *Dugesia bengalensis*, Kawaskatsu

Shyamal Mitra & A.K. Aditya
Deptt. of Zoology, DSA-UGC
Visva-Bharati, Santiniketan

Turbellarians constitute first evolutionary step among animals that have well defined organ system. The planarians are simple organisms and many works have been undertaken to unearth the interesting complicated phenomena underlying the process of regeneration in them. *Dugesia bengalensis*, Kawakatsu is free living aquatic planarian species. It is endowed with high regenerative efficiency associated with multipotential neoblast cell participation and proliferation, the bilataral symmetry and with novel axial gradient system.

Indiscriminate disposal of heavy metal pollutants from different industries causes a great threat to all aquatic biota. Cadmium being one of the black listed metal is known to be highly toxic to most organisms, even in small concentration (Kopp and Koner, 1970; Loverti *et al.* 1972, Suresh *et al*, 1991). It is a common hazardous nondegradable and potentially harmful aquatic pollutants (Babich and Stoxky, 1978). Because of its chemical similarity it occurs naturally in association with zinc minerals. Cadmium intoxification is responsible for decreased growth rate, it is also reported to inhibit glycogen storage (Lowe-Jinde and Niimi, 1984) blocking m-RNA synthesis and protein synthesis (Vijayram, *et al*, 1989). Cadmium enters into the planarians through their body surface and with nutrients which result in accumulation of cadmium in the tissues, causing impairment in the functioning of normal physiological processes, retarding the growth and reproduction and may ultimately lead to death.

The induction of abnormal differentiation and neoplasia in planarians by action of carcinogens and mutagens have been studied by many investigators (Goldsmith 1939; 1941; Lange, 1966; Hall et al 1986, Best and Morita, 1990, Mahapatra and Aditya, 1992) heavy metal can also cause damage to the intricate system of these animals (Aditya and Bandopadhyay, 1995). Benzo(a)pyrene is a well known carcinogen occuring in crystalline form and thus easily manipulated in insertion into tissues. Foster (1969) reported that *D. dorotocephala* treated with the carcinogen formed malignant tumors at high frequency.

Material and Methods

Free living adult aquatic planarians *Dugesia bengalensis* constituted the experimental material for this work (Kawakatsu et al., 1983). The specimens were of approximately of same age and size group of 10 mm to 12 mm long and 1.5 mm to 2.5 mm diameter, collected from local ponds of Santiniketan deer park and kept in glass petridishes with culture medium in the laboratory in a BOD incubator at a temperature of 20±1°C. The planarians were fed with goat liver twice a week and acclimated to laboratory conditions at least for one month before they were exposed to cadmium chloride and Benzo(a)pyrene.

The specimens selected for the experiment were in 72 hours of fasting condition and were divided into different sets. The first set contained 10 animals as control. In the second set (four petridishes of 250 ml water capacity contained 10 planarians each) specimens were exposed to different concentrations of $CdCl_2$. The 24, 48, 72 96H LC_{50} values of cadmium chloride had been determined by Probit-Analysis method. Having determined the 24th LC_{50} value the third set of experiment was conducted with the exposure to 1/5th the sublethal concentration. In these set the intact individuals were kept in petridish and supplied with $CdCl_2$ at a concentration of 3 mg/l and were exposed for a period of 96 hours. The specimens were sacrificed at different post treatment periods as well as at 24, 48, 72 and 96 hours during treatments. Paraffin sections of 6 μ thickness were cut and stained to note the cellular changes in them.

Regeneration was stimulated in experimental animals by removal of the pharynx. To accomplish this, animals were immbolized by placing them on damp filter paper resting on the cover of icefilled petridish and the pharynx was cut off at its base with a glass needle inserted from the dorsal side; the pharynx was then extricated through the mouth. Two days after the wound healed a microcrystal of approximately 5 μg benzo(a)pyrene was inserted into pharyngeal cavity from the dorsal side with forceps. The worms were then maintained for 3-4 weeks in culture medium at 20°C without feeding. Control experiments were performed on 10 animals who were subjected to the same operation except that a piece of glass the size of the benzo(a)pyrene crystals was placed in the pharyngeal cavity instead of the carcinogen.

Results and Discussion

The specimens which have been subjected to $CdCl_2$ for determination of their LC_{50} value for different hours are brought back to the culture medium for further observations. It is noticed that the rate of mortality is directly related to the time and concentrations used.

During the determination of LC_{50} values and later on for exposure to 96 hours at the sublethal dose, the planarians are noted to be initially very uneasy and gradually become very lethargic. Sometimes they adhere to the glass surface and contract their bodies, the lateral part of the body becomes wrinckled and later free the body from the petridish and move vigorously up and down showing the medium is very unsuitable and unhealthy to them.

The animals are exposed to a dual mode fo action of the toxic substances : (i) the polluting agents react into their body in a considerable manner through direct surface contact and (ii) accumulation of persistent xenobiotic materials entering through the nutrients. The physico chemical parameters like temperature, pH, dissolved O_2 and Co_2 concentration, hardness of water have tremendous influence on TLm value of various xenobiotics. The differences in the celluler response to the toxic impact may be pharynx but the control of morphological differentiation is lost. The abnormally intense staining of these masses, their abnormal form and the amitosis in these cells show that these masses are neoplasias; this view is supported by the work of Foster (1963). He is also of the opinion that tumors in planarians consist of neoplastic cells formed by the transformation of neoblasts that appear during regeneration (Foster, 1969).

In most animals, tumors rarely form in highly regereative zones (Hoshina and Teshirogi, 1991). The low rate of neoplasia production in this region despite the treatment with the carcinogen is probably related to that high pharyngeal regenerative capacity. However, for the characterisation of these neoplastic cells and to understand their origin and nature, the detail study is required.

References

Aditya, A.K. and M.P. Bandyopadhyay, (1995). Mercuric chloride induced changes in *Dugesia bengalensis* Kawakatsu, and aquatic plansian from Santiniketan, West Bengal. *J. Environ. Biol.* 16(3).

Babich, H. and Stozky, G. (1978). Effects of Cadmium on the biota : influence of environmental factors. *Adv. Apl. Micro. Biol.* 23-55.

Best, J.B. and Morita, M. (1991). Toxicology of planarians, *Hydrobiologia,* 227.

Foster, J.A. (1963). Induction of neoplasms in planarians with carcinogens *Cancer Res.* 23.

Foster, J.A. (1969). Malformations and lethal growths in planarians treated with carcinogens. *Nat. Cancer Inst.* Monogr. 31.

Ghosh, P. (1985). Preliminary observations on a new species of aquatic planaria, *Dugesia bengalensis* Kawakatsu, from West Bengal M. Phill Dissertation. Burdwan University.

Goldsmith, E.D. (1941). Further observations of supernumerary structures in individuals of an artificially produced clone of *Dugesia tigrina Anat Rec.* 81.

Goldsmith, E.D. (1939). Spontaneous outgrowths in *Dugesia tigrina. An. at. Rec.* (suppl.) 75.

Hall, F., M. Morita and J.B. Best (1986). Neoplastic transformation in the planarian I cocarcinogenesis and histopathology. II Ultrastructure and malignant reticuloma *J. Exp. Zool.* 240.

Kawakatsu, M., Oki, I:, Tamura, S. Yamayoshi, T. and Aditya A. (1983). A new fresh water planarian from West Bengal *Bull Biogeogr. Soc.* Japan 38.

Kopp, J.F. and R.C. Koner, (1970). Trace metal in water of united states, *Fed. Water pollut. controll. Admins.,* U.S. Dept. intrior cincinnati, Ohio.

Lange, C.S. (1966). Observations of some tumers found in two species of planaria – *Dugesia etrusca* and *D ilvana J. Embryol exp Morph.* 15.

Loveti, R.J., W.H. Gutenman,, L.S. Pakkala, W.D. Youngs, D.J. Lisk, G.E. Burdick and E.J. Haris, (1982). A survey of the total cadmium content 406 fish from 49 New York State fresh waters. *J. Fish. Res. Board. Can* 29.

Lowe, Jinde, R. and A.J. Niimi, (1984). Short term and long term effect of cadmium of glycogen reserves and liver size in rainbow trout (Salmo gairdneri Richardson) *Arch. Environ. Contam. Toxicol,* 13.

Mahapatra, S. and A.K. Aditya, (1992) Effect of carcinogens and mutagens on phospholipid and protein in aquatic planaria *Dugesia bengalensis,* Kawakatsu, indian Journal of Experimental Biology Vol. 30, October.

Suresh, A.B. Sivaramakrishna, P.C. Victoriamma, and K. Radhakrishaiah (1991). Shifts in protein metabolism in some organs of fresh water fish *Cyprinus carpio* under mercury stress. *Biochem. Int.* 24(4).

Proceedings of the National Seminar on Environmental Biology, April 03-05, 1998
***Edited By :* A.K. Aditya & F. Haldar**
DAYA PUBLISHING HOUSE, DELHI

The Macroinvertebrate Diversity of Some Urban Wetlands of Calcutta

Manika Mukherji, Sujit Pal & N.C. Nandi
Zoological Survey of India
M-Block, New Alipur
Calcutta - 700 053

Abstract

Distribution, diversity and density of macroinvertebrates of three urban wetlands, namely, Rabindra Sarovar (RS). Subhas Sarovar (SS) and Brace Bridge Nature Park (BBNP) in Calcutta metropolis, studied for period of two years, 1995-97, reveals a total of 104 species belonging to 11 major groups, mainly Oligochaeta (2 species), Hirudinea (3 species), Crustacea (7 species), Arachnida (14 species), Ephemeroptera (1 species), Odonata (9 species), Hemiptera (24 species), Coleoptera (19 species), Diptera (6 species), Gastropoda (14 species) and Bivalvia (9 species). The population density of Oligochaeta, Crustacea, Diptera and Gastropoda was highest in post monsoon but density of Bivalvia and Coleoptera was more in premonsoon in three wetlands surveyed. However, in RS and SS gatropod density was highest followed by Oligochaetes and Bivalves. In BBNP, Oligochaete density was found to be higher than gastropods in all seasons indicating organic pollution.

Introduction

The macroinvertebrates of fresh water wetlands primarily annelids, molluscs and arthropods provide significant support to the aquatic food web and are responsible for the sustenance of cultivable fishes, water snakes, aquatic birds and other wildlife (Singh and Roy, 1991). Their composition, abundance and distribution provide an index of the ecosystem emphasising on the trophic structure, water quality and eutrophication of the ecosystem (Wiederholm, 1980 and Lang, 1985).

In India, in recent years, though there has been a large number of investigations on the diversity, both qualitative and quantitative aspects of aquatic life specially macroinvertebrates, by authors like Govind (1969); Munawar (1970); Gupta (1981; 1983); Gupta and Pant (1986); Rai *et al.* (1981); Kulshrestha *et al.* (1988) and Singh and Roy (1991); but unfortunately little work has been done on the macroinvertebrate fauna of the wetlands of West Bengal specially those located in and around Calcutta, except Datta and Sarangi (1980), Ghosh and Chattopadhyay (1990; 1994) and Nandi *et al.* (1993). Hence the present study on the species composition, diversity and abundance of macroinvertebrates of three urban lakes of Calcutta viz., Rabindra Sarovar (RS), Subhas Sarovar (SS) and the wetlands of Brace Bridge Nature Park (BBNP) has been undertaken.

Materials and Methods

Study Area : The Calcutta Metropolitan Area (CMA) is a part of mature delta of the river Ganga, having a large number of natural wetlands scattered over the urban, industrial and rural areas of this region lying approximately between latitudes 22°28' and 22°33' North and longitudes 88°17' and 88°23' East. Of the three wetlands surveyed, RS and SS are two artificial rainfed reservoirs situated in South and North Eastern part of Calcutta respectively. Both the lakes were excavated by Calcutta Improvement Trust (CIT), the governing body of the two lakes, from the second decade of the 20th century for extending the city through acquisition of marshy land. RS has recently attained the status of 'National Lake' under the National Lake Conservation Programme (NLCP) set up by the Union Ministry, Government of India, and is larger than SS (39.5 acres), having land and water area of about 119 and 73 acres respectively. The maximum length of RS is about 1770 m and maximum width at its broadest part is 286 m. SS has maximum length of 533.3 m and width is about 366 m.

BBNP is located in the Garden Reach Area of the South Western industrial part of Calcutta. The wetlands and the adjoining area belong to the Calcutta Port Trust (CPT), a statutory body of the Government of India. Formerly a swampy bed, it has now been converted into, sewage-fed fisheries by the Mudially Fishermen's Co-operative Society (MFCS). In 1985, MFCS established 50 ha of this water area and 20 ha of the surrounding embankment as an 'Ecological Park' or 'Nature Park'. The waterbodies receives untreated effluent from more than 152 industrial units (Deb *et al.*, 1996).

Qualitative studies were made seasonally to inventorise the faunal diversity of these reservoirs. Quantitative studies were done using a metal box type sampler (15 × 15 cm^2) for benthic organisms and a perplex box type sampler (20 × 20 × 40 cm^3; Pal and Nandi, in press) for macrophyte associated fauna. The collected samples were washed in sieves of 40 mesh size in the field so as to retain the macroinvertebrates for calculating their abundance and population density as number of organism/sq.m.

Results

Qualitative study reveals the presence of 11 major groups of macroinvertebrates (Table 1) with a total of 104 species (Oligochaeta 2, 1.92%; Hirudinea 3, 2.88%; Crustacea 7, 6.73%; Arachnida 14, 13.46%; Ephemeroptera 1, 0.96%; Odonata 9, 8.65%; Hemiptera 24, 23.07%; Coleoptera 19, 18.26%; Diptera 6, 5.76%; Gastropoda 14, 13.46% and Bivalvia 5, 4.80%).

The population (number/m^2) of macroinvertebrate along with percentage composition presented in Table-2 showed considerable seasonal variations in abundance and percentage composition with maximum population in postmonsoon (October-January) and minimum in monsoon (June-September) in all the waterbodies (RS, SS and BBNP). The number of macroinvertebrates/m^2 in RS was always found to be more than SS and BBNP, which had the least density.

Annelida – Annelids were represented by 2 species of Oligochaeta, *Branchiura sowerbyi* and *Limnodrilus hoffmeisteri* and 3 species of Hirudinea, *Hemiclepsis marginata asiatica, Hirudinaria manillensis* and *Placobdella emydal*. Oligochaetes showed greatest abundance in postomonsoon and least in monsoon seasons in all the wetlands. In RS its number varied from 1592.85 – 2255.76/m^2 (36.87% – 43.74%), in SS from 1359.13 – 2028.58/m^2 (35.09% – 44.67%) while in BBNP from 1176.60 – 2997/m^2, but its percentage composition (55.25% – 63.64%) was comparatively much higher than RS and SS. On the other hand, percentage of hirudineans was very low ranging from 0 – 0.65% (0–32.54%/m^2), but they showed a considerable increase in number in RS (2.48/m^2), SS (32.54/m^2) and BBNP (3.25/m^2) during postmonsoon.

Table 1 : Qualitative Collection of Macroinvertebrate Fauna of RS, SS and BBNP

		Rabindra Sarovar	*Subhas Sarovar*	*BBNP*
	1	*2*	*3*	*4*
I.	**ANNELIDA**			
	Oligochaeta (2 species)			
	Branchiura sowerbyi	+	+	+
	Limnodrilus hoffmeisteri	+	+	+
	Hirudinea (3 species)			
	Hemiclepsis marginata asiatica	+	+	+
	Hirudinaria manillensis	+	+	–
	Placobdella emydal	–	+	–
II.	**ARTHROPODA**			
	Crustacea (7 species)			
	Amphipoda (unidentified)	–	+	+
	Caridina sp.	+	+	–
	Isopoda (unidentified)	+	+	+
	Macrobrachium dayanum	+	+	+
	M. lamarrei	+	+	+
	Sartoriana spinigera	–	+	+
	Varuna litterata	–	+	–
	Arachnida (14 species)			
	Arrenurus sp.	+	–	–
	Evippa shivajii	+	–	+
	Larinia sp.	+	–	+
	Lycosa sp.	+	+	–
	Myrmarachne sp.	+	–	+
	Pardosa alii	+	+	–
	P. annandalei	+	+	+
	P. birmanica	+	+	+
	P. pusiota	+	+	+
	Salticus sp.	+	–	–
	Tetragnatha sp.	+	+	–
	Theridion sp.	–	+	+
	Trochosa punctipes	–	+	–
	Unionicola sp.	+	–	–
	Insecta			
(A)	*Ephemeroptera* (1 species)			
	Cloeon sp.	+	+	–
(B)	*Odonata* (9 species)			
	Agriocnemis pygmaea	–	+	–
	Brachythemis sp.	+	+	+
	Ceriagrion coromandelianum	+	+	+
	Crocothemis sp.	+	+	–
	Ictinogamphus rapak	+	+	–
	Ischnura spp. (2 species)	+	+	+
	Orthetrum sp.	–	+	–
	Pseudagrion sp.	+	+	+

Contd...

Table 1 – *contd...*

	1	2	3	4
(C)	*Hemiptera* (24 species)			
	Anisops spp. (2 species)	+	+	+
	Diplonychus annulatum	+	+	+
	D. rusticum	+	+	+
	D. sp.	+	+	+
	Gerris spp. (2 species)	+	+	+
	Helotrephes indicus	–	–	+
	Hydrometra spp. (2 species)	+	+	+
	Laccotrephes spp. (2 species)	+	+	+
	Limnogonus spp. (2 species)	+	+	+
	Micronecta quadristrigata	+	+	+
	M. thyesta	+	+	+
	Microvelia albomaculata	–	–	+
	M. diluta	–	–	+
	Noboandelus signatus	–	–	+
	Plea sp.	+	+	+
	Ranatra elongata	–	–	+
	R. filiformis	–	+	+
	R. sordidula	+	+	+
	Rhagadotarsus kraepelini	–	–	+
(D)	*Coleoptera* (19 species)			
	Amphiops pedestris	+	+	+
	Berosus indicus	+	+	+
	Canthydrus laetabilis	+	+	+
	Cassida sp.	+	+	+
	Clypeodytes sp.	–	+	–
	Dicladispa armigera	–	+	+
	Dineutus unidentatus	+	+	–
	Enochrus sp.	+	+	+
	Haliplus sp.	+	+	–
	Helochares ancholaris	+	+	+
	Hydaticus fabricii	–	–	+
	Hydrocoptus subvittulus	+	+	+
	Hydrophilus olivaceus	+	–	–
	Hydrovatus sp.	+	+	+
	Laccophilus spp. (2 species)	+	+	+
	Regimbertia sp.	+	+	–
	Sternolophus rufipes	+	+	+
	Uvarus quadrilineatus	+	+	–
(E)	*Diptera* (6 species)			
	Chironomids (3 species)	+	+	+
	Mosquito larva (2 species)	+	+	+
	Odontomyia dorsoangulata	+	+	+
III.	**MOLLUSCA**			
	Gastropoda (14 species)			
	Bellamya bengalensis	+	+	+

Contd...

Table 1 – *contd...*

1	2	3	4
Brotia costula	+	+	+
Digoniostoma cerameopoma	+	+	+
Gobbia orcula	+	+	+
Gyraulus convexiusculus	+	+	+
G. labiatus	+	+	+
Indoplanorbis exustus	+	+	+
Lymnaea acuminata	+	+	+
L. luteola	+	+	+
Pila globosa	+	+	+
Thiara granifera	+	+	+
T. lineata	+	+	–
T. tuberculata	+	+	+
T. scabra	+	+	–
Bivalvia (5 species)			
Lamellidens correanus	+	–	+
L. marginalis	+	+	+
Parreysia caerulea	+	–	–
P. pachysoma	+	–	–
Pisidium clarkaenum	+	+	+

Crustacea : Custaceans were represented by 7 species, *Macrobrachium lamarrei, M. dayanum, Caridina* sp., *Varuna litterata, Sartoriana spinigera,* an isopod and an amphipod. Their population decreases in monsoon and tends to increase during post and premonsoon seasons. In RS its density varied from 22.51–38.01/m^2 (0.56%–0.73%), in SS from 13.50–51.17/m^2 (0.35%–1.45%) and in BBNP from 28.34–62.14/m^2 (1.31%–1.80%).

Arachnida : This group is represented by 14 species, some common ones are *Pardosa annandalei, P. pusiota* and *P. alii.* The percentage composition of arachnids was insignificant ranging from 0.002%–0.10% and did not show any definite seasonal pattern.

Ephemeroptera : It is represented by only 1 species *Cloeon sp.* with the density fluctuating between 0–9.18/m^2 (0–0.26%), increasing in premonsoon.

Odonata : Total 9 species comprise the odonate group. Among these *Pseudagrion microcephalum, Brachythemis contaminata* and *Ischnura* sp. are most dominant. In RS and SS, the abundance of odonates was highest in monsoon, but in BBNP it was more in postmonsoon. In RS the odonate density varied from 7.09–12.09/m^2 (0.14–0.27%), in SS from 7.21–10.28/m^2 (0.14%–0.26%) and in BBNP from 1.20–5.80/m^2 (0.05%–0.12%).

Hemiptera : It is the largest macroinvertebrate group consisting of 24 species of which *Diplonychus annulatum, Ranatra sordidula, Anisops* sp., *Limnogonus* sp. and *Micronecta* sp., are most dominant, but its density as well as percentage was very less ranging from 3.21–18.12/m^2 and 0.15%–0.59% respectively, showing no definite seasonal fluctuations.

Coleoptera : This group has rich species diversity comprising of a total of 19 species. Of the 19 species *Amphiops pedestris, Hydrocoptus subvittulus, Laccophilus* sp., *Enochrus* sp. and *Canthydrus laetabilis* are most common. The maxium number of coleopterans was observed in premonsoon in RS (7.55/m^2, 0.18%) and SS (25.51/m^2, 0.72%). In BBNP its density was highest in postmonsoon (11.33/m^2), but the percentage composition was greatest in premonsoon (0.33%). In general the population density is higher in SS than RS and BBNP.

Table 2 : Seasonal Abundance (Number/m²) and Percentage (%) Distribution of Different Macroinvertebrate Groups in RS, SS and BBNP

Group	Rabindra Sarovar			Subhas Sarovar			Brace Bridge Nature Park		
	Premonsoon No./m²	Monsoon No./m²	Postmonsoon No./m²	Premonsoon No./m²	Monsoon No./m²	Postmonsoon No./m²	Premonsoon No./m²	Monsoon No./m²	Postmonsoon No./m²
Oligochaeta	1630.13 (40.58%)	1592.82 (36.87%)	2255.76 (43.74%)	1575.76 (44.67%)	1359.13 (35.09%)	2028.58 (40.74%)	1687.20 (61.75%)	1176.60 (55.25%)	2997.00 (63.64%)
Hirudinea	0.30 (0.007%)	0.82 (0.01%)	2.8 (0.05%)	0.24 (0.006%)	– –	32.54 (0.65%)	0.15 (0.005%)	1.10 (0.05%)	3.25 (0.06%)
Crustacea	22.51 (0.56%)	24.78 (0.57%)	38.01 (0.73%)	51.17 (1.45%)	13.50 (0.35%)	48.39 (0.97%)	49.32 (1.80%)	28.34 (1.33%)	61.14 (1.31%)
Arachnida	2.72 (0.06%)	0.52 (0.01%)	4.02 (0.07%)	0.01 (0.002%)	2.88 (0.07%)	3.57 (0.07%)	1.09 (0.03%)	2.31 (0.10%)	3.21 (0.06%)
Ephemeroptera	0.89 (0.02%)	0.76 (0.02%)	0.26 (0.005%)	9.18 (0.26%)	0.01 (0.002%)	– –	1.12 (0.04%)	0.34 (0.01%)	0.89 (0.01%)
Odonata	9.74 (0.24%)	12.09 (0.27%)	7.09 (0.14%)	8.29 (0.23%)	10.28 (0.26%)	7.21 (0.14%)	2.31 (0.08%)	1.20 (0.05%)	5.80 (0.12%)
Hemiptera	9.04 (0.22%)	16.43 (0.38%)	16.63 (0.32%)	11.02 (0.31%)	10.41 (0.26%)	16.50 (0.33%)	16.23 (0.59%)	3.21 (0.15%)	18.12 (0.38%)
Coleoptera	7.55 (0.18%)	0.52 (0.01%)	3.52 (0.06%)	25.51 (0.72%)	9.87 (0.25%)	20.00 (0.40%)	9.14 (0.33%)	6.22 (0.29%)	11.33 (0.24%)
Diptera	5.72 (0.14%)	1.50 (0.03%)	11.76 (0.22%)	15.03 (0.42%)	0.24 (0.006%)	28.96 (0.58%)	222.00 (8.12%)	44.40 (2.08%)	488.40 (10.37%)
Gastropoda	2127.72 (53.10%)	2531.78 (58.61%)	2637.84 (51.14%)	1720.22 (48.76%)	2404.81 (62.08%)	2708.28 (54.39%)	632.60 (23.15%)	821.40 (38.57%)	1042.40 (22.13%)
Bivalvia	190.23 (4.74%)	137.20 (3.18%)	179.75 (3.48%)	110.81 (3.14%)	62.09 (1.60%)	84.78 (1.70%)	111.00 (4.06%)	44.40 (2.08%)	76.60 (1.62%)

Diptera : Among the 6 dipteran species, chironomid larvae were recorded as abundant. Its number was observed maximum in post monsoon (RS–11.76/m², 0.22%; SS–28.96/m², 0.58% and BBNP–488.4/m², 10.37%) and minimum in monsoon (RS–1.50/m², 0.03%; SS–0.24/m², 0.006% and BBNP–44.4/m², 2.08%). Chironomid population was much higher in BBNP than RS and SS.

Gastropoda : Gastropods contribute to the major portion of the macroinvertebrate community in respect of density and percentage composition in all the wetlands under study. This group comprises of 14 species of which *Bellamya bengalensis, Gabbia orcula, Digoniostoma cerameopoma, Thiara granifera, T. tuberculata, Brotia costula, Indoplanorbis exustus* and *Gyraulus labiatus* are most dominant. Gastropod population is higher in RS (2127.72–2637.84/m²) than SS (1720.22–2708.28/m²) least in BBNP (632.6–1042.4/m²). It speak was observed in postmonsoon and least in premonsoon.

Bivalvia : This group was represented by 5 species, *Lamellidens marginalis, L. correanus, Pisidium clarkaenum, Parreysia caerulea* and *P. pachysoma*. Bivalve population was always highest in RS reaching its peak in premonsoon (RS–190.23/m², 4.74%; SS–110.81/m², 3.14% and

BBNP–111/m^2, 4.06%) and lowest in monsoon (RS–137.20/m^2, 3.18%; SS–62.09/m^2, 1.60% and BBNP–44.4/m^2, 2.08%) in all the three wetlands.

Discussion

The present study on the macroinvertebrate diversity and seasonal abundance of the three urban wetlands in Calcutta reveals the rich insect diversity comprised of 59 species representing 56.70% of the total macroinvertebrates (104 species). But its percent composition in terms of density (no./m^2) is very small signifying that the 5 insect groups (Ephemeroptera, Odonata, Hemiptera, Coleoptera and Diptera) have low population density (0–488.40/m^2), often being >1%, when compared to gastropod (632.6–2708.28/m^2) and oligochaeta (1176.60–2997/m^2) density. Gastropoda with 14 species (13.46%), Oligochaeta with 2 species (1.92%) and Bivalvia with 5 species (4.8%) are the most dominant groups in all the three wetlands. Such dominance of gastropods and oligochaetes in respect of density has been earlier reported by Gupta (1976), Barbhuyan and Khan (1992) and Singh *et al.* (1994). The lentic condition of these wetlands infested with more than 20 submerged and emergent macrophytes provides excellent diverse niches for several insect larvae/adults which are adopted for mining into stems and leaves.

Mention may be made that the macrophytes provide suitable food and shelter for insects and gastropods (McLachlan, 1975; Soszka, 1975 and Maitland, 1978). At the same time, the soft clay soil with decaying leaves and other organic matter in the bottom soil influences the growth and propogation of various benthic animals (Bose and Lakra, 1994). Bose and Lakra (1994) also reported that the maximum number of most of the species of organism in aquatic ecosystem occurred after rain, which also corroborates our investigation showing the highest density of macroinvertebrates during postmonsoon season. During the monsoon months, the inflowing rain water from the catchment area, bearing nutrients and organic matter settle at the bottom of the wetlans, which would have decomposed in postmonsoon favouring the growth of luxuriant vegetation and aquatic organism. During premonsoon macroinvertebrate population showed a decreasing trend due to their reduced rate of multiplication. Majority of the macroinvertebrates except gastropods and odonates showed least popultion in monsoon probably because of frequent and heavy rainfall between June and September. Heavy rainfall and high water discharge displace or destroy the microhabitats of these macrofauna, causing a drastic reduction in their number. Gastropods which showed a good presence throughout the year, but unlike the other macroinvertebrate, this benthic group was found to show an increasing trend in density from monsoon, reaching the peak in postmonsoon and lowest in premonsoon. Srivastava (1956), Mandal and Moitra (1975) and Singh *et al.* (1994) also recorded the same finding.

In BBNP, oligochaete density was always found to be greater than gastropod and its chironomid population was comparatively higher than RS and SS. The predominance of oligochaete in BBNP could be attributed to the inflow of sewage as well as the availability of food in the form of decaying organic matter. Brinkhurst (1966) stated that the occurrance of *Limodrilus hoffmeisteri* nearly alone and in great abundance would indicate pollution. Das (1978) suggested that some oligochaetes such as *Tubifex* sp. and *Limnodrilus* sp. are biological indicators of pollution. Loden (1974) suggested that chironomid larvae can be used as pollution indicator. Mahadevan and Krishnaswamy (1984) have argued that relatively high population size of *Chironomus* sp. is the best biological indicator of pollution. In the present investigation *L. hoffmeisteri*, *Branchiura sowerbyi* and *chironomid* larvae were observed in large number in BBNP which indicate higher pollution level of this sewage fed wetland in comparison to RS and SS.

Conclusion

In these three urban wetlands of Calcutta, gastropods, oligochaetes and bivalves have higher percentage composition than other macroinvertebrates though the species diversity of different aquatic insect groups were much higher. In general, the macroinvertebrate density was found to be highest in post monsoon and lowest in monsoon season. The macroinvertebrate abundance is presumbly directly related to the seasonal appearances and disappearances of macrovegetation and also with the nature of water, bottom soil and rain as reported by Roy *et al.* (1988). The presence of higher number of oligochaetes and chironomids in BBNP, indicates relatively high pollution level than RS and SS among the three urban wetlands surveyed.

Acknowledgement

The authors are thankful to Director, Zoological Survey of India for providing laboratory facilities and to Ministry of Environment and Forests, New Delhi for Senior Research Fellowship to one of us (M.M.).

References

Barbhuyan S.I., Khan, A.A. (1992). Studies on the structure and function of benthic ecosystem in an eutrophic body of water : Temporal and Spatial distribution of benthos. *J. Freshwat. Biol.* 4(4) : 239-247.

Bose S.K., Lakra M.P. (1994). Studies on the Macrozoo-benthos of two fresh water ponds of Ranchi. *J. Freshwat. Biol.* 6(2) : 135-142.

Brinkhurst R.O. (1966). Taxonomical studies on tubificidae (Annelida, Oligochaeta). *Int. Rev. Ges. Hydrobiol. suppl.* 51 : 727-742.

Das S.M. (1978). High pollution in lake Nainital, U.P. as evidenced by biological indicator. *Sci and Cult.* 44 : 236-237.

Datta N.C., Sarangi N. (1980). Preliminary study on the macrobenthos of a brackish water bheri at Taldi, West Bengal. *J. Inland. Fish. Soc. India.* 12 : 81-88.

Deb S.C., Pandey J.S., Santra, S.C. (1996). A practical approach to water pollution control through ecologically balanced waste water management : A case study from Calcutta, India. *In :* S.R. Mishra (ed.) *Assessment of Water Pollution.* APH Publishing Corporation, Delhi, pp. 63-79.

Ghosh A.K., Chattopadhyay S. (1990). Biological resources of Brace Bridge Wetlands. *Indian J. Landscape System & Ecological Studies,* 13(2) : 189-207.

Ghosh A.K. Chattopadhyay S. (1994). Biological resources of periurban wetlands : Santragachi jheel, Howrah district, West Bengal. *Indian J. Landscape system & Ecological Studies,* 17(1) : 1-7.

Govind B.V. (1969). Bottom fauna and macrovegetation in the Tungabhadra reservoir and their role in the food chain of fish communities. *In :* Seminar on "Ecology and Fisheries of Freshwater Reservoirs", sponsered by ICAR at CIFRI, Barrackpore; 27-29, Nov. 1969; 99-128.

Gupta, S.D. (1976). Macrobenthic fauna of Loni reservoir. *J. Inland Fish. Soc. India.* 8 : 49-59.

Gupta, P.K. (1981). An ecological study of the macrobenthic communities in Kumaun Lakes. *Ph.D. Thesis, Kamaun Univ.,* Nainital, p. 303.

Gupta, P.K. (1983). Macrobenthos of Lake Nainital with particular reference to pollution, *Water, Air and Soil Pollution.* 19 : 397-405.

Gupta, P.K., Pant M.C. (1986). Analysis of the inshore macrozoobenthic community in Lake Nainital, U.P., India. *International Revue. Ges. Hydrobiol.* 71(1) : 115-125.

Kulshrestha S.K., Adholia U.N., Bhatnagar A., Khan A.A., Saxena M. (1988). Past, present and future of Bhopal lakes. *Proc. Nat. Sym.* Dept. of Zool. Motilal Vigyan Mahavidyalaya, Bhopal, India, i-v; 1-257.

Lang C. (1985). The oligochaete community of the sub-littoral as indicator of Lake Geneva eutrophication. *Arch. Hydrobiol.* 103 : 325-340.

Loden M.S. (1974). Predation of Chironomid (Diptera) larvae on oligochaetes. *Limnol. and Oceanog.* 19 : 156.

Mahadevan A., Krishnaswamy (1984). Chironomid larval population size an index of pollution in river Vaigai. *Poll. Res.* 3 : 35-38.

Maitland P.S. (1978). Biology of freshwater. Blackie, London.

Mandal B.K., Moitra S.K. (1975). Studies on the bottom fauna of a freshwater fish pond at Burdwan. *J. Inland. Fish. Soc. India.* VIII; 43-48.

McLachlan A.J. (1975). The role of aquatic macrophytes in the recovery of the benthic fauna of a tropical lake after a dry phase. *Limnol. Oceanogr.* 20 : 54-63.

Munawar M. (1970). Limnological studies on freshwater ponds of Hyderabad, India I. The Biotype. *Hydrobiol.* 35(1) : 127-162.

Nandi N.C., Das S.R., Bhuniya S., Dasgupta J.M. (1993). Wetland Faunal Resources of West Bengal. 1. North and South 24–Parganas District. *Rec. Zool. Surv. India. Occ. Paper No.* 150 : 1-50.

Pal S., Nandi N.C. (in press). A simple device for quantitative sampling of macrofauna from littoral macrophytes. *J. Freshwater. Biol.* (Accepted).

Rai D.N., Roy S.P., Sharma U.P. (1981). Freshwater gastropods (mollusca) community structure in relation to the macrophytes of littoral zone of a fish pond at Bhagalpur (Bihar). *Indian J. Ecol.* 8(1) : 88-95.

Roy, S.P., Kumar V., Pathak H.S. (1988). Faunistic composition of aquatic insects of eastern Bihar with notes on their some aspects of ecology. *Rec. Zool. Surv. India.* 85(1) : 49-57.

Singh J.P., Roy S.P. (1991). Seasonal changes in the standing crop of some macroinvertebrate population of the Kawar Lake, Begusarai (Bihar), *J. Freshwater Biol.* 3(1) : 59-64.

Singh U.N., Kumari V., Singh G.P. (1994). Qualitative and quantitative study of benthic fauna of Kararia Lakes of Motihari, Bihar (India). *J. Freshwater Biol.* 6(1) : 21-25.

Soszka G.J. (1975). The invertebrates on submerged macrophytes in three Masurian Lakes. *Ecol. Pol.* 23(3) : 371-391.

Srivastava V.K. (1956). Benthic organism of a freshwater fish tank : *Cur. Sci.* 250 : 158-159.

Wiederholm T. (1980). Use of benthos in lake monitoring. *J. Water Poll. Cont. Fed.*, 52 : 537-547.

Proceedings of the National Seminar on Environmental Biology, April 03-05, 1998
***Edited By :* A.K. Aditya & P. Haldar**
DAYA PUBLISHING HOUSE, DELHI

Insect Fauna Associated with Large Waterhyacinth in Fresh Water Wetlands of West Bengal

D.K. Bhattacharya
Department of Zoology, University of Kalyani
Kalyani - 741235, Nadia, West Bengal, India.

Abstract

The large waterhyacinth (*Eichhornia crassipes*) is a mat forming aquatic plant native to Brazil, now occurs in nearly all countries of the World. It is considered to be one of the World's most serious aquatic weed. In India it infests more than 200,000 hectres of water surfaces. The fresh water wetlands of West Bengal exhibit similar picture, where this plant often completely covers the ponds, lakes and streams. This weed interferes wth fish growth, fish operation and other hazards. *E. crrassipes* makes water courses more favourable to insect fauna. Searching of insects attacking this particular weed in fresh water wetlands of West Bengal revealed the presence of 21 species of insects comprising 45.65% of the total insect fauna occurring in the same microhabitat. Insects use this weed for shelter, food and egg laying substratum.

Introduction

The large waterhyacinth *Eichhornia crassipes* (Martius) is a perenial, mat forming aquatic plant of South American (Brazil) origin and now occurs in nearly all countries with suitable climates, specially in most tropical areas of the world. It is one of the most economically important weeds in the World and most important aquatic weed (Holm *et al.*, 1977). This weed was introduced into different countries as ornamentals as aquarium plants or by accident.

In India this weed occurs commonly in fresh water wetlands. It infests more than 200,000 hectres of water surfaces (Anonymous, 1979). In many cases it completely covers the streams, canals, lakes of the wetlands by its high growth rate. Fresh water wetlands of West Bengal are almost completely covered by this weed.

From the economic point of view this weed contributes to the low water level, causing more water loss by evapotranspiration than would by evaporation alone (Brezny *et al.* 1973). It also interfere with fish growth by reducing oxygen content of the water (Sculthrope, 1967). Recreational use of large wetlands such as boating, swimming and fishing is practically impossible because of the day-night movement of the large waterhyacinth mat.

Searching of insects attacking *E. crassipes* by Bennett, 1970; Bennett and Zwölfer, 1968; Blackwelder, 1947 recorded at least 30 species from different countries. In India informations on the insect faunal composition associated with the vegetation of fresh water wetlands is poor and scattered except the works of Srivastava, 1959; Bhattacharya and Gupta, 1991 and Saxena *et al.*

1989 recorded number of insect species infecting on large waterhyacinth. Informations of the insects associated with the vegetation in fresh water wetlands of West Bengal can only be available from the works of Bhattacharya and Gupta, *op.cit.* and De and Sengupta, 1993. But knowledge on the insect faunal diversity associated with *E. crassipes* in fresh waste wetlands of West Bengal is practically absent.

Considering these, attempt has been made herein to find out a comprehensive account of the insect community associated with the serious weed *E. crassipes* of the fresh water wetlands of West Bengal with a view to provide a base line data for future studies.

Materials and Methods

Surveys were conducted in different fresh water wetlands of West Bengal particularly in the districts of Nadia, North 24-Paraganas, Malda and Murshidabad. Following methods of random sampling from the selected wetlands insect samples were collected using scientific insect nets and hand collection. After collection, specimens were brought to the laboratory. Adult specimens were made dry mount and the immature forms were preserved in 70% alcohol. Specimens were determined by the experts.

Table 1 : List of insect species and their site of occurrence on *Eichhornia crassipes*

Insect species	*Site of occurrence on host*
Order Hemiptera	
1. *Diplonychus rusticus* (Fabricious)	Submerged root.
2. *D. molestum* (Duf.)	Submerged root.
3. *D. annulatum* (Fabricious)	Submerged root.
4. *Limnogonus fossarum* Fab.	On leaf above water level.
5. *L. nitidus* (Mayr)	On leaf above water level.
6. *Micronecta proba* Distant	Submerged root.
7. *Lithocerus indicus* Lep. & Serv.	Below leaf sheath.
8. *Plea* sp.	Submerged root.
Order Coleoptera	
9. *Berosus indicus* Motchulsky	Submerged root.
10. *Cantrydus laetabilis* (Walker)	Within leaf sheath.
11. *Coelostoma* sp.	Around leaf sheath.
12. *Halochares* sp.	Submerged root.
13. *Enocrus escuriens* (Walker)	Submerged root.
Order Odonata	
14. *Inchnura senegalensis* (Rambur)	Within leaf sheath and root.
15. *Brachythemis contaminata* (Fab.)	Within leaf sheath and root.
16. *Anax* sp.	Within leaf sheath and root.
Order Diptera	
17. *Chironomus circumdatus* Kieffer	Around petiole.
18. *C. filitans* Kieffer	Around petiole.
19. *C. javanus* Kieffer	Around petiole.
Order Ephemeroptera	
20. *Caenis* sp.	Below leaf sheath.
21. *Cloeon* sp.	Below leaf sheath.

Results and Discussion

Weed plant *E. crassipes* offers adequate shelter to a wide assemblage of insect species. Altogether 21 species of insects belonging to five orders viz. Hemiptera, Coleoptera, Odonata, Diptera and Ephemeroptera were identified associated with this weed. The list of insect species along with their respective site of occurrence on the host plant has been shown in the Table. Among these five orders of insects recorded, Hemiptera ranks one of the major group containing largest number of species (8) followed by Coleoptera containing 5 species. The orders Odonata and Diptera are represented by 3 species each, while Ephemeroptera contains only 2 species. The quantitative analysis of the insect faunal assemblage on this particular weed of the fresh water wetland comprising 45.65% of the total insects occurring in fresh water wetlands of West Bengal (Bhattacharya and Gupta, *op. cit.*).

The insects were found to occupy almost all parts of the weed host. They were found on different parts such as leaf sheath, petiole, stem and also roots. The members of the Hemiptera and Coleoptera were obtained in adult stages while members of the remaining orders were obtained only in immature forms. It appears that these insect species make associated with *E. crassipes* to exploit this weed in various ways. Majority of the species viz. *Diplonychus rusticus, D. molestum, D. annulatum, Plea* sp., *Limnogonus fossarum, L. nitidus, Lithocerus indicus, Canthrydus laetabilis* and *Micronecta proba* take their shelter for different biological need. The Coleopteran species viz. *Berosus indicus, Coelostoma* sp., *Halochares* sp. presumably use this plant as food. The members of the orders Odonata, Diptera and Ephemeroptera viz. *Inschnura senegalensis, Brachythemis contaminata, Anax* sp., *Chironomus circumdatus, C. filitens, C.J. javanus, Caenis* sp. and *Cloeon* sp. are supposed to be associated with this weed to the emergence of their last instar larval stages into adult. The species *Lithocerus indicus* exploit this weed to egg laying site, as egg masses of the said species were seen attached with this plant. All these insect species are always encountered with this hyacinth plant. Most of them partially complete their life cycle, fulfil thier habit and habitat in association with this weed. A few species viz. *Plea* sp. *Cloeon* sp. are supposed to be the casual visitor for the fulfilment of their biological activity.

Acknowledgement

The author is thankful to Head of the Department, Department of Zoology, University of Kalyani, for providing laboratory facilities.

References

Anonymous (1979). Recommendations of the task force of waterhyacinth. Govt. of India. Ministry of Agriculture and Irrigation, New Delhi, 18 pp.

Bennett, F.D. (1970). Insects attacking waterhyacinth in Northern South America, British Honduras and the USA. *Hyacinth Contr. J.*, 8 : 10-13.

Bennett, F.D. and Zwölfer, H. (1963). Exploration for the natural enemies of the waterhyacinth in northern South America and Trinidad. *Hyacinth Contr. J.*, 7 : 44-52.

Bhattacharya, D.K. and Gupta, B. (1991). Fresh water wet-land inhabiting Insects of West Bengal. *Eny. & Ecol.*, 9(4) : 995-998.

Blackwelder, R.E. (1947). Checklist of the Coleopteran Insects of Mexico Central America, The West Indies and South America, Part 5. *U.S. National Museum Bull.*, 185, p. 825-827.

Brezny, O. Mehta, I. and Sharma, R.K. (1973). Studies on evapotranspiration of some aquatic weeds. *Weed Sci.*, 21 : 197-204.

De, M. and Sengupta, T. (1993). Beetles (Coleoptera : Insecta) of wetlands of Calcutta and its surrounding. *Rec. Zool. Surv. India*, 93 (1-2) : 103-138.

Holm, L.G., Plucknet, D.L., Pancho, J.V. and Herbert, J.P. (1977). The World's worst weed : distribution and biology. University Press, Hawaii, 609 pp.

Saxena, M.N., Sharma, S. and Kaushik, S. (1989). Studies on aquatic bodies communities and their distribution in some aquatic bodies of Gwalior, M.P. India. *J. Hydrobiol.*, 5(1) : 39-42.

Sculthrope, C.D. (1957). The biology of aquatic vascular plants. Edward Arnold Ltd. London, pp. 610.

Srivastava, V.K. (1959). Studies on the aquatic vegetation and their role in breeding and development of some insects. *Nat. Acad. Sci. India (Allahabad).* Ann.n. I, 69 pp.

Proceedings of the National Seminar on Environmental Biology, April 03-05, 1998
***Edited By :* A.K. Aditya & P. Haldar**
DAYA PUBLISHING HOUSE, DELHI

Diversity in Population Composition of a Medically Important Freshwater Snail Species *Lymnaea (Radix) acuminata* (Lamarck)

T.K. Misra and S.K. Raut
Ecology and Ethology Laboratory
Department of Zoology, University of Calcutta
35, Ballygunge Circular Road, Calcutta - 700 019

Abstract

The freshwater snails *Lymnaea (Radix) acuminata* are involved with the spread of some worm diseases in man and domestic animals. Various attempts are being made to control them. Though sporadic reports on the biology and ecology of this snail species are on record nothing is known on its population compostion so as to evaluate the rate of recruitment is respect to time and age of the snails concerned. Accordingly, is an attempt, by culturing 100 zero-day old individuals separately, in each month of a year, for a period of 372 days it is revealed that the density of reproductive population varied from 2% (March) to 19.88% (August) with practically no existance of post-reproductive population. Marked variations have been noted in the age-pyramids in snails in respect to the batches considered for experimentation against the months of a year.

Introduction

The freshwater snails *Lymnaea (Radix) acuminata* (Lamarck) are involved with the spread of some worm diseases in humans and domestic animals (Raut, 1986; Subba Rao, 1989). Various attempts are being made to control them. Though sporadic reports on the bioecology of this snail species are available (Raut, 1981; Raut and Das, 1997; Misra and Raut, 1997; Misra and Raut; In Press) nothing is known on its population composition in respect to dominant age group. Since the dominant age group of a population determines the population size of the species concerned (Hjort, 1926) in the present programme an attempt was made to ascertain the distribution patterns of age groups in the population with a view to throw some light on the influence of dominant age group in regulating the population size.

Materials and Methods

A good number of gravid snails *L. (R.) accuminata* were collected from a pond located at Bagha Jatin, Calcutta. They were released into a plastic container (26 cm in diameter) containing 4 *l* of pond water. A few examples of *Chara, Vallisneria* and *Ipomoea* were also released into the water of the container to provide resting as well as egg laying sites for the snails. The snails were supplied with lettuce (*Lactuca sativa*) and mustard (*Brassica nigra*) leaves regularly as food. The water of the container was changed regularly by fresh pond water. The unconsumed part of leaves, dead snails if any, and faecal pellets were also removed regularly, at an interval of 24-

hr. The snails started egg laying from the very next day. The egg capsules were collected daily. They were then kept in small glass beakers and dated. The eggs hatched within a few days. The newly hatched (zero-day old) snails thus obtained were considered for experimentations.

A total of 12 sets of experiments each with 100 newly hatched (zero-day old) *L. (R.) acminata* individuals were designed to study the natality, mortality, hatching success of the eggs, and the age of attainment of sexual maturity of the individuals concerned. In all cases, for the convenience of studies, at the initial state, for a period of two weeks they were maintained in a glass beaker containing 750 ml pond water. After that, that is, on and from the third week, the snails were transferred to a plastic container (26 cm in diameter and 14 cm in depth) containing 4 *l* pond water. The 12 sets of experiments were designed to record the variability, if any, in the said aspects of *L. (R.) acuminata* borned during the first week of each month for February 1992 to January 1993 and then to find out the diversity in population composition of the snail batches, if any. The snails considered for experimental studies were maintained carefully until the last individual of each experiment was died. Throughout the experiment pond water was used and, lettuce and mustard leaves were supplied *ad libitum* to the snails as food. A strict hygienic condition was maintained throughout by removing the unconsumed food materials, the dead snail individuals if any, the feacal pellets and by changing water regularly, at an interval of 24 hr.

Data on the number of egg capsules laid by the snails of each and all containers were collected, and the number of eggs present in each capsule was counted regularly. The birth rate was calculated by using the formula $\beta i = \frac{niH}{Ni}$ (Saha & Raut, 1992), where niH is the number of eggs hatched and the *Ni* is the number of snails present in the container at an interval of 12 days. Using the data obtained on birth rate and death rete the population projection is done for a period of 372 days in respect to experiments, separately. Then the number of individuals of different age groups (i.e. 0-30 days, 31-60 days old, and so on) was calculated on the basis of the total number of individuals obtained at the end of 372 days. Campbell (1989) was followed for statistical interpretations of the data.

Results

Population size of *L. (R.) acuminata* varied with the snail batches. On the basis of the data obtained on natality and mortality it is revealed that 100 zero-day old individuals would give rise 859918, 1251844, 1477861, 1555072, 1710166, 1717763, 2047290, 2563003, 2937188, 4653791, 5706256 and 6676955 individuals in 372 days if they were considered for studies by the first week of June, December, October, February, January, May, November, August, March, September, July and April of the study year respectively. Out of the total population, in these cases 80.12, 82.31, 85.46, 87.21, 88.88, 90.24, 90.63, 90.92, 91.61, 95.3, 96.61 and 98 per cent were prereproductive individuals in August, September, June, November, May, October, December, April, July, February, January and March batches respectively. The distribution pattern of individuals belonged to different age groups is remarkably different in different experimental sets as could be seen from the age-pyramids (Fig. 1). It is to be mentioned here that, in April-batch there were only few individuals in 211-240 days and 241-270 days old age-groups though there was no trace of any individual of relatively lower age group viz. 181-210 days old in respect to the next higher age group. But on the basis of the total population obtained in each batch by the end of 372 days it is quite clear that higher the age group lower the density (Fig. 1). Though the density of the

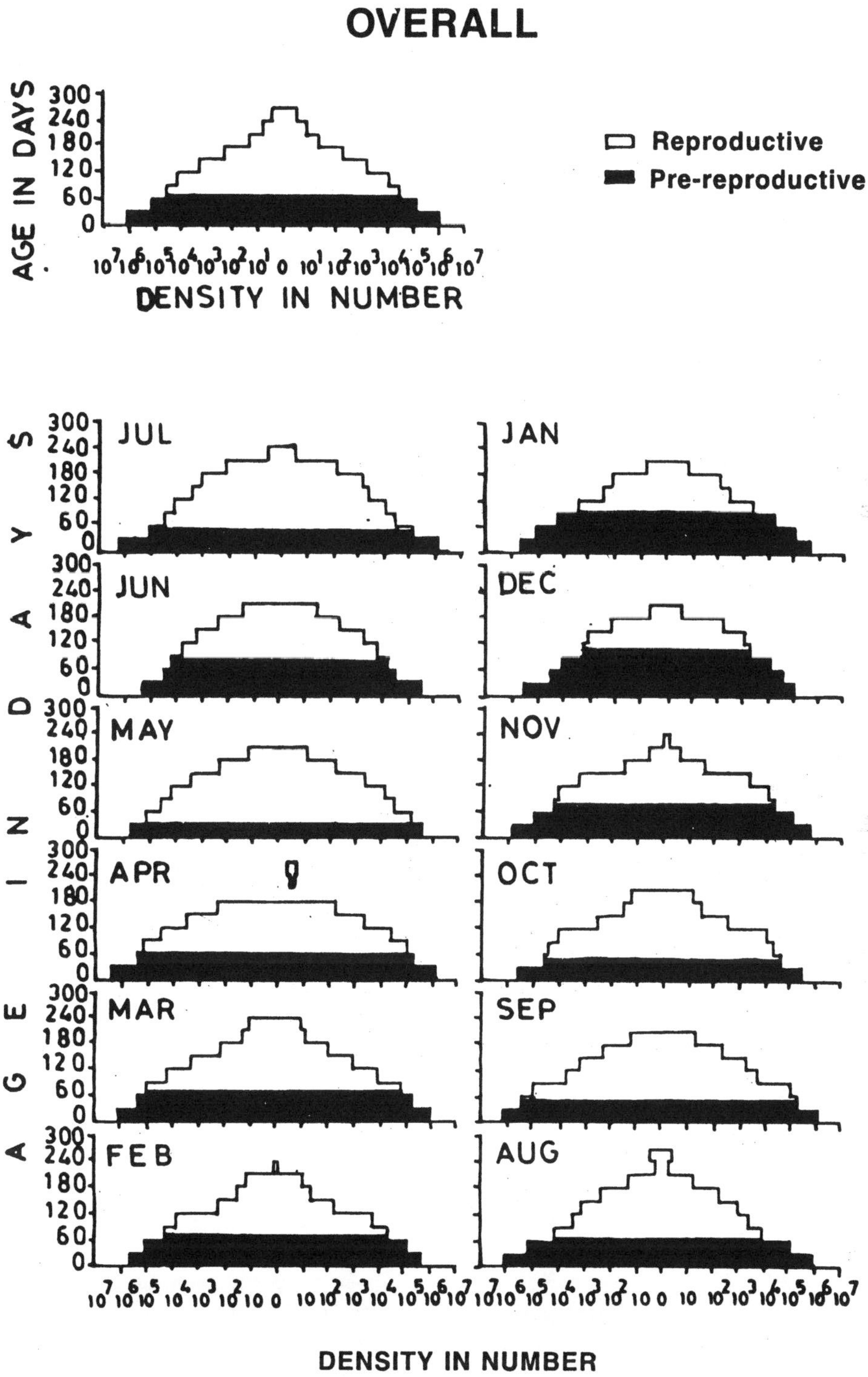

Fig. 1 : Age-pyramids in *L. (R.) acuminata*

snails belonged to the same age group varies with the months in respect to the batches of the individuals considered for experimentations it is well evident from the ANOVA (Table 1) studies ($P < 0.001$) that the size of a population in respect to parental stock, is largely governed by the density of individuals belonged to different age/size groups with the progress of time.

Table 1 : Two-way ANOVA to determine the interaction of snail batches considered during different months of a year with the different age-group distribution pattern within the population of *L. (R.) acuminata*

Source of Variation	SS	df	MS	F	Remarks
Between rows (Month/Batch)	4372937430000	11	397539766400	1.32	Not Significant
Between columns (Age group)	55180794680000	8	6897599335000	22.86	Significant*
Residual	26556967540000	88	301783722000	–	
Total	**86110699650000**	**107**			

* $P < 0.001$

Discussion

Irrespective of months and batches of *L. (R.) acuminata* considered for experimentations the constructed age-pyramids are indication of an expanding population (Odum, 1971). This is really dangerous as the snails *L. (R.) acuminata* are intermediate hosts of certain helminth parasites causing diseases in man and his domestic animals. From the results it is evident that, of the total population 70.15-93.26 (average 85.11), 4.54-18.91 (average 11.86), 0.97-6.38 (average 3.71), 0.09-2.7 (average 0.96), 0.004-0.048 (average 0.17), 0.005-0.08 (average 0.02) and 0.00004-0.005 (average 0.002) per cent individuals are distributed in 0-30, 31-60, 61-90, 91-120, 121-150, 151-180 and 181-210 days old age groups respectively. It is also seen that 2.00, 3.39, 4.70, 8.39, 9.08, 9.37, 9.76, 11.12, 12.79, 14.54, 17.69 and 19.88 per cent individuals of the total population were distributed into the reproductive age groups in the snail batches of March, January, February, July, April, December, October, May, November, June, September and August respectively.

From the results it is seen that the maximum snail individuals is obtained from the April batch. So, obviously one can presume that, either the prereproductive population or the reproduction population concerned to April-batch would be maximum or minimum that, ultimately, be considered as the key factor determining the popultion size. But this idea can not be substantiated in all cases. The result of experiment started during June though revealed the minimum size of population there exists hardly any relationship with minimum/or maximum relative percentage of either prereproductive or reproductive population.

Since the present report is based only on the individuals obtained from the survivors of the individuals taken at zero-day the interacting effects of different size groups were confined among the descendents of the same parental stock. Under such a situation birth and death would be entirely at different rates (Misra and Raut, 1993; In Press) in respect of population having individuals of different snail batches. Such studies are essential to determine the actual distribution patern of the age/size-groups of *L. (R.) acuminata* in natural population at least to identify the dominant age class as has been done by Hjort (1926) in case of food fishes. Of course environmental limitations such as unfavourable temperature (Misra and Raut, 1997) and availability of resources play significant role in determining the dominant age class in different seasons of a year (Pretorius *et. al*, 1992). Since, in *L. (R.) acuminata* only 10.23 per cent population take part in reproduction it is sure that the dominant age class would belong to the said group. Thus, recognition of the dominant age class is urgently needed so as to manage the snail population with a view to control the snail-borne diseases.

Acknowledgements

We are thankful to the Head of the Department of Zoology, Calcutta University for the facilities provided.

References

Campbell, R.C. (1989). *Statistics for biologists,* 3rd ed. Cambridge Univ. Press, Cambridge. XVII + 445 p.

Hjort, J. (1926). Fluctuations in the year classes of important food fishes. *J. du Conseil Parmanent Internatinale pour L `Exploration de la Mer.* 1 : 1-38.

Misra, T.K. & S.K. Raut. (1993). Dynamic of population-dynamics in a medically snail *Lymnaea (Radix) luteola* (Lamarck). *Mem. Inst. Oswaldo Cruz.* 88 : 469-485.

Misra, T.K. & S.K. Raut. (1997). Thermal effects on a medically important snail species *Lymnaea (Radix) acuminata* Lamarck (Gastropoda : Lymnaeidae). *Apex* 12 : 101-108.

Misra, T.K. & S.K. Raut. In Press. Birth rate int eh disease transmitting snails *Lymnaea (Radix) acuminata* Lamarck. *Proc. Nat. Acad. Sci. India.*

Odum, E.P. (1971). *Fundamentals of ecology,* 3rd ed. W.B. Saunders Company, London. xiv + 574 p.

Pretorius, S.J., K.N. de Kock & P.H. Joubert. (1992). Evidence that population growth of the freshwater snail *Bulinus africanus* is density limited in a natural habitat. *J. Med. & Appl. Malacol.* 4 : 113-120.

Rao, N.V. Subba. (1989). *Handbook, Freshwater molluscs of India.* Zoological Survey of India, Calcutta, XXIII + 289 p.

Raut, S.K. (1981). Populaton interaction among freshwater snails. *Proc. Symp. Ecol. Anim. Popul. zool. Surv. India,* Pt.2 : 163-173.

Raut, S.K. (1986). Snails and slugs in relation to huma diseases. *Environ. and Ecol.* 4 : 130-137.

Raut, S.K. & Das, S. (1997). Effects of water depth on hatching of eggs of three disease transmitting freshwater snails. *Apex* 12 : 9-14.

Saha, T.C. & S.K. Raut. (1992). Bioecology of the water-bug *Sphaerodema annulatum* Fabricius (Heteroptera : Belostomatidae). *Arch. Hydrobiol.* 124 : 239-253.

Proceedings of the National Seminar on Environmental Biology, April 03-05, 1998
Edited By **: A.K. Aditya & P. Haldar**
DAYA PUBLISHING HOUSE, DELHI

Ecology and Diversity of Cladocerans in Some Calcutta Wetlands

Chandrima Sinha & R.A. Khan
Zoological Survey of India
234/4, A.J.C. Bose Road
Nizam Palace, Calcutta-700020

Abstract

The species diversity and some aspect of the ecology of Caldocera (Crustacea : Branchipoda) of Calcutta Metropolitan area were studied during the period April 1995 - March 1996. Regular monthly surveys were conducted in four selected wetlands, -two large man made recreational lakes (Rabindra Sarovar and Subhas Sarovar), a complex of Fish culture ponds (*bheries*), situated in Brace Bride Nature Park (BBNP) and a medium sized pond, situated in the heart of the city. A total of 26 species of Cladocera belonging to 18 genera and 6 families were recorded during the period of investigation. The diversity varied considerably from one water body to other. Rabindra Sarovar, which is characterized by dense macrophyte vegetation (mainly belonging to *Ceratophyllum* sp.) and with no fishing activity, harboured a total number of 19 species. Highest number of species (21) was recorded from Subhas Sarovar, a semi-managed lake from angling viewpoint and lowest number of species (8) was recorded from BBNP.

Like species richness, density of cladocerans also varied remarkably in different wetlands and in different seasons. Comparatively high density was observed in Subhas Sarovar thoughout the year with the annual mean of 195.99/litre. This was closely followed by Rabindra Sarovar (187.96/litre). Lowest density was recorded from BBNP (39.67/litre) which was probably due to heavy grazing pressure.

Post-monsoon season (October - January) was characterized by abundance of cladocerans both in density and diversity followed by monsoon (June - September). Lowest richness and density was observed during Pre-monsoon (February - May). A comparison of the species richness of Cladocerans in different tropical water bodies has also been made.

Introduction

The importance of cladocera in the trophic dynamic of freshwater ecosystem, being the main component of zooplankton, have long been recognised. They are the consumers of first order, directly drawing energy from primary producer of the ecosystem viz. phytoplankton. In turn, they form the food of a large number of planktivorous fishes and other invertebrates, thus transferring energy to higher trophic level. Besides, they have also been reported to be reliable indicators of eutrophic nature of water bodies.

The members of the order cladocera, under sub-class Branchipoda, are aquatic microcrustacean. These are highly diverse group and distributed widely throughout the world. Baring a few marine / esturine genera, all others are freshwater inhabitants.

Whereas considerable amount of work have been carried out on taxonomy and ecology of cladocerans in temperate waters, works from tropical region are meagre. Only recently the diversity of cladoceran fauna in tropical water boides have been debated in some details (Dumont 1994, Fernando 1994, Egborge et al., 1994). In India not much specific attention has been paid to this group.

Taxonmic studies on Indian cladocera began during mid 19th century when Baird (1860) published the description of two new species. This was followed by Gurney (1906, 1907), Sewel (1935), Brehm (1936, 1953, 1963), Brehm and Woltereck (1939), and Biswas (1964, 1971). Thereafter, the existing knowledge of cladocera was consolidated by Michael & Sharma (1987). During recent years Venkataraman and Das (1993) reported few new species and several new records from South Bengal. Some notable contribution on the biology and ecology of cladocera in past include the works of Michael (1962), Vijayaraghavan (1970), Murugan (1975), Yousuf and Quadri (1977) and Khan (1983a, b) etc.

Few taxonomic studies on cladocera in West Bengal have been carried out in past (Sharma 1978, Venkataraman and Das 1993). However, these works were based on general faunistic survey of few water-bodies without making any effort to study their diversity, spatial or seasonal variation or their relationship with environmetal factors.

Realising the importance of the group and lack of specific knowledge in the region, a detailed programme on their taxonomy, population structure and dynamics and interrelationship with environmental factors have been initiated few years ago. The present paper is a part of the detailed study and is related to the work carried out on their diversity and density and their seasonal and spatial variations during 1995-96, in few urban lakes/ ponds, situated in thickly populated aglomerate of Calcutta.

Description of the Study Area

The studies were carried out in following four wetlands.

1. *Rabindra Sarovar :* This is a large man made lake, situated in the southern part of the city and is chiefly meant for recreational purposes. The lake is elongated in shape covering an area of 72 acres with depth varying between 9.3 and 10.5 meters. The lake has been characterised by dense growth of macrophytes, both rooted and submerged. *Ceratophyllum* sp. was the main component of macrophytes, constituting nearly 80 per cent of total vegetation. This was followed by *Vallisneria* sp., *Chara* sp., and *Hydrilla* sp. etc. Water hyacynth (*Eichhornia* sp. also grew in localised area of the lake during postmonsoon season). However, these were removed quickly and were not allowed to grow. Other submerged macrophytes were also removed periodically, specially from littoral zone.
2. *Subhas Sarovar :* Subhas sarovar is also an artificial lake situated in northeastern part of the city and covers an area of 39.5 acres. This is also a recreational lake meant mainly for aesthetic purposes. Currently it is also used for angling / sport fishing. This lake is mainly covered by monospecific vegetation of *Vallisneria* sp.
3. *Generals Tank :* This is almost a rectangular small pond of 2.4 acre area and situated in the heart of the city (Maidan area). It is also meant for recreation purposes and is of comparatively recent contruction. The pond is almost devoid of any submerged vegetation. However certain floating strands of *Eichhornia* sp. and *Pistia* sp. were found to develope periodically, but these were not been able to form dense strands because of their periodical removal.

4. ***Ponds of Brace Bridge Nature Park :*** These ponds, situated in south western part of the city, belong to a complex of fish culture ponds (bheries) where commercial pisciculture activities are being carried out on a very large scale. Water is drawn from a nearby canal which carries sewage from domestic establishment and several industries located in the area. The entire systems covers an area of 48 hectres which is divided into 6 *Jheels*. There is also no submerged or floating vegetation in these ponds. Sampling were done in four Jheels.

Materials and Methods

Qualitative and quantitative samples of zooplankton were collected from littoral zone of all wetlands at regular interval, with the help of a plankton net made of bloting nylon cloth (no. 21). Qualitative samplings for taxonomic studies were done by sweeping the net several times at each occasion. For quantitative analysis, 50 lit of water was filtered through the net. Samples were preserved in 4 per cent formalin. Detailed taxonomic identification was carried out following Edmondson (1959), Pennak (1978) and Michael and Sharma (1987). For quantitative analysis of relative abundance and density, identification and enumeration were done simultaneously in a Segwick Rafter counter by taking 1ml. sub sample and then raised to total volume of water filtered. Only cladoceran were taken into consideration ignoring other groups.

Similarity index of Sorensen (1948) was applied in order to compare the similarities of different wetland in respect of cladoceran diversity.

Seasons have been defined as Pre-monsoon (Feb - May), Monsoon (June - Sept.) and Post-monsoon (Oct - Jan).

Results

(a) ***Species richness :*** This study revealed diverse cladoceran occurrence in these four wetlands. These were mainly represented by 26 species belonging to 20 genera under 6 families (Table 1). Highest number of species belonged to family chydoridae (9 species) followed by Daphnidae (6 speices) and Sididae (4 species). Macrothricidae. Moinidae and Bosminidae were represented by two species each. Species richness was greater in Subhas Sarovar (21 species) followed closely by Rabindra Sarovar (19 species). General Tank harboured 13 species and the lowest richness was recorded from the ponds of Brace Bridge Nature Park (8 species) (Table 1 Fig. 1).

While several species like *Moina micrura, Pleuroxus similis. Alona davidi* and *Oxyurella signalensis* were common to all wetlands. Many others were confined to one or two wetland only. *Ceridaphnia cornuta. Simocehalus expinosus. Simocephalus vetulus. Chydorus sphaeicus* etc. occurred only in the large water bodies like Rabindra Sarovar and Subhas Sarovar.

Similarity index analysis (Sorensen 1948) revealed that the two lakes viz. Rabindra Sarovar and Subhas Sarovar were most similar while Rabindra Sarovar and Generals Tank were most dissimilar in respect of cladoceran community (Table 2).

(b) ***Density :*** Highest cladoceran density was recorded from Subhas Sarovar where annual mean was 195.99/L. This was followed by Rabindra Sarovar (187.96) and lowest density was recorded from Brace Bridge Nature Park (39.67/L) Fig. 2, 3 and 4 shows the seasonal fluctuation in the density of total cladocera in four wetlands. The patterns were almost similar in all the wetlands as peaks occurred during postmonsoon season.

Table 1 : Occurrence of Cladoceran Species in Different Wetlands of Calcutta

	Rabindra Sarovar	Subhas Sarovar	General Tank	Ponds of Brace Bridge Nature Park
I. FAMILY : SIDIDAE				
1. *Sida Crystallina* (O.F. Muller, 1776)	●	●	–	–
2. *Diaphanosoma sarsi* Richard, 1894	●	●	–	–
3. *Diaphanosoma excisum* Sars, 1885	●	●	–	●
4. *Pseudosida bidentata* Herrick, 1884	–	●	–	–
II. FAMILY : DAPHNIDAE				
5. *Ceriodaphnia cornutu* Sars, 1885	●	●	●	–
6. *Daphnia carinata* King, 1853	–	●	●	–
7. *Daphnia lumholtzi* Sars, 1885	–	●	–	–
8. *Scupholeberis kingi* Sars, 1903	●	●	–	●
9. *Simocephalus expinosus* (Koch, 1841)	●	●	●	–
10. *Simocephalus vetulus* (O.F. Muller, 1776)	●	●	●	–
III. FAMILY : MOINIDAE				
11. *Moina micrura* Kurz, 1874	●	●	●	●
12. *Moinodaphnia macleayi* (King, 1853)	–	●	–	●
IV. FAMILY : BOSMINIDAE				
13. *Bosmina longirostris* (O.F. Muller, 1776)	●	●	–	–
V. FAMILY : MACROTHRICIDAE				
14. *Macrothrix goeldi Richard*, 1897	●	●	–	–
15. *Macrothrix triserials* (Bruty, 1886)	–	●	●	–
FAMILY : CHYDORIDAE				
16. *Chydorus barroisi* Richard, 1894	●	●	–	–
17. *Chydorus sphacricus* (O.F. Muller, 1776)	●	●	●	–
18. *Dunhevedia crassa crassa* (King, 1853)	●	●	–	–
19. *Pleuroxus similis* Vavra, 1900	●	●	●	●
20. *Pseudochydorus globosus* (Baird, 1843)	●	–	●	–
SUB-FAMILY : ALONINAE				
21. *Alona davidi* Richard, 1895	●	●	●	●
22. *Alona pulchella* King, 1853	●	●	●	–
23. *Alona rectangula*, Sars, 1862	–	–	–	●
24. *Oxyurella singulensis* (Daddy, 1898)	●	–	●	●
25. *Camptocercus australis* Sars, 1896	–	–	●	–
26. *Acropus harpae* (Baird, 1834)	●	–	–	–
Total Cladocera species	**19**	**21**	**13**	**8**

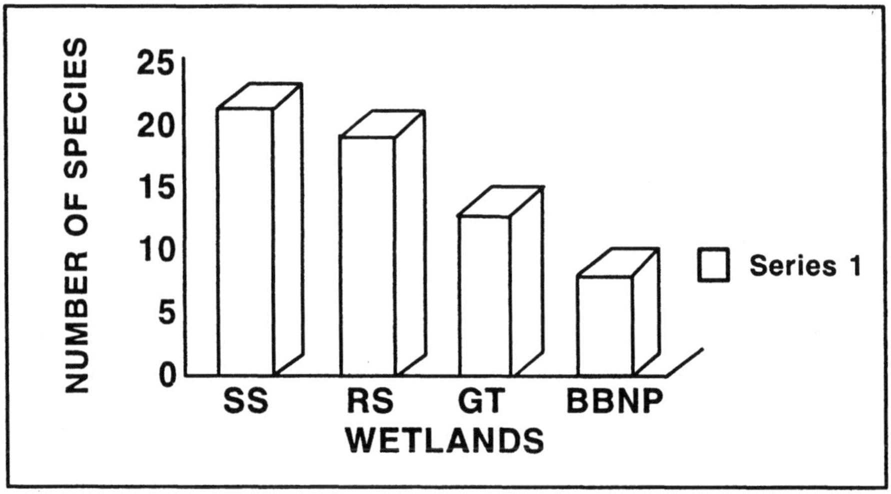

Fig. 1 : Number of Cladoceran Species in Some Wetlands of Calcutta

Table 2 : Quotient of Similarity (Sorensen 1978) of Cladocerans in Four Wetlands of Calcutta

	Brace Bridge Nature Park	General Tank	Subhas Sarovar	Rabindra Sarovar
Rabindra Sarovar	0.44	0.375	0.85	–
Subhas Sarovar	0.48	0.59	–	–
General Tank	0.38	–	–	–
Brace Bridge Nature Park	–	–	–	–

This was followed by a sudden drop in Pre-monsoon season, when minimum numbers were recorded from all the wetlands. The density again started to increase following the onset of monsoon.

(c) *Numerically abundant species :* Out of all cladoceran species recorded from these wetlands, few species in each wetland formed the bulk of density almost throughout the year. Table 3 gives the contribution of such numerically abundant species of each wetlands during different seasons.

Individual species densities revealed significant difference among different water bodies. *Diaphanosoma sarsi, Bosmina longirostris, Ceriodaphnia cornuta and Moina micrura* accounted for a large portion of total cladoceran density throughout the year in Subhas Sarovar, whereas in Rabindra Sarovar, besides these species, *Chydorus sphaericus, Simocephalus expinosus. Simocephalus vetulus,* and *Scapholeberis kingi* also contributed significantly. The pattern was different in General Tank and ponds of Brace Bridge Nature Park, former with the dominance of *Alona pulchela. Alona davidi* and *Pseudochydorus globosus* and later with *Diaphanosoma excisum, Moina micrura, Pleuroxus similis* and *Scapholeveris-kingi* as dominant species.

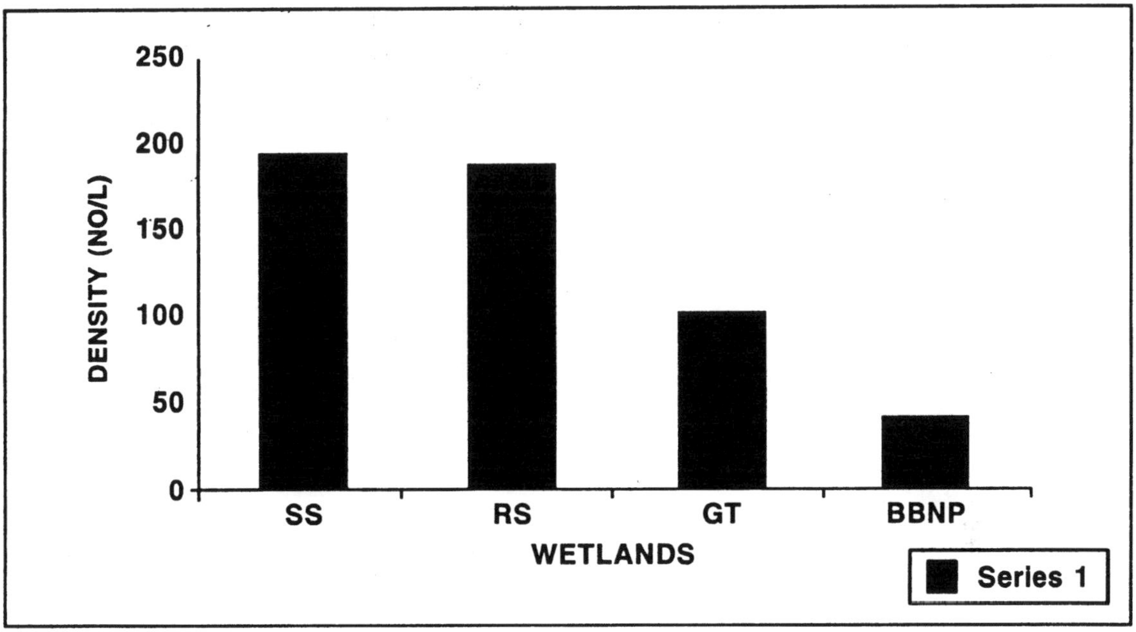

Fig. 2 : Average Annual Density of Cladocerans in Some Wetlands of Calcutta

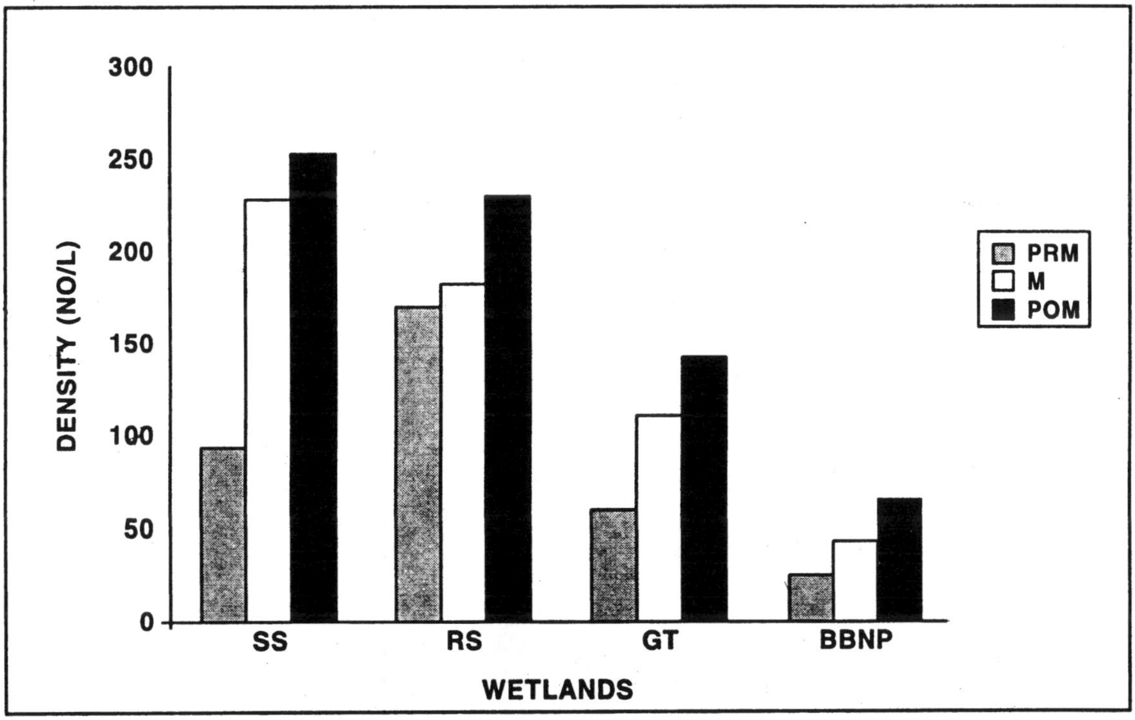

Fig. 3 : Seasonal Fluctuation in Numerical Density of Cladocerans in Some Wetlands of Calcutta

Table 3 : Seasonal fluctuation in the density of numerically abundant cladoceran species

Name of Species	SUBHAS SAROVAR						RABINDRA SAROVAR					
	Pre-monsoon		Monsoon		Post-monsoon		Pre-monsoon		Monsoon		Post-monsoon	
	Density	%	Density	%	Density	%	Density	%	Density	%	Density	%
1. *Alona pulchela*	16.25	17.58	16.67	7.36	–	–	14.48	9.07	28.34	16.23	5.42	2.36
2. *Alona davidi*	–	–	–	–	2.50	0.93	0.72	0.45	–	–	15.00	6.53
3. *Bosmina longirostris*	11.35	12.17	19.58	8.64	14.75	5.48	17.00	10.64	25.00	14.32	18.75	8.17
4. *Ceriodaphnia cornuta*	7.50	8.10	43.96	19.40	53.13	19.75	23.84	14.93	37.71	21.60	34.58	15.06
5. *Chydorus sphaericus*	5.83	6.31	21.04	9.90	2.08	0.77	41.02	25.69	68.96	39.50	28.33	12.34
6. *Diaphanosoma sarsi*	33.30	36.01	109.79	48.46	118.33	43.99	18.96	11.87	3.13	1.83	11.25	4.90
7. *Diaphanosoma excisum*	5.00	5.41	–	–	–	–	–	–	–	–	–	–
8. *Moina micrura*	12.50	13.52	10.73	4.74	31.87	11.85	14.26	8.93	4.38	2.51	10.42	4.53
9. *Macrothrix goeldi*	–	–	–	–	4.58	1.70	–	–	–	–	–	–
10. *Oxyurella singalensis*	–	–	–	–	–	–	–	–	–	–	–	–
11. *Pleuroxus similis*	–	–	–	–	–	–	–	–	–	–	8.33	3.63
12. *Pseudochydorus globosus*	–	–	–	–	–	–	–	–	–	–	1.25	0.54
13. *Simocephalus vetulus*	0.83	0.90	2.71	1.20	–	–	5.98	3.74	7.08	4.05	33.33	14.52
14. *Simocephalus expinosus*	–	–	–	–	15.83	5.88	23.46	14.69	–	–	33.83	14.74
15. *Scapholeberis kingi*	–	–	–	–	12.50	4.65	–	–	–	–	2.50	1.09
16. *Sida sp.*	–	–	2.08	0.92	9.83	3.65	–	–	–	–	8.33	3.63

It has also been noted that while total cladoceran densities followed almost similar seasonal patterns in all water bodies being higher in post-monsoon, some individual species behaved differently (Table 2). For example the density of *Diaphanosoma saris* in Subhas Sarovar was highest in monsoon and *Diaphamosoma excisum* was highest during pre-monsoon in Rabindra Sarovar. Highest density of *Bosmina longirostris* was recorded in post-monsoon period from Subhas Sarovar, while monsoon was the period of its highest abundance in Rabindra Sarovar. Similarly, the period of abundance of some other species was also different in different water bodies (Table 2).

Discussion

Diversity and species richness of cladocerans have been discussed by several workers both in temperate as well as tropical water bodies. As far as tropical wetlands are concerned, there are two divergent view points. Earlier concept, advocated by a number of worker (Fernando and Kanduru 1984, Green 1990), asserted that the diversity of cladoceran fauna in tropical water bodies is considerably lesser than those of temperate water bodies. The reasons cited for such impoverishment of species are intense grazing pressure by planktivours fishes throughout the year resulting in elimination of large sized species (Fernando 1994) and eutrophic nature of tropical waterbodies restricting species richness (Green, 1976; Ravera, 1996). Refuting this viewpoints Dumont (1994) mentioned that diversity of cladocera in tropical waters is in no way lesser than temperate waters. To prove this he cited that out of 290 cladoceran species known from from world, almost 150 species are exclusively confined to tropical water besides 40 species being common to both. During present investigations, it was noted that richness of cladoceran in two large man made lakes was quite moderate, where 19 and 21 species was observed in Rabindra Sarovar and Subhas Sarovar respectively. Although these numbers from individual lakes are lesser than those generalised by Dumont (1994) for species rich water body, these can be considered quite sufficient in Indian context. Dumont (1994) classfied a water body as rich if there are 30-50 species. This range appears to be extremely high as far as Indian water bodies are concerned. A survey of the literature revealed that the number of the cladoceran species reported from any single water body of the country generally varied between 5-10 species (Singh & Pandey 1991, Sinha *et al* 1992, Pati & Sahu 1993, Revti 1993) excepting few cases (Ganapati 1943, Kumar and Dutta 1994). Such small number reported by these workers also appears to be unrealistic. Since most of the studies were under taken from general limnological view points, where proper taxonomic attention was not given to cladocera as well as other planktonic species, there is always a possibility of many species being over looked. Many of the workers have casually mentioned certain taxa only upto generic level. It is quite certain that detailed taxonomic examination of cladoceran fauna of those water bodies will increase the numbers of species several times. This can also be judged from the detailed taxonomic work of Venkataraman & Das (1994) where they reported a total 57 cladoceran species including 22 new records from few selected water bodies of south Bengal, mostly from single qualitative samplings.

The richness and high species diversity of Chydoridae (11) in comparison with Daphnidae (6) Siddae (4), Moinidae (2), Bosminidae (2) Macrothricidae (2), in the water bodies studied supports the view of earlier workers (Green 1962, Mamaril and Fernando 1978, Dumont 1980, Dumont et al. 1981, Sharma & Michael 1987, Egborge et al. 1994) that such combination, specially the dominance of chydorids is the characteristics of tropical water bodies.

However, the diversity of cladocean, like any other faunal group is affected considerably by various local factors operating in the particular water body. Increased eutrophication due to extensive use by surrounding floating human population have resulted in the impoverishment

Table 4 : Seasonal fluctuation in the density of numerically abundant cladoceran species

Name of Species	GENERAL TANK						BRACE BRIDGE NATURE PARK					
	Pre-monsoon		Monsoon		Post-monsoon		Pre-monsoon		Monsoon		Post-monsoon	
	Density	%	Density	%	Density	%	Density	%	Density	%	Density	%
1. *Alona pulchela*	40	66.67	8.0	7.69	37.0	26.27	–	–	–	–	–	–
2. *Alona davidi*	–	–	2.0	1.92	12.0	9.18	5.0	25	–	–	3.0	4.84
3. *Bosmina longirostris*	–	–	–	–	–	–	–	–	–	–	–	–
4. *Ceriodaphnia cornuta*	–	–	–	–	15.4	11.22	–	–	–	–	–	–
5. *Chydorus sphaericus*	–	–	4.0	3.85	–	–	–	–	–	–	–	–
6. *Diaphanosoma sarsi*	–	–	2.0	1.92	–	–	–	–	–	–	–	–
7. *Diaphanosoma excisum*	–	–	–	–	7.60	5.54	5.0	25	25.0	67.57	5.0	8.06
8. *Moina micrura*	–	–	4.0	3.85	4.0	2.92	2.5	12.5	5.0	13.51	15.0	24.19
9. *Macrothrix goeldi*	–	–	–	–	–	–	–	–	–	–	–	–
10. *Oxyurella singalensis*	–	–	–	–	–	–	–	–	1.0	2.70	2.0	3.23
11. *Pleuroxus similis*	20	33.33	–	–	–	–	5.0	25	1.0	2.70	2.0	3.23
12. *Pseudochydorus globosus*	–	–	80	76.92	35	25.31	–	–	–	–	–	–
13. *Simocephalus vetulus*	–	–	–	–	25	18.22	–	–	–	–	–	–
14. *Simocephalus expinosus*	–	–	2.0	1.92	–	–	–	–	–	–	–	–
15. *Scapholeberis kingi*	–	–	2.0	1.92	–	–	2.5	12.5	5.0	13.51	35.0	56.45
16. *Sida sp.*	–	–	–	–	–	–	–	–	–	–	–	–

of many susceptable species from Generals Tank. Further the increased grazing pressure by planktivorus fishes in fish ponds of Brace Bridge Nature Park has also limited the species richness. These ponds are extensively used for pisciculture.

The composition of cladoceran fauna was different in two large lakes. In Rabindra Sarovar, chydorids were represented by 9 species as compared to Subhas Sarovar where only 5 species occurred. The factor responsible for the richness of chydorid in Rabindra Sarovar appears to be the abundance of littoral macrophytes mainly belonging to *Ceratophyllum demersum*. The entire littoral zone of this lake upto nearby 10-12 ft. from shore was covered by macrophytes.

The analysis of numerically abundant species revealed that different water bodies have different set of abundant species. In Subhas Sarovar while *Diaphanosma sarsi, Ceridaphmia cornuta, Bosmina longirostris, Moina micrura* etc., contributed maximum to mean annual density, in Rabindra Saravor such dominant contribution was formed by *Chydorus sphaericus, Ceriodapnnia cornuta, Bosmina longirositis, Diaphanosma excisum, Simocephalus expinosus* etc. At present it is very difficult to ascertain core cladoceran fauna as reported by some earlier workers for tropical water bodies (Egborge et al 1994). This requires further detailed studies covering a large number of water bodies.

Acknowledgement

We wish to thank Director, Zoological Survey of India, for providing necessary laboratory facilities, constant encouragement and providing fellowship to one of us (C.S.)

References

Baird, W. (1860). Description of two new species of Entomostracous curstacea from India. *Proc. Zool. Soc. London*, 213-234.

Biswas., S. (1964). Species of Daphnidae (Crustacea : cladocera) from Simla hills in India, with a new record of *Alona costata* sars, from NEPHA *J. Zool. Soc. India.*, 16 : 92.

Biswas, S. (1971). Fauna of Rajasthan, India. Part II. (Crustacea : cladocera). *Rec. Zool. Surv. India.* 63 : 95-141.

Brehm, V. (1936). Yale North India Expedition. Report on Cladocera Article XVI. Mem. Cann. Acad., Arts. Sci., 10 : 282-297. Brehm. V. 1952. Indische Diaptomides Pseudodiaptomiden and cladoceren. *Ost. Zool. Zeit.*, 4 : 241-345.

Brehm. V. (1953). Indische Diaptomides Pseudodiaptomiden and cladoceran. *Ost. Zool. Zeit*, 4:241-345.

Brehm, V. (1963). Einige Bemerkugen Zu vier Indischer Entomostraken *Int. Rev. Ges. Hydrobiol.* 48 : 159-179.

Brehm, V. and Woltereck, R. (1939). Die Daphniden der Yale - North India Expedition. *Ibid.*, 39 : 1-19.

Dumont. H.J., J. Pensaert and I.van de velde, (1981). The crustacean zooplankton of Mali (West Africa). *Hydrobiologia* 80 : 161-187.

Dumont, H.J., (1980). Zooplankton and science of biogeography : the example fo Africa. In W.C. Kerfoot (ed.), *Evolution and ecology of zooplankton communities.* University Press of New England, Hanover, New Hampshire : 685-696.

Dumont, H.J. (1994). On the diversity of the Cladocera in the tropics *Hydrobiol.*, 272 : 27-38.

Edmondson, W.T. (1959). *Freshwater Biology.* John Wiley and Sons, New York, 757pp.

Egborge, A.B.M., Onwudinjo C.C. and Chigbu, P.C. (1994). Cladocera of coastal river of Western Africa *Hydrobiol.*, 39-46.

Fernando, C.H. (1994). Zooplankton, fish and fisheries in the tropical fresh water *Hydrobiol*, 272 : 95-104.

Fernando, C.H. and Kanduru, A., (1984). Some remarks on the latitudinal distribution of cladocera on the Indian subcontinent *Hydrobiol.* 113 : 69-76.

Ganapati, S V. (1943). An ecological study of a garden pond containing abundant Zooplankton. Proc. Acad. Sci., 17 : 41-58.

Green, J.E., (1962). Zooplankton of the river Sokoto. The Crustacea. *Proc. Zool. Soc. London* 138 : 415-453.

Green J. (1976). Changes in the zooplankton of lakes Mutanda Bunyonyi and Mulehe (Uganda). *Fresh Wat. Biol.* 6 : 433-436.

Green, J. (1990). Zooplankton associations in Zimbabwe, *J. Zool. Lond.* 222 ; 259-283.

Gurney, R. (1906). On some freshwater Entomostraca in the collection of the Indian Museum, Calcutta *J. Asiatic Soc. Bengal.*, 2 : 273-281.

Gurney, R. (1907). Further notes on Indian freshwater Entmostraca. *Rec. Indian. Mus.*, 1 : 21-33.

Khan, R.A. (1983). Pupulation dynamics biology and production ecology of a littoral zooplankton, *chydorus sphaercus* (O.F. Muller) (cladocera : chydoridate) *Bull. Zool. Sury. India.* 5(1) : 83-93.

Khan, R.A. (1983). The effects of food on growth, life span and reproduction of a cladoceran zooplankteer *Ceriodaphnia cournuta* (sars). *Bull. Zool. Surv. India,* 5(1) : 99-106.

Kumar, S. and Dutta, S.P.S. (1994). Population dynamics of cladocera in a subtropical pond, Jammu. *Indian J. Environ. Hlth.*, 36(1) : 19-23.

Mamaril, A.C. & C.H. Fernando, (1978). Fresh water zooplankton of the Phillipines (Rotifera, cladocera, copepoda). *Natural and Applied Science Bulletin* 30 : 106-221.

Michael, R.G. (1962). Seasonal events in a natural population of the cladoceran *Ceriodaphnia cournuta* sars and observation on its life cycle. *J. Zool. Soc. India.*, 14 : 211-218.

Michael, R.G. and Sharma, B.K. (1987). *Fauna of India and Adjacent Countries* - Indian cladocera (Crustacea : Branchipoda : Cladocera) Zoological Survey of India.

Murugan, N. (1975). The biology of *Ceridaphnia cornuta* sars (Cladocera : Daphinidae) *J. Inland Fish Soc. India.* 7 : 80-87.

Pati, S. and Sah, B.K. (1993). Hydrology and Plankton of the recently impounded Rengali reservoir of Orissa (India). *J. Fresh Water Biol.* 5(4) : 283-292.

Pennak, R.W. (1978). *Freshwater invertebrates* of the United States. 2nd Ed. Johnwiley & Sons., New York, 803 pp.

Ravera, O. (1996). Zooplankton and trophic state relationship in temperate lakes. *Mem. 1st Ital. Idrobiol.*, 54 : 195-212.

Revti, K. (1993). Biology of two temple tanks at Kovur and Tiruninrawur. *Geobios* 20 : 100-104.

Sewel, R.B.S. (1935). Studies on the bionomics of freshwater in India II. On the fauna of the tank in the India Museum compound and the seasonal changes observed. *Int. Rev. Hydrobio.*, 31 : 203-238.

Sharma, B.K. (1978). A note on Freshwater cladocerans from West Bengal. *Bangladesh. J. Zool.* 6 : 149-151.

Singh, U.N., and Pandey, Shakuntala (1991). Water quality of stagnant water bodies of North Bihar. *Env. and Eco.* 9(3) : 770-774.

Sinha, M.K., Pandit, N., Saha, L.C. and Datta Munsi J. (1992). Comparative limnological study of two religious ponds, Papharni and Sankhkund of Mandar Hill, Bounsi, Bhagalpur (Bihar). *J. Freshwater Biol.* 4(4) : 255-260.

Sorensen, T., (1948). A method of establishing groups of equal amplitude in plant sociology based on similarity of species content. *Det. Kong. Danske Vidensk. Selsk. Biol. Skr (copenhagen)* 5(4) : 1-34.

Vankataraman, K. and Das, S.R. (1993). Freshwater cladocera (Crustacea : Branchiopoda) of Southern West Bengal. *J. Andaman, Sci Associ,* 9(162) ; 19-24.

Vijayraghban, S. (1970). Seasonal events in a natural population of *Daphnia cariniata* King. *Proc. Ind. Acad. Sci.*, 71 : 193-203.

Yousuf, A.R. and Quadri, M.Y. (1977). Cladocera of Malpur sar, Kashmir. *J. Sci. Univ. Kash.*, 3 : 87-92.

Proceedings of the National Seminar on Environmental Biology, April 03-05, 1998
***Edited By* : A.K. Aditya & P. Haldar**
DAYA PUBLISHING HOUSE, DELHI

Macrophyte Preference and Insect Diversity of Freshwater Wetlands in Southeastern Bengal

Sujit Pal, S.R. Dey & D.K. Bhattacharya
Deptt. of Zoology, University of Kalyani
Kalyani - 741 235
and
S.K. Das & N.C. Nandi
Zoological Survey of India
M-Block, New Alipore, Calcutta - 53

Abstract

The present paper deals with insect diversity occuring in freshwater wetland of Southeastern Bengal viz., Calcutta, Nadia, North 24-Parganas and South 24-Parganas districts. Field works conducted so far from these districts have yielded a total of 73 species of insects belonging to five major groups (1 Ephemeroptera, 9 Odonata, 25 Hemiptera, 32 Coleoptera and 6 Diptera) associated with 39 species of macrophytes. Further studies on the assoication of insects with six different species of macrophytes, two from each of floating, submerged and emergent vegetation, have revealed highest insect diversity (39 spp.) associated with a species of submerged macrophyte viz., *Ceratophyllum demersum*. The degree and diversity of association of insects with six different speices of macrophytes have been documented and discussed.

Introduction

The shallow depth of freshwater wetlands supports suitable growth of macrophytes and offers ample micro-niches to diverse insect groups. From the economic point of view the macrophyte associated insects of freshwater wetland are important for their significant role in aquaculture (Jhingran, 1997) and serve as a reliable indicator of water quality. The macrophyte associated insects are also important component of food web in the wetland ecosystem. Information on wetland inhabiting insects with reference to their association with macorphyte being limited in India (Srivastava, 1959; Roy and Munshi, 1978; Mitra and Kumar 1988, Sexena *et al.*, 1989; Bhattacharya and Gupta, 1991; Rai and Sharma, 1991; Mishra *et al.*, 1992 and De and Sengupta, 1993), an attempt is made herein to highlight the macrophyte associated insect faunal diversity of Southeastern Bengal and especially to find out the insect community associated with six different species of macrophytes, selecting two from each of the floating, submerged and emergent categories.

Material and Methods

Several local faunistic surveys were made from a large number of wetlands of Southeastern

Bengal located in Calcutta, Nadia, North 24-Parganas and South 24-Parganas districts (Bhattacharya and Gupta, 1991; Nandi *et al.*, 1993) using a lift net (0.5 m^2) made of nylon cloth (75-80 mesh/cm^2), D-shaped hauling net and a box type of sampler (20x20x40 cm; Pal and Nandi, in press) and following methods of random sampling from some selected wetlands of this region. The collected materials were assorted after filtering though standard sieve no. 40. The insect specimens werw reserved in 70 per cent alcohol for laboratory studies and a few herbarium sheets were made for identification of macrophytes.

Results

A total of 73 insect species belonging to five orders were found to be associated with macrophytes viz. Ephemeroptera (1 species), Odonata (9 species), Hemiptera (25 species), Coleoptera (32 species) and Diptera (6 species). The equatic macrophte species are listed in Table 1. These macrophytes are represented by 39 species of 23 plant families and 34 plant genera. Similarly insects associated with these aquatic macrophytes are listed in Table 2. These insects represent 73 species distributed under 19 families and 45 genera. Among these, Coleopteran diversity tops the list comprising the largest number of 32 species belonging to 4 families and

Table 1 : List of the macrophytes collected from the freshwater wetlands of Southeastern Bengal

A. FREE FLOATING.

Family : Aroideae
1. *Pistia stratiotes*

Family : Convolvulaceae
2. *Ipomea aquatica*

Family : Gramineae
3. *Coix aquatica*

Family : Hydrocharideae
4. *Hydrocharis cellulosa*

Family : Lemnaceae
5. *Lemna perpusilla*
6. *Spirodella polyrrhiza*
7. *Wolffia arrhiza*

Family : Pontederiaceae
8. *Eichhornia crassipes*

Family : Salviniaceae
9. *Azolla pinnata*

B. SUBMERGED

(a) Rooted to the bottom

Family : Characeae
1. *Chara branchypus*

Family : Hydrochorideae
2. *Blyxa roxburghii*
3. *Hydrilla dentata*
4. *Hydrilla verticillata*
5. *Otellia alismoides*
6. *Vallisneria spiralis*

Family : Naiadaceae
7. *Potamogeton crispus*

Family : Najadaceae
8. *Najas indica*

Family : Scrophularineae
9. *Limnophila heterophylla*

(b) Not rooted to the bottom

Family : Ceratophyllaceae
10. *Ceratophyllum demersum*

Family : Lentibulariaceae
11. *Utricularia stellaris*

C. EMERGENT

Family : Amarantaceae
1. *Alternanthera philoxeroides*

Family : Cyperaceae
2. *Elocharis dulcis*
3. *Elocharis spiralis*
4. *Scirpus grossus*
5. *Scirpus articulatus*

Family : Leguminosae
6. *Aeschynomene indica*

Family : Nymphaceae
7. *Nelumbium specipsum*
8. *Nelumbo nucifera*
9. *Nymphaea* lotus
10. *Nymphaea pubiscens*

Family : Onagraceae
11. *Trapa bispinosa*

D. MARGINAL

Family : Alismaceae
1. *Saggitaria sagittifolia*

Family : Cyperaceae
2. *Cyperus exaltatus*
3. *Cyperus tegetiformis*

Family : Hydrophyllaceae
4. *Hydrolea zeylanica*

Family : Marsiliaceae
5. *Marsilea quadrifoliata*

Family : Onagraceae
6. *Jussiaea repens*

Family : Polygonaceae
7. *Polygonum barbatum*

Family : Typhaceae
8. *Typha angustata*

19 genera followed by hemipterans representing 25 species of 8 families and 12 genera. The odonates and dipterans are represented by 9 and 6 species distributed under 3 families and 8 genera and 3 families and 5 genera respectively. Only one species of ephemeropterans has so far been recorded. The insect species collected from six different species of macrophytes are listed in Table 3. It is evident from Table 3 that maximum numbr of insect species (39 species) was recorded from *Ceratophyllum demersum,* a submerged plant and lowest (13 species) from *Alternanthera philoxeroides,* and emergent species. Among the three broad categories taken into consideration, the number of insect species associated with different species of macrophytes of each category differs significantly ($0.001<p<0.05$) from one category to the other. It is also apparent from Table 3 that suberged vegetation is the most preferred macrophyte type by the insects. Furthermore, of the five groups of insects, ephemeropterans were recorded from all categories of macrophytes.

Table 2 : List of the insect species associated with the macrophytes in freshwater wetlands in Southeastern Bengal

A. EPHEMEROPTERA

Family : Baetidae

1. *Cloeon kimminsi*

B. ODONATA

Family : Libellulidae

1. *Brachythemis contaminata*
2. *Crocothemis servilia*
3. *Orthetrum sobina*

Family : Coenagrionidae

4. *Agrioenemis pygmaea*
5. *Ceriagrion cormondelianum*
6. *Ischunura aurora*
7. *Ischnura senegalensis*
8. *Pseudagrion microcephallum*

Family : Gomphidae

9. *Ictinogamphus rapak*

C. HEMIPTERA

Family : Belostomidae

1. *Diplonychus annulatus*
2. *Diplonychus rusticus*
3. *Diplonychus* sp.
4. *Lethocercum indicum*

Family : Gerridae

5. *Limnogonus fossarum*
6. *Limnogonus nitidus*
7. *Limnogonus parvulus*
8. *Limnogonus* sp.
9. *Noboandelus signatus*
10. *Rhagadotarsus kraepelini*

Family : Hydrometridae

11. *Hydrometra-greeni*
12. *Hydrometra* sp.

Family : Nepidae

13. *Laccotrephes griseus*
14. *Ranatra sordidula*
15. *Ronatra filiformes*

Family : Notonectidae

16. *Anisops breddini*
17. *Anisops bouvieri*
18. *Anisops sardea*
19. *Anisops batilliformis*
20. *Anisops* sp.

Family : Pleidae

21. *Plea liturata*
22. *Plea* sp.

Family : Velidae

23. *Mesovelia vittigera*

Family : Corixidae

24. *Micronecta scultellaris*
25. *Micionecta* sp.

D. COLEOPTERA

Family : Dytiscidae

1. *Clypeodytes* sp.
2. *Canthydrus laetabilis*
3. *Canthyarus luctuosus*
4. *Canthydrus morsbachi*
5. *Conthydrus ritsemai*
6. *Cybister limbatus*
7. *Cybister* sp.
8. *Eretes sticticus*
9. *Hydrocoptus subvittulus*
10. *Hydraticus fabricii*
11. *Hydrovatus* sp.
12. *Laccophilus sharpi*
13. *Laccophilus parvulus*
14. *Laccophilus anticotus*
15. *Laecophilus* sp.
16. *Uvarus quadrilineatus*

Family : Hydrophilidae

17. *Amphiops pedestris*
18. *Amphiops* sp.
19. *Berosus indicus*
20. *Enorcus* sp.
21. *Globeria* sp.
22. *Hydrophilus olivaceus*
23. *Helochares anchoralis*
24. *Helochares crenatus*
25. *Helochares* sp.
26. *Regimbartia attenuata*
27. *Regimbartia* sp.
28. *Sternolophus rufipes*
29. *Sternolophus* sp.

Family : Haliplidae

30. *Haliplus* sp.

Family : Gyrinidae

31. *Dineutus indicus*
32. *Dincutus unidentatus*

F. DIPTERA

Family : Chironomidae

1. *Chironomus circumdatus*
2. *Chironomus* sp.

Family : Culicidae

3. *Aedes* sp.
4. *Anopheles* sp.
5. *Culex* sp.

Family : Stratiomyidae

6. *Odontomyia dorsoangulata*

They showed marked preference for emergent plants. The emergent vegetation is also preferred by odonates while submerged vegetation is preferred by insects like Hemiptera, Coleoptera and Diptera.

Table 3 : List of the insect species associated with six different species of macrophytes

Category	Macrophyte species	Insect Order: Ephemeroptera	Odonata	Hemiptera	Coleoptera	Diptera	Total
Floating	*Eichhornia crassipes*	1 (1)	2 (2, 8)	5 (2, 13, 14, 21, 22)	12 (2, 3, 9, 12-18, 22, 23)	2 (1,7)	22
	Lemna purpusilla	1 (1)	1 (2)	7 (1-3, 21, 22, 24, 25)	6 (2, 8, 9, 27, 28, 32)	4 (1, 5-7)	19
Submerged	*Ceratophyllum demersum*	1 (1)	4 (1, 3, 8, 9)	12 (1-3, 13-17, 21, 22, 24 25)	17 (1-4, 6, 11-17, 23, 25, 26 30, 32)	5 (1, 5-8)	39
	Vallisneria spiralis	1 (1)	3 (1, 7, 8)	9 (3, 6, 10, 13-15, 21, 22, 25)	12 (2-4, 9, 11-13, 15-17, 20, 23)	4 (1, 4, 6, 7)	29
Emergent	*Alternahthera philoxeroides*	1 (1)	7 (1, 2, 4-8)	4 (1, 2, 14, 14)	1 (9)		13
	Nelumbo nucifera	1 (1)	5 (1, 2, 5, 8, 9)	6 (1-3, 11, 22, 25)	6 (17-19, 22, 23)	3 (4, 6, 7)	21

Note : Total number of species along with the serial no. of the insect species shown in Table 2 is given in parentheses.

Discussion

There are more than hundred species of macrophytic plants occurring in freshwater wetlands of West Bengal (Prain, 1903; Subramanyam, 1962; Deb, 1976). In freshwater wetlands insects often represent dominant life forms and are prevalent in shallow ponds and streams. There are 10 orders of class Insecta which are truly aquatic and several others may be treated as semi-aquatic (Usinger, 1956). In the present study only 39 species of macrophytes have been recognised harbouring 73 insect species belonging to five orders. The representation of the species of insect fauna, both larvae and adults, reveals that coleopteran and hemipteran species inhabit the wetland environment throughout their life, while species belonging to the rest insect orders living in the wetland environment during their immature stages are in a sense amphibious.

Among six different macrophytes, insects, in general, preferred submerged vegetation *Ceratophyllum demersum* with dissected leaf surface that also offers adequate shelter to a wide assemblage of insects. The majority of the insect species make associated with macrophytes for their shelter, while few coleopterans use them as food and few species exploit them as egg laying site (Mishra *et al.*, 1992). The occurrence and abundance of ephemeroptera and odonate larvae associated with the emergent vegetation may be related to their emergence of last instar nymphal stages into adult phase of aerial existence (Srivastava, 1986 and Singh and Roy, 1991).

Acknowledgement

The authors are thankful to the Head of the Department of Zoology, University of Kalyani, Kalyani and to the Director, Zoological Survey of India, Calcutta for necessary facilities.

References

Bhattacharya D.K., Gupta B., (1991). Freshwater wetland inhabiting insects of West Bengal. *Env. & Ecol.* 9(4) : 995-998.

De (Pal) M., Sengupta T. (1993). Beetles (Coleoptera : Insecta) of Wetlands of Calcutta and its surroundings. *Rec. Zool. Surv.* India 93 (1-2) : 103-138.

Deb D.B., (1976). Aquatic Plants of India. *Bull. Bot. Soc. Bengal* 29 : 155-170.

Jhingran V.G., (1997). Fish and Fisheries of India. Hindustan Publishing Corporation (India) Delhi, 3rd Ed. (corrected reprints) pp. xxiii + 728.

Mishra R.K., Jha B.P., Jha V., Singh S.K., Mahato A. (1992). Insect associations of *Euryale ferox* Salisb. in the ponds of Darbhanga North Bihar. *J. Freshwater. Biol.* 4(3) : 199-208.

Mitra K., Kumar K. (1998). Common insects of freshwater ponds and their control. *Central Inland Capture fish, Res. Instt. Barrackpore Bull.* 54 : 12-38.

Nandi N.C., Das S.R., Bhunia S., Dasgupta J.M., (1993). Wetland faunal resources of West Bengal 1. North and South 24-Parganas Districts. *Rec. Zool. Surv. India. Occ. paper no.* 150 : 1-50.

Pal S., Nandi N.C. in press. A simple device for quantitative sampling of macrofauna from littoral macrophytes. *J. Freshwater Biol.*

Prain D. (1903). Bengal Plants. 2 vols. London (Regr. ed. 1963).

Rai, D.N., Sharma U.P. (1991). Co-relation between macrophytic biomass and macroinvertebrates community structure in wetlands of North Bihar. *Int. J. Ecol. & Envt. Soc.* 17 : 27-36.

Roy S.P., Munshi J.S.D., (1978). Freshwater aquatic beetle (coleoptera) of Bhagalpur (Bihar). *Proc. IVth All India Zool. Congr.* Bodh-Gaya.

Saxena M.N., Sharma S., Kaushik S. (1989). Studies on Aquatic bodies communities and their distribution in some aqautic bodies of Gwalior, M.P. India. *J. Hydrobiol.* 5(1) : 39-42.

Singh J.P., Roy S.P. (1991). Interaction between macrophytic biomass and macroinvertebrate abundance in Kawar Lake. Begusari (Bihar) *J. Freshwater Biol.* 3 (3) : 229-234.

Srivastava V.D., (1986). Ephemeroptera larva as component of aquatic environment and their role as bio-indicator of pollution. *Proc. Symp. Pest Resid & Env. Pollu.* pp. 206-218.

Srivastava, V.K. (1959). Studies on the aquatic vegetation and their role in breeding and development of some insect. *Nat. Acad. Sci. India* (Allahabad) Annual No. 69.

Subramanyam K. (1962). Aquatic Angiosperms : A Systematic Account of Common Indian Aquatic Angiosperms. New Delhi. CSIR.

Usinger R.L. (1962). Aquatic insects of California. California University Press, Berkely.

Section IV

INVERTEBRATE BIODIVERSITY : TERRESTRIAL ECOSYSTEM

Proceedings of the National Seminar on Environmental Biology, April 03-05, 1998
***Edited By :* A.K. Aditya & P. Haldar**
DAYA PUBLISHING HOUSE, DELHI

Diversity in Soil Mites (Acari) of West Bengal

A.K. Sanyal & A.K. Bhaduri*
Zoological Survey of India,
Calcutta - 700 053
* Vidyasagar College, 6 Sankar Ghosh Lane
Calcutta-700 006

Abstract

The study on soil dwelling mite fauna of West Bengal was initiated by Bhattacharya in 1965. Later, many workers studied the taxonomy and ecology of soil mites and till date 264 species under 162 genera and 75 families are known from the state. Among the four suborders, the cryptostigmatids have been studied extensively. The range of diversity, endemicity and Zoogeographical relationship of the species are discussed in the paper.

Introduction

The mites inhabiting soil are the most dominant and important group of Acarina and surpass the number of all other groups of microarthropods living in soil. They are tiny creatures having sizes range from 100-2000 mm. Their somatic segments are generally inconspicuous or absent. The legs are five segmented with variously modified claws at the tip. Most of these mites are oviparous. The life cycle passes through larval and nymphal stages.

The soil mites have drawn special attention because many of these play a vital role in the process of humification of organic matter resulting in increase of soil fertility. Few species also act as predators of soil nematode pests. Various helminth diseases carried by anoplocephaline cestodes in cattle and domesticated animals are transmitted by different species of soil cryptostigmatid mites. These mites are also injurious to plants and producing alergy in man. They are sometimes used as agents of biological control and are also considered as a homogenous material for the study of Zoogegraphy.

Though the study on soil mites in India was initiated by Oudemans in 1906, the work in West Bengal was started very late. Bhattacharyya (1965) and Deb and Roychaudhury 1965) first recorded mesostigmatid mites from the soils of West Bengal. Bhaduri and Raychaudhuri (1967) first reported soil crytostigmatid species from Calcutta. Later, serval workers described and recorded a good number of genera and species of soil mites from the State. The detailed accounts of those studies have been summarised by Sanyal and Bhaduri (1986), Sanyal and Bhaduri (1988), Sanyal (1991a, 1991b), Sanyal (1992) and Bhattacharyya, Sanyal and Sengupta (1996).

Range of Diversity

The total number of taxa in different suborders of mites so far known from Indian soils is shown in Figure 1. It indicates that soil mites belonging to all the four suborders are found to occur in India. It shows that cryptostigmatid taxa occupy the highest position in number. The

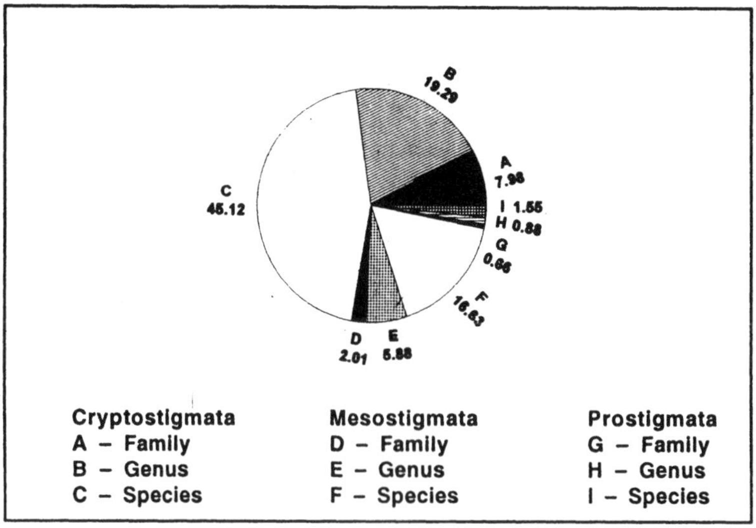

Fig. 1 : Total No. (%) of taxa of soil mites in India

soil mites belonging to other subordes i.e. Mesostigmata, Prostigmata and Astigmata occupy the second, third and fourth position, respectively.

The number of soil mite species of West Bengal (Figure 2) when compared with the total number of mite species known from Indian soils, it is noted that West Bengal alone represents 47.4 per cent of Indian fauna and occupies highest position among the other States. The total number of taxa of mites so far known from the soils of West Bengal is 75 families 162 genera and 246 species. The data when considered sub order-wise, it is observed that cryptostigmatids

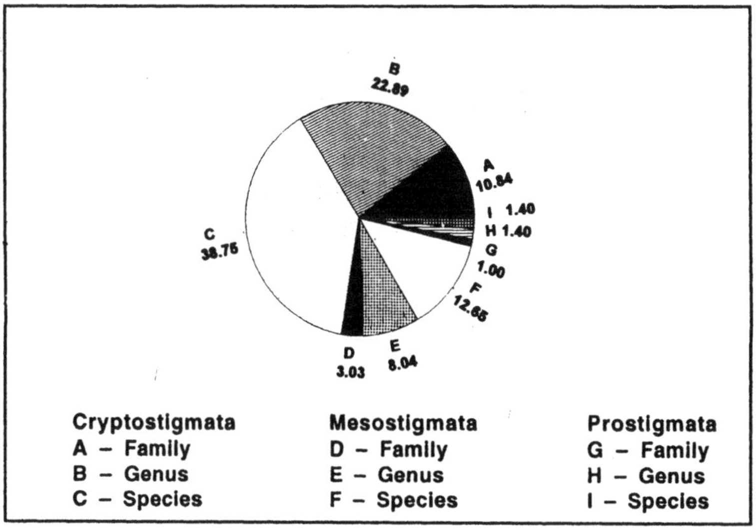

Fig. 2 : Total No. (%) of taxa of soil mites in West Bengal

comprise largest number of species (193), genera (114) and families (54). The mesostigmatids occupy the second position with 15 families, 40 genera and 63 species. The prostigmatid (15 families, 7 genera and 7 species) and astigmatid (1 family, 1 genus and 1 species) taxa are very few in number because of little attention paid to these groups by the acarologists.

While analysing the status of taxa of soil mites in West Bengal, it is noted that 2 genera and 59 species of Cryptotigmata and genea and 49 species of Mesostigmata have have described as new to science from this state. It is also recorded that among the species so far known from West Bengal, only a few are to be considered as domiant soil mite species from the state (Table 1).

Table 1 : Dominant Soil Mite Species Known from West Bengal

Cryptostigmata	Mesostigmata	Prostigmata	Astigmata
Scheloribates albialatus	*Asca garmani*	*Tydens* sp.	*Tyrophagus* spp.
S. parvus	*Gamasellodes bicolor*	*Cunaxa* spp.	
S. huancayensis	*Lasioseius perberlesei*	*Chelacaropsis* sp.	
S. natalensis	*Hypoaspis bengalensis*	*Allothrombium* sp.	
Oppia yodai	*Uroobovella villosella*	*Coccotydeus* sp.	
Galumna flabellifera orientalis	*Veigaia serrata*	*Dinothrombium gigas*	
Lamellobates palustris			
Hoplophorella africana			
Rhysotritia ardua			
Haplochthonius clavatus			

Thus, the distribution pattern of soil mites in West Bengal indicates that they have adapted themselves to live in different soil types, ecosystems and habitats, from mangrove forests to higher elevation of the Himalayas.

The Zoogeographical relationship of the soil mites so far known from West Bengal is shown in Figure 3. From the data presented in the figure it would be premature to make any specific comment of the relationship as the knowledge on soil mesostigmatid, prostigmatid and astigmatid mites of West Bengal are still very scanty. However, it indicates that cryptostigmatid species recorded from West Bengal show maximum similarity with Neotropical species. The other regions in order of degree of similarity are palearctic, pacific, ethiopian, nearctic and antarctic. It is also noted from the figure that maximum number of palearctic species of soil mesostigmatid mites have been recorded from West Bengal. The other regions in order of degree of similarity in case of soil mesostigmatids are neotropical, ethiopian, nearctic and pacific.

Endemicity in Soil Mite Taxa

The studies so far done on soil mites in West Bengal show that most of the areas in the State are not yet extensively surveyed particularly for mesostigmatid, prostigmatid and astigmatid mites, hence their distributional data are insufficient and it is rather difficult to make any comment on the degree of endemism in soil mite taxa known from West Bengal. However, on the basis of data so far available, it may be noted that the soil mite fauna in the state has a number of endemic genera and species. The endemicity is more pronounced in Cryptostigmata, in which one genera and 55 species are endemic to West Bengal. In the suborder Mesostigmata, endamicity is restricted to one genus and 38 species. There is no endemic species in prostigmatid and astigmatid mites known from the state.

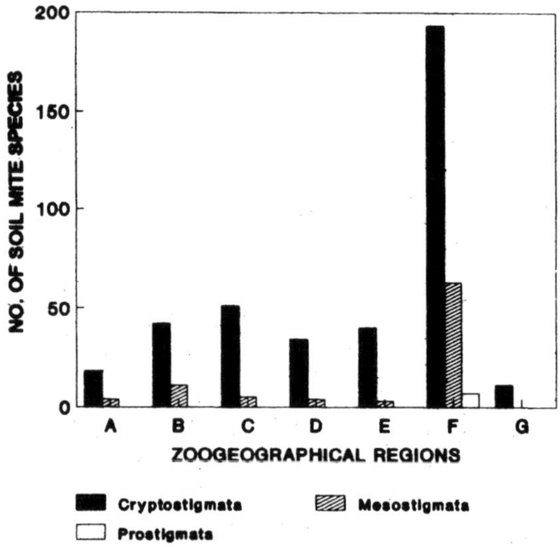

A–Nearctic; B–Palaearctic; C–Neotropical; D–Ethiopian
E–Pacific; F–Oriental; G–Antarctica

Fig. 3 : Zoogeographical Distribution of Soil Mite Species of West Bengal

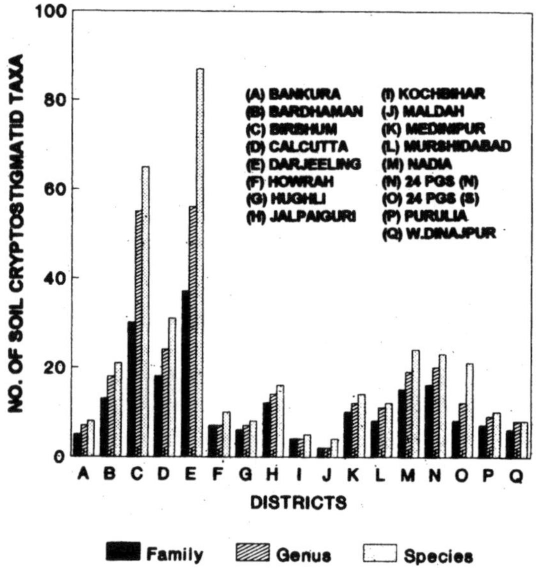

Fig. 4 : Total Number of Soil Cryptostigmatid Taxa in West Bengal

Distribution of Soil Mites in West Bengal

The distribution of soil mite taxa in different districts of West Bengal is shown in Table 2 and Figures 4-6. These indicate that the soil inhabiting cryptostigmatids are known from all the districts, whereas mesostigmatids, prostigmatids and astigmatids are described and or recorded from eight, two and one districts respectively. It also shows that the soil mite fauna of 9 districts *viz.*, Bankura, Darjeeling, Jalpaiguri, Koch Bihar, Maldah, Medinipur, Murshidabad, Purulia and West Dinajpur are known only by Cryptostigmata. The absence of any record of other groups of soil mites in these districts is not due to the fact that these mites are absent there, but only because of very scanty works done on these groups. The table also records that Darjeeling district alone represents 49.33% of families, 34.56% of genera and 32.45% of species out of the total soil mite taxa known from West Bengal and it occupies highest position among the other districts. The district Birbhum occupies the second highest position. The other districts in order of number of total soil mite taxa are South 24-Parganas, Calcutta, North 24-Parganas, Nadia, Hughli, Baddhaman, Howrah, Jalpaiguri, Medinipur, Murshidabad, Purulia, West Dinajpur, Bankura, Koch Bihar and Maldah. The analysis of available data records that the distribution of one genera each and 59 and 49 species of Cryptostigmata and Mesostigmata respectively are restricted to West Bengal, though they are known from outside India. Further, it is recorded that 23 families, 31 genera and 41 species of Cryptostigmata and 7 families, 12 genera and 17 species of Mesostigmata are known West Bengal as well as from outside the State.

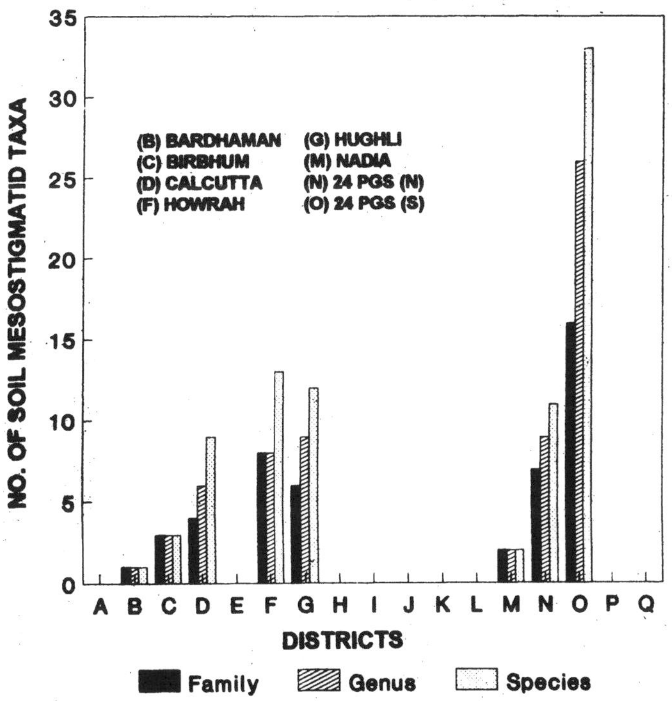

Fig. 5 : Total Number of Soil Mesostigmatid Taxa in West Bengal

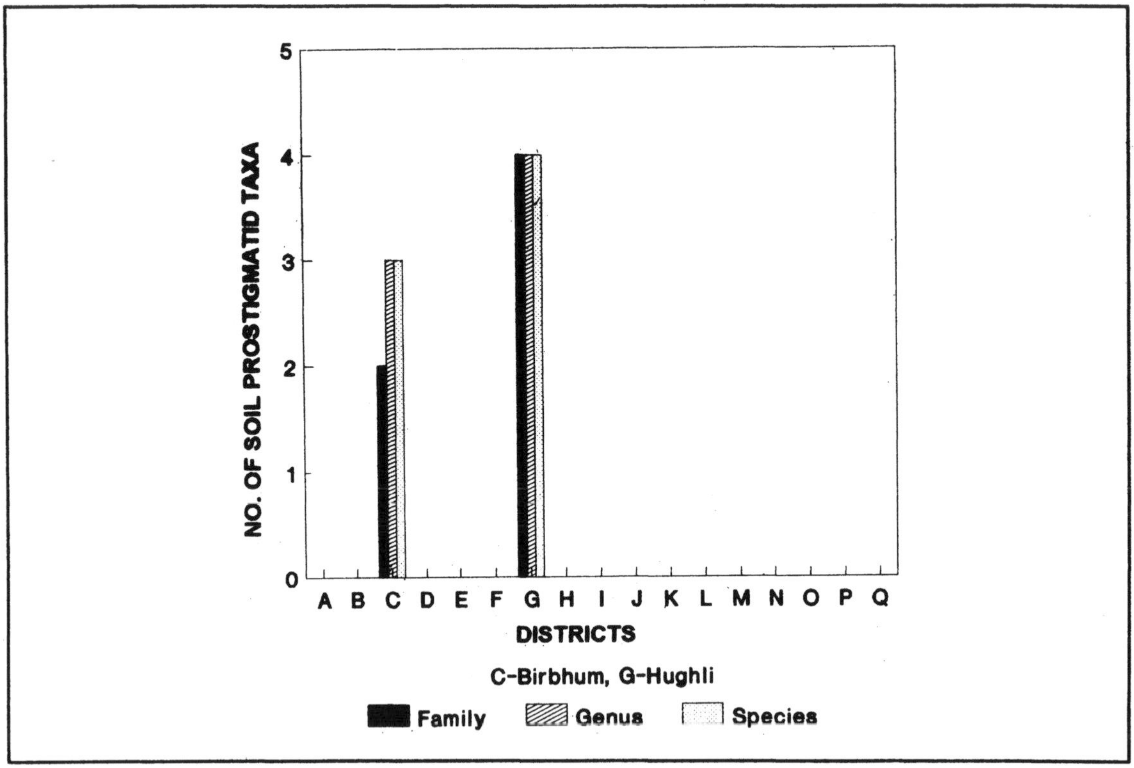

Fig. 6 : Total Number of Soil Prostigmatid Taxa in West Bengal

Table 2 : Total Number of Taxa of Soil Mites in India and West Bengal

Districts	Cryptostigmata			Mesostigmata			Prostigmata			Astigmata		
	Fam.	Gen.	Spp./ SSpp.	Fam.	Gen.	Spp./ SSpp.	Fam.	Gen.	Spp./ SSpp.	Fam.	Gen.	Spp./ SSpp.
India	72	174	407	18	53	150	6	8	14	03	03	03
West Bengal	54	114	193	15	40	63	5	7	7	01	01	01
Bankura	5	7	8	–	–	–	–	–	–	–	–	–
Baddhaman	13	18	21	01	01	01	–	–	–	–	–	–
Birbhum	30	55	65	03	03	03	02	03	03	–	–	–
Calcutta	18	24	31	04	06	09	–	–	–	–	–	–
Darjeeling	37	56	87	–	–	–	–	–	–	–	–	–
Hawra	07	07	10	08	08	13	–	–	–	–	–	–
Hoogli	06	07	08	06	09	12	04	04	04	–	–	–
Jalpaiguri	12	14	16	–	–	–	–	–	–	–	–	–
Koch Bihar	04	04	05	–	–	–	–	–	–	–	–	–
Maldah	02	02	04	–	–	–	–	–	–	–	–	–
Medinipur	10	12	14	–	–	–	–	–	–	–	–	–
Murshidabad	08	11	12	–	–	–	–	–	–	–	–	–
Nadia	15	19	24	02	02	02	–	–	–	–	–	–
24-Parganas (N)	16	20	23	07	09	11	–	–	–	–	–	–
24-Parganas (S)	08	12	21	14	26	33	–	–	–	–	–	–
Purulia	07	09	10	–	–	–	–	–	–	–	–	–
West Dinajpur	06	08	08	–	–	–	–	–	–	–	–	–

The suborder-wise distribution of species shows that highest number of soil cryptostigmatid mites is known from Darjeeling district. Similarly maximum number of mesostigmatid mites is recorded from South 24-Parganas district. The occurrence of maximum number of mite taxa in these two districts is due to the fact that these districts have been surveyed extensively.

Acknowledgements

The authors express gratefulness and thanks to the Director, Zoological Survey of India for extending working facilities.

References

Bhattacharyya, A.K., Sanyal, A.K. and Sengupta, D. 1996. Present state of knowledge of taxonomy of soil mesostigmata (Acari) in India. *Environment & Ecology*, 14(4) : 850-857.

Sanyal, A.K. 1991a. ACARI; Soil Mesostigmata. Animal Resources of India (ed. Director, ZSI) : 501-504.

Sanyal, A.K. 1992b. Oribatid mites (Acari). *States Fauna Series* 3 : Fauna of West Bengal, part 3 : 213-356.

Sanyal, A.K. and Bhaduri, A.K. 1986. Check list of Oribatid mites (Acari) of India. *Rec. zool. surv. India, Occ. Paper* no. 83, 79 pp.

Sanyal, A.K. and Bhaduri, A.K. 1988. The present state of knowledge of Oribatid (Acari) taxonomy in India. *Progress in Acarology* (eds. G.P. Channa Basavanna and C.A. Viraktamath), 1 : 295-300.

Proceedings of the National Seminar on Environmental Biology, April 03-05, 1998
***Edited By :* A.K. Aditya & P. Haldar**
DAYA PUBLISHING HOUSE, DELHI

Insect Biodiversity in Agroecosystem : Consequences of Insecticide Use and Remedial Role of Integrated Pest Management

G.T. Gujar
Division of Entomology,
Indian Agricultural Research Institute
New Delhi-11 0012

Abstract

There are about 1.6 million species living on the earth. Of these, about 0.25 million are plants and rest animals. Of the animals, nearly 75% are insects. Insects constitute about 1 million species. However, there are many that remain unidentified. They have adapted themselves over the evolutionary period of about 250-500 million years to the diverse environmental conditions.

The dominance of insects over the other animal species may be attributed to their variability in their biological and ecological attributes. Only a fraction of insect species is of direct importance to human. They provide prducts like honey, lac, silk for human consumption. They help in crop production by pollination. On the contrary, some are vectors of human and animal diseases, affecting millions of people by causing diseases and creating nuisance. Insects of agricultural importance damage the crop to extent of about Rs 29,000 Crore every year in India alone.

The growing needs of human have led to the regular interference with modification of insect habitat so much so that conflict between insects and human has become a vicious cycle of dominance of one over other. The excessive reliance on pesticides has led to the development of pesticide resistance, resurgence of minor pests, pest outbreaks, destruction of beneficial and useful insects and also contamination of the environment. This has short-lived pesticide usefulness.

In order to preserve insct biodiversity, it is necessary that agroecological diversity be maintained along with simultaneous reduction in indiscriminate insecticide use. Integrated pest management is an essential part of integrated cropping system that aims at preserving insect biodiversity with judicious and compatible use of insect management agents. This paper therefore aims to discuss consequences of insecticide use on insect biodiversity and remedial role of integrated pest management.

Introduction

Insects comprising of about 950,000 described species constitute about 60% of the total biota and about 75% of the animals (Samways, 1994). With the numerical advantage gained over evolution spanning 250-500 million years, they have adapted themselves to various ecological niches and are therfore found practically everywhere. Similarly, they have varied feeding and behavioural attributes. The dominance of insects over other animals is attributed to their morphological, physiological and ecological adaptability. The study of insect biodiversity therefore represents a classic example of their adaptability to the wide range of environmental conditions. Agroecosystem is one of the well-developed habitats that constitute nearly 45% of the total area in this country and therfore any change in agroecosystem affects biodiversity of insects closely.

Insect Biodiversity: Insect biodiversity in India is restricted to about 51,000 species. These constitute about 6% of the world insect fauna. Insects of agricultural importance are found in high endemism in India (table 1).

Table 1 : Insects of Agricultural Importance in India

Insect order	Total species	Endemic species	% endemism
Orthoptera	900	330	37
Isoptera	310	145	47
Hemiptera	5216	2421	46
Thysanoptera	702	493	70
Coleoptera	15289	5198	34
Diptera	6093	2135	35
Lepidoptera	15000	1500	10
Hymenoptera	4264	2920	68
Neuroptera	315	256	80
Mantodea	162	88	54
Odonata	494	115	21

Cf. Ghosh (1998)

Except for Lepidoptera, the endemism of insect in various orders in India is very high. Even in Lepidoptera where endemism is only 10% the economic importance of lepidopterans in terms of thir damage to crop and their role in pollination is very high. About 155.5 insect species are found for every million hectare of land in India while these estimates vary with countries; USA 9.5, Britain 91.4 and Southern Africa 97.2. There are about 2.83 insect species for every flowering plant species in India. This ratio is low when compared with the other countries viz., USA 4.45, Britain 14.9, and Southern Africa 10.9. The low insect to plant ratio in India is largely due to comparatively higher plant biodiversity in the country.

Intensive Agriculture-effect on Insect Biodiversity

India, with its area of about 328 million hectare, has baout 142 million hectares under agriculture. Of the biodiversity of about 45,000 plant species native to the country, about 160 plant species are domesticated, with 20 plants as food crops. The intensive farming has over the years reduced genetic variability within the food crops significantly. Of the total area of 96 million hectare under the cultivation of 5 cereals viz., rice, wheat, sorghum, pearl millet and maize; 75.8 million hectares are under high yielding varieties, thereby, eroding varietal diversity that has been a hallmark of traditional Indian farming. This erosion of plant genetic diversity concurrent with intensive farm practices like extensive use of inputs like fertilizers, pesticides, irrigation and farm mechanization has brought about changes in insect pest complex in agriculture, particularly towards the dominance of single or oligo pest species complex. This has reduced pest enemy population and increased chances of pest outbreaks (Ferro, 1987). The monocropping in India has led to pest outbreaks like those of American bollworm and whitefly in cotton, gallmidge, stem borer and gundhy bug in paddy in recent years. Regupathy *et al.*,

(1994) listed about 808 insect pests in the country damaging various agro-horticultural crops. Considering the fact that about 1% of pest species numbering about 10,000 are reported from all over the world (Sailer, 1969), insect pests of agricultural importance alone constitute about 1.6% of insect fauna in India. The mean insect pest species per million hectare of crop area and a domesticated crop works out to be 5.7 and 5.05 respectively. Obviously, for the tropical country like India, this number may appear to be relatively low; since crops like rice are reportedly attacked by as many as 74 insect pests. Nevertheless, considering favourable tropical conditions, the damage potential of even few insect species is high.

Irrigation facilities have over the years led to the increase in cropping intensity to the tune of 1.36 for the country while for state like Punjab, it is as high as 1.7. Cropping intensity coupled with rotatiosn like rice-rice cropping system in the South and rice-wheat in the North India has accentuated pest problem. Even crop rotations like cotton in kharif followed by pulses like gram or extensive cultiveation of pigeon pea as monocrop have led to increase in pest problems, since such cropping systems have provided alternate hosts to the insect throughout the year. Thus, American bollworm has become a serious pest in the most of the South India. Zero tillage or minimum tillage concept has further led to increase in pest problems especially in tropical and subtropical countries where quick plant growth provides necessary food to the insect pests round the year. Although fertilizer use has increased crop yields, their biased use in favour of nitrogenous fertilizers, has led to increase in succulency of leaves/plants that in turn increased pests like whitefly, brown planthopper in some crops. Alongwith fertilizes, other chemicals like pesticides, plant growth regulators have become a part of intensive farming. Of these, insecticides are the main chemicals used for pest control in the country.

Consequences of Use of Insecticides in the Country

In India, pesticides are used for crop protection and public health to the extent of about 85,000 tons annually. Of these, about 77.8% of pesticides are insecticides, rest fungicides, herbicides, rodenticides, fumigants and miscellaneous pesticides. The use of insecticides is mostly in agriculture both in terms of quantity and quality. Insecticides being provider of immediate relief to the crops are often used indiscriminately as the only means of pest control. The use of insecticides over the last four decades had peaked up to about 85,000 tons in 1980s, but has been consistently declining from about 50,700 tons in 1990-91 to about 40,230 tons in 1994-95. The decline is mainly due to phasing out of DDT, hexachlorocyclohexane and cyclodienes. But in terms of toxicity, this decline in tonnage is concurrently compensated with gradual increase in consumption of more toxic organophosphates and synthetic pyrethroids. Further the use of insecticides for crop protection has become biased based upon the economic importance of crop. Thus cash crop like cotton which occupies only about 7.8 million hectare of area (5.5% of total crop area) consumes about 52% of pesticides. Rice with 24% of crop area consumes about 17.2% of pesticides, vegetables and fruits with 3% of crop area 13.2% of pesticides while cereals with 58% of crop area consumes only 6.7% of pesticides. Pesticide consumption within the country varies widely. Andhra Pradesh is a leading state in terms of pesticide consumption with 33.6%, Karnataka 16.2%, Gujarat 15.2%, Punjab 11.4%, Maharashtra 5.1% and rest 18.5%. This biased consumption pattern has partly contributed to the pesticide resistance, pest resurgence and pest outbreaks. Maximum pesticide resistance in the American bollworm and pest outbreaks has been reported from Andhra Pradesh. Increased use of pesticides have led to a) erosion of insect diversity leading to insect resurgence and pest outbreads, b) development of insecticides resistance, c) adverse effects on beneficials and natural enemies of insect pests and d) environmental contamination and health hazards to the higher animals including humans. Inview of dominant use of pesticides in cotton and rice agroecosystems, these two crops present interesting case studies for changes in their ecosystem, their effect on insect diversity and success stories of integrated pest management (IPM).

Insect Diversity in Rice Agroecosystem

Many insect pests attack various stages of growth of rice (*Oryza sativa*) that is cultivated over an area of 42 million hectares with annual production of about 82 million tons in India. Nayer *et al.* (1976) listed about 221 insect pests of rice while Dale (1994) reported only 74 insect pests belonging to 6 orders, 20 families and 55 genera (table 2). Although some of genera have more than one species, dominating insect populations often belong to single species and as diverse as possible. The pest complex of rice has undergone a change over years. Datta (1997) reported the changing pest dynamics of rice over three decades at Sriniketan in West Bengal. Almost all common pests of rice were used to be recorded on a single rice variety (TN-65) in a single plot in the 1960s (Datta *et al.*, 1967); but in recent years, only limited species, 8 in number, viz., whorl maggot, leaf roller, leaf folder, green leafhopper, brown planthopper, yellow stem borer, gundhy bug and earhead cutting leaf caterpillar cause most of the damage to rice. Thus, an increase in intensification of rice cultivation has narrowed down insect diversity.

Table 2 : Insect Diversity in Rice Agroecosystem

Order	Family	Genera	Species
Order	Family	Genera	Species
Orthoptera	Acrididae	11	16
	Gryllotalpidae	1	1
Hemiptera	Alydidae	1	1
	Aphididae	7	11
	Cicadellidae	3	3
	Delphacidae	2	2
	Pentatomidae	2	4
	Pseudococcidae	1	1
Thysanoptera	Phlaeothripidae	1	1
	Thripidae	1	1
Lepidoptera	Hesperiidae	2	2
	Noctuidae	5	7
	Pyralidae	6	10
	Satyridae	1	1
Diptera	Cecidomyiidae	1	1
	Ephydridae	1	1
	Muscidae	1	2
Coleoptera	Chrysomelidae	2	2
	Curculionidae	2	2
	Scarabaeidae	4	5

Cf. Dale (1994)

Pest Resurgence and Outbreaks

Concurrent with increase in cultivation of high yielding short duration, non-logging japonica rice cultivars pest complexity has changed to a dominance of some key pests and their outbreaks occasionally (Table 3). About 20 insect pests are considered of economic importance in India (Pathak and Pawar, 1983). Of these, brown planthopper *Nilaparvata lugens*, Green leafhopper *Nephotettix virescens*, white-backed planthopper *Sogatella furcifera*, yellow stem borer *Scirtophaga incertulas*, gallmidge *Orseolia oryzae*, leaf folder *Cnaphalocrocis medinalis*, hispa *Dicladispa armigera* and rice earcutting caterpillar *Mythimna separata* are found to cause serious outbreaks. Pathak *et al.* (1996) reported that more than 55% of total rice production was lost annually due to pests and diseases, with insect pest contributing about 30%.

Table 3 : Chronological Importance of Insect Pest of Rice

Year	Rice Type	Insect pests
Before 1960s	Indica varieties	Stem borer
1960s to 1970s	Japonica varieties	Gallmidge, Green leafhopper, Brown plant hopper, white-backed brown planthopper, hispa, leaf folder, gundhy bug
1980s to 1990s	High yielding varieties, hybirds	Brown planthopper, green leafhopper, leaf folder, Stem borer, gallmidge, hispa

Cf. Pathak and Pawar (1983), Krishnaiah (1993)

With an increase in area under high yielding varieties of japonica type of rice, germplasm diversity for various traits has eroded greatly. Some of these high yielding varieties show resistence to one or more insect pests. This has however led to evolution of biotypes to nullify host resistance. In India, three biotypes of brown planthopper, five biotypes of green leafhopper and five biotypes of gallmidge have been reported so far.

Increase in cropping intensity by cultivation of rice throughout year has led to dramatic increase in pest population especially of monophagous yellow stemborer, *S. incertulas* over polyphagous stripped borer, *Chilo suppressalis* (Loevinsohn, 1994).

Insecticide induced resurgence of insect pests of rice has been mainly due to the extensive use of pesticide in rice ecosystem. About 13% of pesticide consumption are used in rice ecosystem with regional variations. It is reported that organophosphates like chlorpyriphos, fenitrothion, diazinon, quinalphos, and synthetic pyrethroids like cypermethrin, deltamethrin and fenvalerate have induced resurgence of sucking pests like brown planthopper and white-backed planthopper in the southern states of the country. Phorate has caused resurgence of green leafhopper and stemborer in Tamil Nadu while cypermethrin and deltamethrin has caused resurgence of stemborer in Orissa. Carbofuran use in Orissa, Kerala and Andhra Pradesh has led to resurgence of leaf folder in these states. The resurgence of harmful pests as result of excessive use of insecticides is mediated by decimation of beneficaial fauna like parasites, predators, insectivorous fish and other vertebrates found in rice agroecosystem (Teng *et al.*, 1993).

Insect Diversity in Cotton Agroecosystem

Cotton (*Gossypium* spp., espiecially *G. hirsutum*) is important fibres crop grown on 8.9 million hectares. Over the years, cotton ecosystem has undergone a tremendous change with shift in cultivation from old world varieties belonging to *G. arboreum* and *G. herbaceum* to *G. barbadense* and *G. hirsutum;* with only 22 hybrid varieties belonging to the last species occupying more than 40% of area and contributing more than 45% of production. With concurrent erosion of plant diversity, pest problems have accentuated in recent years. Naya *et al.* (1976) listed about 136 insect pests of cotton belonging to six orders viz., Orthoptera, Hemiptera, Thysanoptea, Lepidoptera, Diptera and Coleoptera, and 38 families. Of the families, important ones are noctuidae with 21 species, curculionidae 19, acrididae 13, thripidae 8 cicadellidae and arctiidae 7 each, pseudococcidae 5 and lymantriidae 4 while rest has 1-3 species each. Presently, however, about a dozen insect pest are the key pests (Simwat, 1994). These are jassid *Amrasca bigutula biguttula,* whitefly *Bemisia tabaci,* thrip *Thrips tabaci,* tobacco caterpillar *Spodoptera litura,* leaf roller *Sylepta derogata,* aphid *Aphis gossypii,* stem weevil *Pempherulus affinis,* mealy bug *Ferrisia virgata,* cotton stainer *Dysdercus koenigii* and most importntly, bolloworm complex of pink bollworm *Pectinophora gossypiella,* spotted bollworm *Earias insulana* and *E. vittela,* and American bollworm, *helicoverpa armigera* (Simwat, 1994; Puri, 1997). Over the years, bollworm complex has acquried importance in view of its damage potential and development of resistance to insecticides. *Emergence of American bollworm and whitefly in 1980s-1990s and introduction of new pest in agroecosystem:* With the introduction and subsequent spread of long-steple American cotton varieties, the pest complex has restricted to bollworms, with the predominance of the American bollworm. Being polyphagous in nature, it occurs throughout the year. The larvae damage mainly on cotton flowers and bolls by feeding on them. The outbreak of *H. armigera* in cotton in Andhara Pradesh especially in Guntur and Prakasham districts was first reported in 1988-89 when farmers could not control the crop despite 20-30 sprays of insecticides including synthetic pyrethroids. This was attributed to the development of resistance in the insect pest. Since then pesticide resistance has been widely recorded in this pest. In 1997-98 too, pest outbreak caused a large-scale damage to the crop despite use of insecticides. The pest outbreak has been attributed to prolonged drought in the first 60-70 days of crop growth followed by late rains at flowering and indiscriminate use of insecticides, faulty application techniques of pesticides, monocropping of cotton, and extensive cultivaton of alternate hosts like tomato, pigeon pea during off-season providing food to the pest throughout the year.

The whitefly, *B. tabaci* has become a key pest of cotton that damages crop by affecting vitality and transmitting leaf curl virus disease. In some parts of the country, the tobacco caterpillar, *S. litura* has become a major pest defoliating plants. Besides, new pests like stem borer and leaf miner are emerging as key pest of cotton in some regions in the country.

Remedial Role of Integrated Pest Management

The failure of insecticide to control insect pests realized in early 1950s, which was attributed to the development of resistance. Early strategy therefore included a combination of biological control with that of chemical control. However, the concept of integrated pest management (IPM) began to take shape in 1960s. Even international organization like Food and Agriculture Organization of the United Nations advocated IPM as a rationale policy of crop and environmental protection. Kogan (1998) comprehensively defined it as a "descision support system for the selection and use of pest control tactics, singly or harmoniously coordinated into a management strategy, based upon cost benefit analysis that take into account the interest of and impact on producers, society and the environment." With such an approach, IPM aims to preserve insect biodiversity and maintain parasites and predators in an agroecosystem.

Integrated Pest Management in Rice Agroecosystem

In view of high infestation of insect pests in the tropical conditions of the country, use of insecticides has become imperative. Use of insecticides should however be need-based and selective so to make their use cost-effective, simultaneously reducing environmental pollution This also helps in encouraging the effectivity and biodivesity of parasites and predatores in an agroecosystem. The ecofriendly safe insecticides are rated high on IPM rating based upon their effectiveness, safety towards higher animals and natural enemies and their persistence. Gujar (1993) found that most insect growth regulators have higher IPM rating than many of the neurotoxic insecticides. It is therefore necessary to promote insecticides of high IPM rating termed as green insecticides. Although some of the insecticides are with low IPM rating, these could be selectively applied at the target sites to achieve their judicious use and minimize environmental pollution. The judicious use of insecticides is also dependent upon their use only on insect infestation attaining economic threshold level (ETL) that justifies the cost of intervention of pest control tactics. In rice ecosystem, (ETL) that justifies tha cost of intervention of pest control tactics. In rice ecosystem, ETL that justifies the cost of intervention of pest control tactics. In rice ecosystem, ETL levels are worked out for differnt pest viz., stem borer, 5% dead hearts, gallmidge 5-10% silver shoots, brown planthopper 10 hoppers/clump, green leafhopper 10-15 hoppers/5 net sweepings. In studies carried out on the effect of insecticides on phytophages and natural enemies in rice ecosystem, judicious use of toxic insecticides like carbofuran @0.75 kg/ha or phorate @ 1.0 kg/ha at the base of plants did not affect parasites and predators. However, a spray of quainalphos @0.5 kg/ha affected parasites of stem borer and gallmidge, also predators like mirid bug, coccinellids and spiders; as a result of wide exposure of insecticide (DRR, 1993). Krishnaiah and Kalode (1992) reported that rice seedling treated with bioinsecticide like neem, maintained mirid bug and spider population similar to check while monocrotophos @ 0.75 kg/ha lowered these natural enemies. Further studies carried out at multilocational trials under the auspices of All India Coordinated Rice Improvement Project (DRR, 1994) have confirmed that carbofuran and phorate, if selectively applied, did not affect egg parasites of stemborer and at the same time gave an effective control by reducing dead hearts in relation to control. However, sprays of quainalphos, monocrotophos and chlorpyriphos affected natural enemies while achieving control of stem borer.

The host plant affects growth and development of insect pests depending upon its nutritional and non-nutritional contents. The susceptible crop varieties offer balanced nutrition to insect pests feeding on them as comapred to resistant varieties. This is turn affect insect susceptibility to insecticide. The insects well fed on the susceptible crop variety tend to show more tolerance to insecticides than the ones on resistant variety. White-backed brown planthopper, *S. furcifera* and brown planthopper, *N. lungens* feeding on the moderately resistent rice varieties like N 22 or ASD 7 were found to be more susceptible to insecticides like carbofuran, acephate than those feeding on susceptible varieties like TN 1 and IR 8 (Heinriches, 1994). In India, many varieties viz., Surekha, Shaktiman, Rasmi, Daya for gallmidge, brown planthopper, white-backed brown planthopper and stemborer resistsnce have been identified. Since level of resistance varies, these varieties support insect pests to some extent, thus helping in build-up of parasites and predators. In studies carried out on varietal and insecticidal effect on pest and natural enemy complex of rice, Panda *et al.* (1997) found that resistant variety, IET 3116 showed less of white earheads and white-backed brown planthopper as comapred to IET 9576 when the crops were treated with insecticides, first root dipping of seedling at 0.02% chlorpyriphos before transplanting and basal dose of carbofuran @ 1 kg/ha 39 days after transplanting. Predators like spiders, mirid bug *Cytorhinus lividipennis* and rove beetle *Paederus fuscipes* were more in IET 3116 than in IET 9576 were same both in treatment as well as control.

Biopesticides, derived from natural sources, show a high safety towards natural enemies due to their selective action. Viral diseases like nuclear polyhedrosis virus are highly specific and hence do not affect pest diversity in agroecosystem. Biological agents like parasites and predators are reported to control insect to a varying extent depending upon various factors. Singh and Dhaliwal (1994) recorded parasitoids like *Telenomus* sp., *Tetrastichus* sp., *Trichogramma* sp., for eggs of lepidopterous pests; *Chelonus* sp., *Apanteles* sp., *Stenobracon, Brachymeria,* for larval and *Xanthopimpla flavolineata, Brachymeria* sp., *Tetrastichus hawardi* for pupal lepidopterous pests giving high parasitization in rice ecosysytem. Gautam (1998a) reported that about 25 species of parasitoids belonging to Diptera and Hymenoptera, and 20 predator species belonging to Coleoptera (9), Diptera (2), Hemiptera (6), Odonata (2), Orthoptera (1) alongwith 8 species of spiders are found in rice agroecosystem in India.

Integrated Pest Management in Cotton Agroecosystem

As cotton cultivation intensifies with high agrotechnological inputs, the managements of pest complex becomes an equally technical and complex. In recent years, IPM tactics in cotton have included more of ecologically biointensive approaches to avoid insecticide resistance and environmental pollution.

The judicious use of insecticides is necessary as last resort when pest species attains ETL despite use of other tactics. ETL levels for various insect pest of cotton have been worked out viz., for bollworm complex 5% damage to reproductive parts/fruiting bodies, whitefly 20 nymphs/ leaf before 10 a.m. or 8-10 adults/leaf, jassids 2/leaf or 50-100 nymphs/50 leaves, aphids 15-20% infested plants.

Use of resistant cotton varieties especially belonging to *G. hirsutum* has been reported by Sundaramurthy and Chitra (1992). Puri *et al.* (1998) advocated use of resistant varieties like Bikaneri Nerma, G.cot 10 & 12, Rs. 875, ABH 466, H-777, RST-9, F-5-5. Fateh for jassids, Supiya, Kanchana, LK 861 for whitefly and LH 900, F-414 and Abadita for bollworms.

Monocultivation of crops has often led to the emergence of major pests, thereby limiting parasitoids and predators. It is therefore recommended that cotton grown, as a main crop should be intercropped or mixcropped with non-leguminous or non-solanaceous crops. Sorghum or Maize should be used as a border crop to avoid dominance of the American bollworm. Cowpea in cotton crop supports predator biodiversity (Natarajan and Seshadri, 1989).

In studies on interaction of crop varieties with insecticidal treatments, Sundaramurthy and Basu (1985) reported that Varalaxmi being long duration and high canopy variety, needed more insecticidal treatments that affected natural enemies, to achieve the same level of control as compared to MCU 5 or 9. IPM tactics in cotton have paid rich dividents amongst farmers and helped in enhancing diversity of natural enemies and reducing insecticide treatments in villages near Coimbatore during trials from 84-85 to 89-90 (Sundaramurthy and Chitra, 1992)

In studies on effect of host plant on natural enemies, Mohite and Uthamasamy (1997) found that trichome density on the leaves of plants is negatively correlated with rates of *H. armigera* egg-predation by *Chrysoperla carnea* and also egg-parasitization by adults of *Trichogramma chilonis,* thus suggesting importance of tritrophic interactions.

Interestingly, biodiversity of natural enemies in cotton ecosystem in very rich. More than 600 parasites and predators have been recorded in cotton ecosystem at Arkensas in USA. Similar attempt by Gautam (1988b) showed presence of 38 natural enemies; 17 parasites belonging to Hymenoptera and amongst predators, Coleoptera 14, Hemiptera 5 and Dictyoptera and Neuroptera one each, and 23 spider species in cotton in India. The important parasitoids of bollworm complex recorded so far are *T. achaeae, T. chilonis, T. chilotraeae, T. brasiliensis, Trichgrammatoidea* sp., *Chelonis* sp., *Apanteles angaleti, Aphelinus* sp., *Elasmus johnstonii, Erythmelus empoascae, Goniozus* sp., *Campoletis chloridae, Goniophthalmus halli* and *Carcelia illota*. IPM of cotton pests includes cultural practices like deep ploughing, use of resistant varieties, avoidance of monocropping along with cotton seed treatment with imidachloprid 70WS @ 5 gg./kg or carbosulafn 20DS @ 2 gg./kg to control preflowering sucking pests, need based application of 5% neem seed kernal suspension, monitoring of infestation of bollworms with pheromones or light traps, release of parasitoids like *T. chilonis* @ 1.5 lakh/ha (35 days after sowing), followed by its second application a week later. Release of green lace wing *C. carnea* @ 50,000/ha (2 releases at fortnightly interval), use of nuclear polyherdrosis virus @ 250 LE/ha, spray of *Bacillus thuringiensis* var. *kurstaki* @ 1 kg/ha in case of heavy infestation and failure of NPV, need based ETL dependent sprays of synthetic insecticides like phosalone 35 EC, monocrotophos 36WSC or quinalphos 25EC, spray of synthetic pyrethroids like cypermethrin 10EC or deltamethrin 25EC between 90 and 120 days after sowing, will help in effective pest management, avoiding dominance of a few pests, maintaining pest diversity and protecting environment (Puri *et al.*, 1998). The case studies dealing with rice and cotton agroecosystems show that insect diversity is adversely affected by modern intensive farming practices including especially, indiscriminate use of insecticides. The rice and cotton agroecosystem in India accounts for about 75% of insecticide consumption in the country. The large-scale use of insecticides has created adverse effects like development of insecticide resistance, resurgence of minor pests, pest outbreaks and loss of natural enemies in these ecosystems. IPM tactics used in recent years on large scale have shown their successes especially leading to reduction in insecticide use and maintenance of biological control agents in ecosystems. IPM is however a knowledge intensive approach that take into consideration socio-economic and environmental aspects of the society. This holistic approach needs further impetus for its popularization amongst the people especially farmers. This appears to be only remedy available to maintain insect biodivesity and sustainability of crop production in Indian agriculture.

Acknowledgement

Author is grateful to Head, Division of Entomology, Director, Indian Agricultural Research Institute, New Delhi and Indian Council of Agricultural Research, New Delhi for funding research reported herein and for providing necessary infrastructural facilities.

References

Dale D. (1994). Insect pests of rice plant-their biology and ecology. Pp. 363-485, In: EA Heinrichs (ed.), *Biology and management of rice insects.* Wiley Eastern Ltd. New Delhi.

Datta AR, Majumdar DK, Das Gupta, D. (1967). Susceptibility of Taichung 65 variety; of paddy to multiple pest infestation. *Indian Journal of Entomology* 29: 224-231.

Datta AR. (1997). Insect pest dynamics and management scenario in West Bengal over three decades with special reference to Birbhum. Pp. 160-166, In: MK Das Gupta (ed.), *Proceedings of symposium on Pest Management in Changing Agricultural Situation.* March 15-16, 1997, Sriniketan, Visva-Bharati, Sriniketan.

DDR, (1993). *All India Coordinated Rice Improvement Programme, Progress Report.* Kharif 1993, Entomology and Pathology, Directorate of Rice Research, Hydreabad

DDR. (1994). *Annual Report of Directorate of Rice Research,* Hyderabad

Ferro DM. (1987). Insect pest outbreaks in agroecosystem. Pp. 195-212, In: P Barbosa and JC Sculz (eds.), *Insect Outbreaks.* Academic Press, New York.

Gautam RD. (1988a). Indonesian model of rice IPM and its prospects in India. Pp. 13-37, In: D Prasad and RD Gautam (eds.), *Potential IPM tactics.* Westvill Publ. Hs., New Delhi.

Gautam RD. (1988b). IPM in cotton crop. Pp. 359-374, In: D Prasad and RD Gautam (eds.), *Potential IPM tactics.* Westvill Publ. Hs., New Dellhi

Ghosh AK. (1998). *Biodiversity and Entomology-Ninth Dr. C.P. Alexander memorial lecture.* Pp. 1-11, March 10, 1998, University of Delhi, Delhi University Press, Delhi

Gujar GT. (1993). Scope of insect growth regulator in insect pest management. Pp. 299-312, In: HC Sharma and M Veerabhadra Rao (eds.), *Pests and Pest Management in India-the changing scenario.* Plant Protection Association of India, Hyderabad.

Heinrichs E.A. (1994). Host plant resistance. Pp. 517-547, In: EA Heinrichs (ed.), *Biology and management of rice insects.* Wiley Eastern Ltd. New Delhi.

Kogan M. (1998). Integrated pest management: Historical perspectives and contemporary developments. *Annual Review of Entromology* 43: 43-70.

Krishnaiah K. (1993). Changing scenario in rice insect pest problems. Pp. 11-18, In: HC Sharma and M Veerabhadra Rao (eds.), *Pests and Pest Management in India-The changing scenario.* Plant Protection Association of India, Hyderabad.

Krishnaiah NV, Kalode MB. (1992). Environmental impact studies of neem. Pp. 50-53, In: *Proceedings of IRRI-ADB Final Workshop on Botanical Pest Control Project Phase* 2. July 28-31, 1992, IRRI, Los Banos, Philippines.

Loevinsohn ME. (1994). Rice pests and agricultural environments, Pp. 487-513, In: EA Heinrichs (ed.), *Biology and management of rice insects.* Wiley Eastern Ltd., New Delhi.

Mohite PB, Uthamasamy S. (1997). Interaction of host-plant resistance and natural enemies in the management of *Helicoverpa armigera* on cotton. Pp. 531-536, In: GM Bharad, RS Bonde, SA Nimbalkar and SV Sarode (eds.), *Integrated Pest Mamagement in Agriculture.* Dr. Panjabrao Deshmukh Krishi Vidyapeeth, Akola

Nayar KK, Ananthakrishnan TN, David BV. (1976). *General and applied entomolgy.* Tata McGraw Hill Publishing Co. ltd., New Delhi.

Natarajan K, Seshadri V. (1989). Abundance of natural enemies of cotton insects under intercropping system. *Journal of Biological Control.* 2: 3-5.

Panda SK, Dash PC, Samalo T. (1997). Effect of varietal resistance and insecticidal protection in IPM of summer rice in western Orissa. *Journal of Applied Zoological Research.* 8: 6-10.

Pathak MD, Pawar AD. (1983). Insect pests of rice. Pp. 67-108, In: PD Srivastava, MG Jotwani, RA Agarwal, SR Wadhi, RK Bhanotar, and RK Bhatnagar (eds.), *Agricultural Entomology.* Vol. 2, All India Scientific Writer's Soceity, New Delhi.

Pathak MD, Jayaraj S, Rao YRVJ, Row KVSRK, Mukhopadhyay SK. (1996). Integrated pest management in rice. Pp. 52-55, In: *International Symposium on Rainfed Rice for sustainable food security.* Sept. 23-25, 1996, CRRI, Cuttack

Puri SN. (1997). Present status of integrated pest management in cotton. Pp. 1-10, In: GM Bharad, RS Bonde, SA Nimbalkar and SV SArode (eds.), *Integrated Pest Management in Agriculture.* Dr. Panjabrao Deshmukh Krishi Vidyapeeth, Akola

Puri SN, Murthy KS, Sharma OP. (1998). Integrated pest management in cotton. *Extension folder 5,* National Centre for Integrated Pest Management, IARI, New Delhi.

Regupathy A, Palanisamy S, Chandramohan N, Gunathilagaraj K. (1994). *A Guide on crop pests.* 2nd edition, Sooriya Destop Publishers, Coimbatore.

Sailer RT. (1969). A taxonomists' view of environmental research and habitat manipulation. *Proceedings of Tall Timber Conference on Ecology Animal Control Habitat Management* 1: 37-45.

Samways MJ. (1994). *Insect conservation biology,* Chapman and Hall, London,

Simwat GS. (1994). Modern concepts in insect pest management in cotton. Pp. 186-237, In: GS Dhaliwal and R Arora (eds.), *Trends in agricultural insect pest management,* Commonwealth Publishers, New Delhi

Singh J, Dhaliwal GS. (1994). Insect pest management in rice – A perspective Pp. 56-112, In: GS Dhaliwal and R Arora (eds.), *Trends in agricultural insect pest management,* Commonwealth Publishers, New Delhi

Sundaramurthy VT, Basu AK. (1985). Management of cotton insect pests in polycrop system in India. *Out look on Agriculture* 14: 79-82

Sundaramurthy VT, Chitra K. (1992). Integrated pest management in cotton. *Indian Journal of Plant Protection* 20: 1-17.

Téng PS, Hoong KL, Moody K. (1993). Advances in tropical rice integrated pest management research. Pp. 241-255, In: R Muralidharan, EA Siddiq (eds.), *New frontiers in rice research.* Directorate of Rice Research, Hyderabad.

Proceedings of the National Seminar on Environmental Biology, April 03-05, 1998
***Edited By :* A.K. Aditya & P. Haldar**
DAYA PUBLISHING HOUSE, DELHI

Biodiversity in Island Environment with Special Reference to Andaman and Nicobar Islands

A.K. Das
Zoological Survey of India
'M' Block, New Alipore, Calcutta - 700053

Introduction

Islands and archipelagoes are the living laboratories on earth for demonstrating the process of organis evolution, adaptive radiation and speciation. Probably therefore, both Charles Darwin and his contemporary naturalist Alfred Russel Wallace could independently conceive and develop the primary concept of organic evolution by natural selection only after studying and analysing the fauna of two island groups – the Galapagos and the Malyan Archipelago respectively, even though these are lying on two opposite sides of the globe. Subsequent studies of island biota in different parts of the world clearly reveal that island harbour simple and impoverish biotic communities with high percentage of endemicity mainly due to varying degree of geographical isolation, topography and simpler ecology. Thus island biotas offer ample opportunity to study various biological and ecological phenomena, more particularly dispersal mechanism, mode of colonisation and, subsequent ecological and evolutionary changes as discussed in details by Das and Maiti (1992). Indian islands, specially Andaman and Nicobar islands also harbour very rich and unique biodiversity which is being dealt with in the present communication.

Biodiversity of Andaman and Nicobar Islands

The Andaman and Nicobar islands are the summits of a submarine mountain range, lying on the great tectonic suture zone that extends from the eastern Himalayas along the Myanmar border to the Arakan and finally Sumatra and the Lesser Sundas. This archipelago consists of more than 500 islands, islets and rocky outcrops running north to south in the Bay of Bengal between 6°45' N and 13°30'N latitudes and 90°20'E and 93°56'E longitudes. It comprises main island chains of Andaman and Nicobars, Ritchie's archipelago and the two outlying volcanic islands – Narcondum and Barren. The northernmost part of these islands (North Andaman) is isolated from Cape Negrais in Southern Myanmar by the North Preparis Channel with 225 m depth and 285 km. width. The southern most part of this archipelago (Great Nicobar) is also separated from the Acheen Head of western Sumatra by the Great Channel having 1600 m depth and 189 km. width. Besides, there are two other deep channels, namely, the Ten Degree Channel with 900 m depth and the Sombero Channel with 250-275 m depth which isolate the Andamans from the Nicobars and the middle Nicobars from the Great Nocobar respectively. Since it is now accepted that Pleistocene sea level lowering never exceeded 160 m (Gascocyne *et al*, 1979) it is quite evident that a permanent water barrier isolated these island chains from the adjacent mainland and some segments of this chain from another such segment long, back, some 100 million years since upper Mesozoic.

This archipelago has a total land area of 8293 sq. km. of which 6340 sq. km. is occupied by the Andaman group and 1953 sq. km. by the Nicobar group. The Andaman group consists of 324 islands of which 25 are inhabited while the Nicobar group comprises 28 islands 13 of which are inhabited.

Habitat Diversity

This archipelago is characterised by undulating terrain with mountains and intervening valleys. There are, however, some more or less flat islands like Car Nocobar and Trinket. Excepting in the Great Nicobar there is no perennial freshwater river in these islands. There are, however, several rain fed streams which dry up during the summer. Being influenced by tropical climates and by both the south-west and north-east monsoons all the major islands support a luxuriant growth of evergreen, semievergreen, moist deciduous and littoral forests from the water edge to the mountain tops, depending on topography and nature of the soil. The forests occupy about 86 per cent of the total land area of this archipelago.

Champion and Seth (1968) classified 7 main forest types in these islands, namely, Giant Evergreen forest, Andaman Tropical Evergreen forest, Southern Hill-top Tropical Evergreen forest, Andaman Semi-evergreen forest, Andaman Moist Deciduous forests, Littoral forest and Tidal Swamp/Mangrove forest. In addition to these, there are also few more forest types like Cane Brakes, Bamboo brakes, etc. Further, grasslands are also present in Car Nicobar and Nancowry island.

The coast line of these islands, occupying about 1962 km is very wavy, with large number of bays, lagoons and serpentine creeks. The sheltered coastal areas, bays and several creeks support rich mangrove forests which is estimated at 1190 sq. km covering about 18 per cent of the total Indian mangales (Das, 1996). At several places tidal creeks penetrate far inside the land and form outlets for freshwater streams. The continental self is narrow with a steep slope. The is a luxuriant coral growth in marine waters around these islands. The corals have grown in the form of fringing reefs on the eastern side of these islands. Whereas on western side they constitue barrier reefs.

Species Diversity

As stated earlier, these islands possess undulating topography with mountains of variable heights and rich tropical rain forests, forming terrestrial ecosystem. Moreover, mountainous parts of the southern islands get around 300 cm. rain fall annually. In insular environment this type of vegetation with undulating topography and abundant rain fall offers a wide gamut of ecologcal habitats from the coast to the mountain top, ideal for supporting rich and diversified terrestrial fauna migrating from mainland source or from other islands (Hamilton *et al*, 1964). Further, such habitats in the island are very much suitable for their subsequent establishment and evolution.

This archipelago supports richest coral formations in its marine ecosystem and luxurious mangrove fringed coasts at the interphase between the terrestrial and marine ecosystems. It is well known that the coral ecosystem itself is one of the richest ecosystems sustaining innumerable variety of life forms. Over and above, mangrove ecosystem is an important reservoir and, feeding, breeding and nursery grounds of a large number of animal species both terrestrial and aquatic. It is, therefore, quite inevitable that this archipelago would harbour a rich faunal biodiversity.

A total of 2513 species of plants have so far been reported from these islands, comprising 2200 species of angiosperms, 130 species of pteridophytes, 50 species of lichens, 15 species of mosses and hepatics and 118 species of marine algae (Table 1). About 10 per cent of the total number of vascular plant taxa of these islands is endemic (Dagar, 1989). Parkinson (1923) compared these islands with Galapagos archipelago and St. Helena for their biological richness.

These islands also harbour very rich faunal biodiversity as detailed in Table 1. So far, 6637 species have been documented from this archipelago (Tikader and Das, 1985; Tikader *et al*, 1986; Das and Dev Roy, 1989; Chandra, K., *personal communication*), of which 2880 species are terrestrial and 3757 species are marine.

Table 1 : Species diversity and endemism in Andaman and Nicobar Islands

Group	No. of species /subspecies	No. of endemics	Endemic %
1	2	3	4
A. Terrestrial biota			
Flora :			
Lichens	50	–	–
Mosses and hepatics	15	–	–
Pteridophyta	130	14	10.8
Angiosperms	2200	232	10.5
Total : Flora (Terrestrial)	2395	246	10.2
Fauna			
Mollusca	110	77	70.0
Annelida	30	9	30.0
Insecta			
Collembola	8	–	–
Odonata	50	7	14.0
Orthoptera	67	12	17.9
Phasmida	4	–	–
Dermaptera	8	2	25.0
Embioptera	2	–	–
Mantodea	5	2	40.0
Blattaria	12	–	–
Isoptera	38	21	55.2
Phthiraptera	4	–	–
Hemiptera	250	36	14.4
Thysanoptera	30	26	6.6
Neuroptera	11	–	–
Coleoptera	541	154	28.3
Siphonoptera	2	–	–
Diptera	231	23	9.9

Contd....

Table 1 – *Contd....*

1	2	3	4
Lepidoptera	795	200	25.1
Trichoptera	20	–	–
Hymenoptera	176	24	13.6
Arachnida	94	38	40.4
Amphibia	18	3	16.7
Reptilia	76	24	31.6
Aves	246	99	40.2
Mammalia	52	33	63.5
Total : Fauna (Terrestrial)	**2880**	**768**	**26.6**
B. Marine Biota			
Flora			
Algae	118	1	0.8
Fauna			
Porifera	72	–	–
Anthozoa	326	2	0.6
Polychaeta	184	4	2.2
Crustacea	586	6	1.0
Mollusca	932	18	1.9
Echinodermata	336	4	1.2
Pisces	820	2	0.2
Reptilia	12	–	–
Mammalia	3	–	–
Meiofauna	486	102	21.0
Total : Fauna (Marine)	**3757**	**138**	

It needs mention here that many islands still remain either unexplored or underexplored. Consequently our knowledge on animal species diversity, more particularly terrestrial invertebrates and coral reef inhabiting animals is far from complete. In view of this and considering habitat diversity as discussed earlier it can safely be inferred that many more animal species would be available in this archeipelago when planned island-wise and ecosystem-wise survey is complete and fauna thus collected is studied and identified.

However, available data on island biota (Table 1) clearly reveals that faunal endemism in terrestrial animals is significantly high in these islands 26.6 per cent as expected in insular environment. Amongst mammals, the Crab-eating Macaque, Wild Pig, Palm Civet, all the Spiny Shrews (*Crocidura*) and the Tree shrews (*Tupaia*) which could reach these island have become endemic (Tikader and Das, 1985). Even considerable number of species and subspecies of bats (11 out of 26) and rats (11 out of 14) which possess high degree of dispersal ability have also achieved endemic status.

Birds of these islands were interesting not only due to their high percentage of endemism but also for their restricrted distribution within the island chains. The Narcondam Hornbill which is found only in Narcondam island is the typical example. Some endemic subspecies of birds are found to inhabit one particular island or few adjacent islands. For example, three endemic subspecies of White-headed Myna, namely, the Andaman White-headed Myna (*Sturnus erythropygius andamanensis*), the Nicobar White-headed Myna (*S. erthropygius erythropygius*) and the Katchal White-headed Myna (*S. erythropygius katchalensis*) occur in these island. Among these, the first one occurs in several adjacent Andaman islands and, the second and the last ones are restricted to Car Nicobar and Katchal islands respectively. Distribution of Megapods, *Megapodius freycinet* is also noteworthy. Its one subspecies, the North Nocobar Megapode (*M. freycinet nicobarensis*) is restricted to Nicobar group of islands lying north of Sombero Channel (except Chowra and Car Nicobar) and the second subspecies, the South Nicobar Megapode (*M. freycinet abbotti*) is found only in the Little and Great Nocobar islands.

The terrestrial fauna of these islands also display several evolutionary changes, such as, dwarfism, gigantism, etc. For example, Serpent Eagles belonging to the genus *Spilornis* display dwarfism whereas Green Imperial Pigeon, *Ducula aenea* and Red-cheeked Parakeet, *Psittacula longicauda* show gigantism in these islands.

Biogeographical Affinity

According to Smith (1930) Andaman contains an impoverished Burmese fauna and fauna of Nocobars approximates to Sumatran type. His above comments might be due to the geographical proximity and geological connections of Andaman islands with Myanmar (Burma) and Nicobar islands with Sumatra. Some workers (Blanford, 1901; Mani, 1974) also stated that biotas of Andaman and Nocobar islands differ substantially from one another. But, the data which is presently available on terrestrial vertebrates, more particularly on avifauna does not support these generalised statements regarding distribution pattern of these island fauna.

After analysing the distributional list of breeding birds of the Andaman and Nicobar island Ripley and Beehler (1989) suggested the major path of colonisation of birds was extended along a single route from South-western Myanmar *via* a series of water barriers which acted as filters. They further stated that avifauna of Nocobars is an impoverish subset of Andaman avifauna, although they observed few species of birds, *e.g. ottus balli, Columba palumboides* and *Macropygia rufipennis* may probably be migrated from Sumatra.

As early as 1870 Stoliczka analysed the distribution pattern of reptiles and commented "Andaman and Nicobar will show a great similarity to each other; several species of lizards and snakes are common to both and the whole fauna greatly resembles the Malayan gradually passing into Burmese fauna". Thirty two species of lizards and forty two species of snakes have so far been reported from these islands (Tikader and Das, 1985). The present data reveals that dispersal of snakes and lizards probably took place both from north and south of this archipelago. Indo-Chinese elements entered the Andamans through Myanmar and the Malayan elements reached the Nicobars *via* Sumatra.

Acknowledgement

The author is grateful to Dr. J.R.B. Alfred, Director, Zoological Survey of India for extending facilities in connection with the preparation of this manuscript.

References

Blanford, W.T. (1901). The distribution of vertebrate animals in India, Ceylon and Burma. *Phil. Trans. Roy. Soc. B, 194* : 201-436.

Champion, H.G. and Seth, S.K. (1968). *A revised survey of forest types of India,* i - xxvii + 1-404.

Dagar, J.C. (1989). Endemic species of Bay Islands. *J. Andaman Sci. Assoc., 5* (2) : 161-168.

Das, A.K. (1996). Faunal diversity, distribution pattern and zonations in the mangrove ecosystem of Andaman and Nicobar islands. *Proc. Zool. Soc., Calcutta, 49* (2) : 57-65.

Das, A.K. and Dev Roy, M.K. (1989). A general account of the mangrove fauana of Andaman and Nicobar islands. *Fauna of Conservation Areas : 4, Zool. Surv. India* : 1-173.

Das, A.K. and Maiti, P.K. (1992). Island fauna : its dispersal, mode of colonisaton and evolutionary pattern. *Proc. Zool. Soc., Calcutta, 45* (Suppl. B) : 41-55.

Gascocyne, M., Benjamine, G.J. and Schwartz, H.P. (1979). Sea level lowering during the Illinian glaciation : evidence from the Bahama 'blue hole'. *Science, 205* : 806-808.

Hamilton, T.H., Barth, R.H. and Rubinoff, I. (1964). The environmental control of insular variation in bird species abundance. *Proc. Natn. Acad. Sci. U.S.A., 52* : 132-140.

Mani, M.S. (Ed.) (1974). *Ecology and Biogeography of India.* W. Junk, Hague, Netherland.

Parkinson, C.E. (1923). *A forest flora of the Andaman islands.* Simla.

Ripley, S.D. and Beehler, B.M. (1989). Ornithographic affinities of Andaman and Nicobar islands. *J. Biogeography, 16* : 328-332.

Smith, M.A. (1930). The reptilia and amphibia of the Malaya Peninsula from the isthumus of the Kra to Singapore, including adjacent islands. *Bull. Raffles Mus.,* 3 : 1-49.

Stoliczka, F. (1970). Observation on some Indian and Malayan amphibia and reptilia. *J. Asiat. Soc. Bengal, 39* : 136-138.

Tikader, B.K. and Das, A.K. (1985). *Glimpses of Animal Life of Andaman and Nicobar Islands. Zool. Surv. India,* Calcutta : i-xi + 1-170.

Tikader, B.K., Daniel, A. and Subba Rao, N.V. (1986). *Seashore Animals of Andaman and Nicobar Islands. Zool. Surv. India,* Calcutta : 1-188.

Proceedings of the National Seminar on Environmental Biology, April 03-05, 1998
***Edited By :* A.K. Aditya & P. Haldar**
DAYA PUBLISHING HOUSE, DELHI

A Contribution to the Diversity of Insects with Reference to Pollination Mechanism in Some Angiosperms

Ashoke Bhattacharya & Sudhendu Mandal
Department of Botany
Visva-Bharati, Santiniketan - 731 235

Abstract

Pollination is an important phenomenon for gene recombination as a result of which genetic and species diversity is promoted. Regular observation on the pollination mechanism of five angiospermic plants belonging to three families namely *Lantana camara* Linn. (Verbenaceae), *Moringa oleifera* Lamk. (Moringaceae), *Syzygium jambos* (Linn.) Alston. (Myrtaceae), *Tectona grandis* Linn. f. (Verbenaceae) and *Vitex negundo* Linn. (Varbenaceae) revealed that different members of Thysanoptera, Hymenoptera, Lepidoptera, etc. visited flowers and enhanced pollinating potential. All the plants showed diurnal anthesis. Pollen-ovule ratio per flower was found to be variable. Netting and bagging of flowers indicated that the external agents might be required for successful fruit setting which some times varied depending upon the availability of effective pollinators.

Introduction

Pollination, a vital phenomenon in sexual reproduction of flowering plants is the basic force for gene recombination by which genetic and species diversity is promoted. Different types of insects belonging to the members of Thysanoptera, Hymenoptera, Lepidoptera, Coleoptera, Diptera, etc. are being interacted with flowers and accelarated pollinating potentialities. Different pollinators and flowers are adapted to cope up with sustainable development of environment for preservation of biodiversity. The functional reproductive units and stable traits of plants i.e. the flowers provide a crucial role in pollination and breeding system. An integrated evolving biological system is established with the help of co-adaptive features of both flowers and insects. Plant-pollinator interactions have been recognised widely since long days. It is assumed that the proliferation of angiospermic species took place in cretaceous period due to rapid adaptation of flowers and pollination insects (Ananthakrishnan, 1982). Blossom-pollinator relationship is established due to the availability of pollen, nectar, odour, colour and of sexual attractants. Besides bees, other insects like thrips, butterflies, moths, etc. used to visit the flowers and are very useful for pollination. It has been shown that the flower-pollinator interactions are essentially a community based. A constant energy flow is being maintained at a minimum level for sustainable development of pollinators community (Faegri and Van der Pijl, 1980; Subba Reddi and Reddi, 1984).

In present investigation, the diversity of different insects in relation to their role in pollination and fruit set of five selected plants, growing in the lateritic belt of Birbhum district, have been studied. This type of study will yield rewarding results and provide necessary information required for the management and improvement of a particular vegetation.

Materials and Methods

Frequent observations were made to study the anthesis and other flowering phenology of five selected plants following the method of Reddi and Janaki Bai (1981) and Mathur and Mohan Ram (1986). The selected plants are: *Lantana camara* (Verbenaceae), *Moringa oleifera* (Moringaceae), *Syzygium jambos* (Myrtaceae), *Tectona grandis* (Verbenaceae) and *Vitex negundo* (Verbenaceae). Pollen-ovule ratio were determined by counting total pollen number per flower following the proposed method of Mandal and Chanda (1981) which were divided by total number of ovules per flower. Ten flowers from different plants were taken into consideration and an average value was considered for each taxon. Stigma receptivity and *in vivo* pollen behaviour were investigated as per the method of Joshi Rao and Saoji (1989). The flower visitors were identified from available reference specimens kept in our laboratory which were already identified from Zoological Survey of India, Calcutta. Contribution of pollinators to fruit setting was noticed by netting and bagging of flowers. Ten different flowers/inflorescence from different individuals were netted and bagged randomly at bud stage with nylon net and butter paper. The net and butter paper were removed after fruit formation and at the end of flowering. The percentages of fruit set in netted and bagged condition were compared with open flowers which were easily allowed to different insects i.e. pollinators.

Results and Discussion

The members of Thysanoptera like *Heterothrips, Microcephalothrips,* etc. form an integral complex with the flowers of *L. camara, M. oleifera* and *T. grandis.* The aphids (Homoptera) have been noticed to inhabit the flowers of *M. oleifera.* All five selected plants except *S. jambos* are visited by some members of Lepidoptera of the families Papilionidae, Nymphalidae and Pieridae.

The flowers of *M. oleifera, S. jambos, T. grandis* and *V. negundo* are visited by different types of bees and wasps like *Apis* sp., *Megachile* sp., *Xylocopa* sp., etc. Even other insects type like Diptera and Coleoptera have also been observed to forage the flowers of *M. oleifera, T. grandis,* and thus an integrated relationship between selected plants and their flower visiting groups has been shown (Fig. 1). Insects visit to *V. negundo* in purpose of collecting nectar only (Table-1 to 5).

Table 1 : Flower visitors of *Lantana camara*

Name of the visitors	Visiting time	Forage substance
Thysanoptera :		
Hetrothrips sp.	Day and Night	Nectar and Pollen
Lepidoptera :		
Catopsilia pomona	Day	Nectar
Danaus chrysippus	Day	Nectar
Euploea sp.	Day	Nectar
Papilio sp.	Day	Nectar
Precis lemonias	Day	Nectar

Table 2 : Flower visitors of *Moringa oleifera*

Name of the visitors	Visiting time	Forage substance
Thysanoptera :		
Heterothrips sp.,	Day and Night	Nectar and pollen
Microcephalothrips abdominalis	Day and Night	Nectar and pollen
Homoptera (Aphids)	Day and Night	Nectar
Hymenoptera :		
Xylocopa sp.	Day	Nectar and pollen
Apis sp.	Day	Nectar and pollen
Vespa sp.	Day	Nectar and pollen
Lepidoptera :		
Papilionidae	Day	Nectar
Pieridae	Day	Nectar
Coleoptera :		
Unidentified	Day	Nectar

Table 3 : Flower visitors of *Syzygium jambos.*

Name of the visitors	Visiting time	Forage substance
Hymenoptera :		
Apis sp.	Day	Pollen and Nectar
Megachile sp.	Day	Pollen and Nectar
Formicidae (Ants)	Day and Night	Nectar

Table 4 : Flower visitors of *Tectona grandis*

Name of the visitors	Visiting time	Forage substance
Thysanoptera (Thrips)	Day and Night	Pollen and Nectar
Hymenoptera :		
Apis sp.	Day	Pollen and Nectar
Megachile sp.	Day	Pollen and Nectar
Xylocopa sp.	Day	Pollen and Nectar
Ceratina viridissima	Day	Pollen and Nectar
Eumenes conica	Day	Pollen
Lepidoptera :		
Papillionidae	Day	Nectar
Nymphalidae	Day	Nectar
Pieridae	Day	Nectar
Diptera :		
Musca sp.	Day	Nectar
Unidentified	Day	Nectar

Table 5 : Flower visitors of *Vitex negundo*

Name of the visitors	Visiting time	Forage substance
Hymenoptera :	Day	Pollen and Nectar
Apis sp.	Day	Pollen and Nectar
Vespa sp.	Day	Pollen and Nectar
Xylocopa sp.	Day	Pollen and Nectar
Megachile sp.	Day	Pollen and Nectar
Ceratina viridissima	Day	Pollen and Nectar
Eumenes conica	Day	Pollen and Nectar
Formicidae (Ants)	Day	Pollen and Nectar
Lepidoptera :		
Papiliondae	Day	Nectar
Nymphalidae	Day	Nectar
Pieridae	Day	Nectar
Diptera	Day	Nectar
Musca sp.	Day	Nectar
Coleoptera :		
Unidentified	Day	Nectar

Different species of thrips inhabit the corolla tube and at the base of ovary of *L. camara*. In *M. oleifera* thrips are noticed to reside at the juncture of petals. Thrips live on pollen grains, nectar and stigmatic exudation and take shelter in the flower (Reddi, 1989). They alight on the stigmas of the flowers after flower opening. After exhausting the food resources they suddenly flight to another flower and carry an appreciable amount of pollen on the bodies. The twigs covered with butter paper bags yielded no fruits thereby indicating that the plants are completely dependent on external agents for their pollination. When covered with nylon nets to avoid large bodied insect, fruits are formed in less quantities in *L. camara, M. oleifera* and *T. grandis* (Table-6). Thus it could be inferred that thrips play a minor role in assisting the pollination of these three plants. Flowers of *M. oleifera, S. jambos. T. grandis* and *V. negundo* are visited by some Hymenopteras like *Apis, Megachile, Xylocopa, Eumenes, Ceratina,* etc. Among these *Apis* and *Megachile* have been found to be dominant over others in relation to forage nature and behaviour. It has been noticed that more time is consumed to forage the flowers by small insects than larger ones, possibly it is due to the difference of energy exploitation for foraging. The pollen grains fo *T. grandis* and *V. negundo* are adherred to ventral part of *Apis* and *Megachile,* whereas the pollen grains of *S. jambos* are dusted over whole body surface and pollen mass of *M. oleifera* are attached with head and abdominal part of *Apis* and *Megachile. Xylocopa* used to forage the flowers of *M. oleifera, T. grandis* and *V. negundo* for collecting forage matters by inserting head within corrolla tube of *V. negundo* and by landing over flowers of *M. oleifera* and *T. grandis.* Other members of Hymenoptera visit the flowers for very short duration and carry the pollen grains through their body parts. Thus an insect-pollen complex is formed to pollinate the other flowers showing geitonogamy or xenogamy. The variation of pollen association with insects may be due to floral architecture.

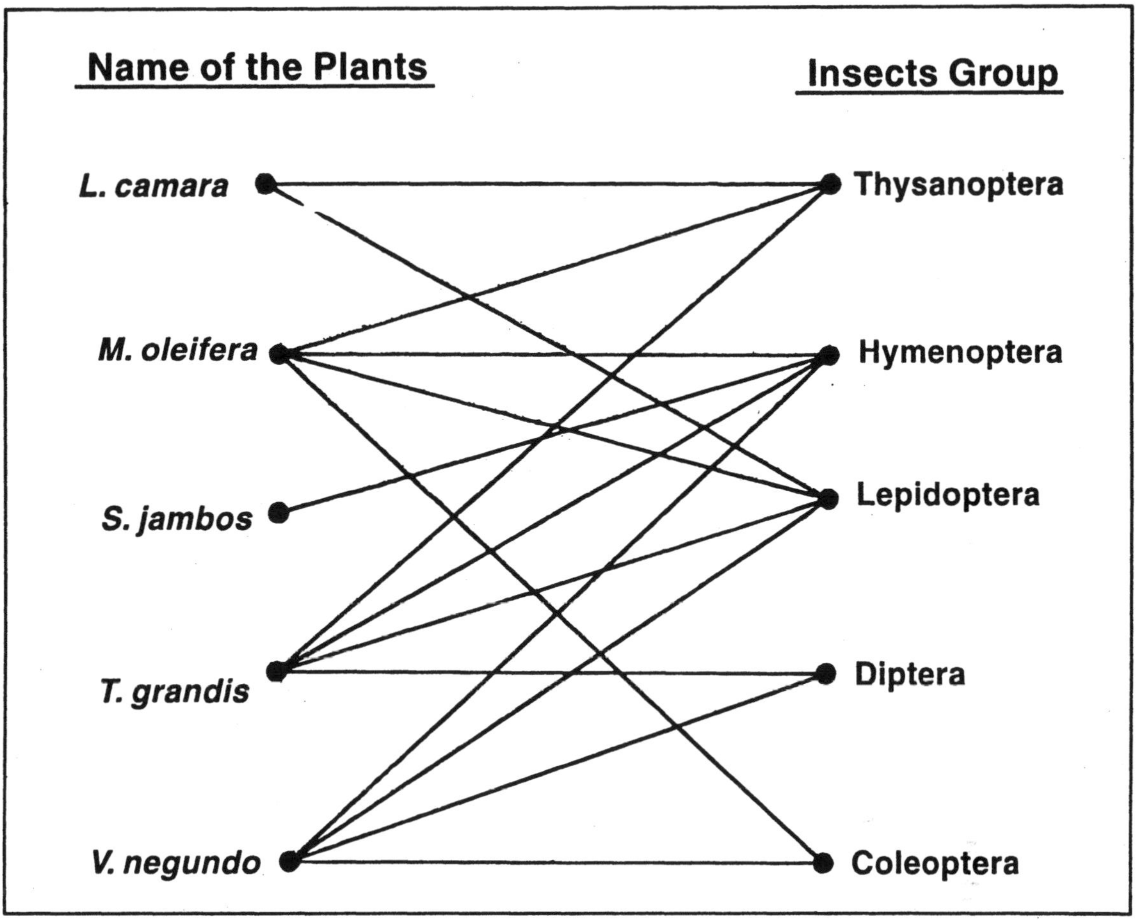

Fig. 1 : Relationship Between Insects and Selected Plants

Different types of bees (Hymenoptera) are attracted by pollen, nectar, colour, etc. of flowers. They require large quantities of pollen to meet the protein requirements. Sometime they reject the pollen. Pollen requirements and consumption may vary from species to species (Reddi and Subba Reddi, 1983). Visiting frequencies of bees over the flowers are dependent on nectar concentration (Rymashevskii, 1976). Bees normally forage during day time although Linsley and Cazier (1970) recognised five temporal categories depending upon forage time. It is believed that these categories are mostly regulated by environmental factors. Andrews (1969) reported that in the arctic summer bees forage during night. Faegri and Pijl (1980) categorised the pollen transferring process by bees in two ways i.e. in sternotribic flowers the pollen grains are deposited over ventral part of insects and in nototribic flowers the pollen grains are deposited on dorsal part of insects. In nototribic flowers it is possible because of the residual pollen on body surface is inaccessible for grooming (Subba Reddi and Reddi, 1984). The type of pollen transfer may depend upon the blossom type and body size of vector. During forage times bees must come in contact with essential whorls of flowers for successful pollination except in *Wrightia tinctoria* (Subba Reddi *et al.*, 1979). Foraging activities are influenced by environmental characteristics and floral resources (Osmar, 1982).

Table 6 : Flowering Phenology and fruit set of selected plants.

Name of the Plants with family	Anthesis pattern	Pollen-Ovule ratio per flower	Maxium stigma recepti-vity period after anthesis	Maximum stigma respect-vity (%)	Maximum in *vivo* germinat-ing pollen (%)	Fruitset (%) in open flowers	Fruitset (%) after netting	Fruitset (%) after bagging
Lantana camara (Verbenaceae)	Early morning	3466:1	First day	60	69.54	84.70	5.30	Nil
Moringa oleifera (Moringaceae)	Forenoon	522.8:1	Third day	40	48.47	10.28	1.00	Nil
Syzygium jambos (Myrtaceae)	Middle day	456462.5:1	Second day	70	48.76	25.21	Nil	Nil
Tectona grandis (Verbenaceae)	Forenoon	1875:1	First day	70	68.75	73.07	6.15	Nil
Vitex negundo (Verbenaceae)	Forenoon	1000:1	First day	50	60	61.29	Nil	Nil

Butterflies (Papilionidae, Nymphalidae and Pieridae) visit the flowers of all selected plants except *S. jambos* where visiting is hindered due to structural complexity of flowers. While visiting the flowers butterflies insert their proboscis in sucking nectar. During this time the pollen grains get attached with body parts like proboscis, head, thorax, legs, wings, etc. and transported to other flowers. Mostly butterflies are diurnal in activities and visit many flowers which open in morning using innate and imprinted "Search images" (Swihart, 1971). They may rely on flower colour for locating and identifying the forage matter. The members of advanced families like Nymphalidae and Pieridae are sensitive to yellow/orange and red colour whereas the primitive Hesperiidae are sensitive to yellow/orange and can not see red (Ilse, 1928; Swihart, 1969, 1970, 1972). Flowers of *S. jambos* and *V. negundo* are also visited by ants (Formicidae). Ants are fond of sugar and protein. They raid the anthers to fullfil their demand of nourishment and thereby help in pollination (Faegri and Pijl, 1980). *Musca* sp. of Diptera has been found to visit the flowers of *T. grandis* and *V. negundo*. Some unidentified memebers of Coleoptera have been observed over flowers of *M. oleifera* and *V. negundo*. Being the oldest group of insects they may help in pollination, though their smooth bodies are badly adapted for pollen transport (Kendall and Solomon, 1973). By bagging process with nylon net it has been noticed that successful fruit set took place in five selected plants with the help of large bodied insects. Fruit formation were low in netting condition compared to open flowers which were easily allowed to large bodied insects like bees, butterflies, wasps and beetles (Tables 1-6). Pollen-ovule ratio were found to be maximum (456462.5 : 1) in *S. jambos* to minimum (522.8:1) in *M. oleifera* (Table-6). Stigma receptivity and *in vivo* pollen performance showed also a variation. Maximum receptive stigmas (70%) were observed in *S. jambos* and *T. grandis* in second day and first day after anthesis respectively but *M. oleifera* showed minimum (40%) receptive stigmas in third day after anthesis (Table-6).

Fruit production in connection with reproductive success rate may vary depending upon the number of pollen per ovule per flower (Namai Ohsawa, 1986). Stigma surface may play a crucial role in pollen recognition and rejection process. Germinating ability of pollen in *in vivo* condition might be dependent on receptive stigmas. Receptivity of stigmas are very critical to determine and it may vary with environmental, physical and physiological factors (Joshi Rao and Saoji, 1989 and Bhattacharya and Mandal, 1997).

Conclusion

From the present findings, it can be concluded that the association of flower-insect complex might be related with specific floral characters and insect types. Insects of different diverse promoted the fruit yielding potentialities of selected plants. Pollen-ovule ratio, stigma receptivity, *in vivo* pollen performance over stigma surface showed variations for fruit set up and yield potential.

Acknowledgement

The authors are thankful to UGC, New Delhi for financial support to carry out this work and grateful to ZSI, Calcutta for identification of insects.

References

Ananthakrishnan, T.N. (1982). Thrips and pollination biology. *Curr. Sci.*, 51: 168-172.

Andrews, C. (1969). The lives of wasps and bees (London: Chatto and windus).

Bhattacharya, A. and Mandal, S. (1997). Studies on *in vivo* pollen germination of some angiospermic plants. *J. Palynol.* 33: 153-156.

Faegri, K. and Van der Pijl, L. (1980). The principles of Pollination Ecology. Pergamon Press Ltd. New York.

Ilse, D. (1928). Uber den Farbensinn der Tagfalter. Z. Vgl. Physiol., 8: 658-692.

Joshi Rao, J.A. and Saoji, A.A. (1989). Studies on *in vivo* germination of pollen of some alkaloid bearing plants. *J. Palynol.*, 25: 45-50.

Kendall, D.A. and Solomon, M.E. (1973). Quantities of pollen on the bodies of insects visiting apple blossoms. *J. Appl. Ecol.*, 10 : 627-634.

Linsley, E.G. and Cazier, M.A. (1970). Some competitive relationships among matinal and late afternoon foraging activities of Caupolicanine bees in Southeastern Arizona. *J. Kans. Entomol. Soc.*, 43: 251-261.

Mandal, S. and Chanda, S. (1981). Aeroallergens of West Bengal in the context of environmental pollution and respiratory allergy. *Biol. Mem.*, 6: 1-61.

Mathur, G. and Mohan Ram, H.Y. (1986). Floral biology and pollination of *Lantana camara*. *Phytomorphology*, 36(1&2): 79-100.

Namai, H. and Ohsawa, R. (1986). Variation of reproductive success rates of ovule and pollen deposited upon stigmas according to different number of pollen on a stigma in angiosperm. In: Biotechnology and Ecology of Pollen (David L. Mulcahy, Gabriella Bergamini Mulcahy and Ercole Ottaviano Eds.). Springer-Verlag, New York, pp. 423-428.

Osmar, M. (1982). Food-gathering behaviour and morphology in Africanized and Caucasian bees and in their hybrids. *Rev. Bras. Genet.*, 5: 641-642.

Reddi, E.U.B. (1989). Thrips-Pollination in Sapodilla (*Manilkara zapota*). *Proc. Indian Acad. Sci.*, B55 (5&6):407-410.

Reddi, E.U.B. and Subba Reddi, C. (1983). Pollination ecology of *Jatropha gossypiifolia* (Euphorbiaceae). *Proc. Indian Acad. Sci.* (Plant Sci.), 92: 215-231.

Rymashevkii, V.K. (1976). Nectar productivity and plant visitation by insects; pollination of entomophilous agricultural crops by bees (ed.) Kozin, trans. from Russian by B.R. Sharma (New Delhi : Amerind Publishing Co.) pp. 114-117.

Subba Reedi, C. and Janaki Bai, A. (1981). Floral biology of *Mimusops elengi* L. *J. Bomb. Nat. Hist.Soc.*, 77: 471-475.

Subba Reddi, C. and Reddi E.U.B. (1984). Bee flower interactions and pollination potential. *Proc. Indian Acad. Sci.* (Anim. Sci.)., 93(4): 373-390.

Subba Reddi, C., Reddi, E.U.B. and Reddi, M.S. (1979). A novel mechanism of pollination in *Wrightia tinctoria* R. Br.*Curr. Sci.*, 48: 746-747.

Swihart, S.L. (1970). The neural basis of colour vision in the butterfly *Papilio triolus*. *J. Insect Physiol.*, 16: 1623-1636.

Swihart, C.A. (1971). Colour discrimination by the butterfly *Heliconius charitonius* Linn. *Anim. Behav.*, 19: 156-164.

Swihart, S.L. (1972). The neural basis of colour vision in the butterfly *Heliconius erato*. *J. Insect Physiol.* 18: 1015-1025.

Proceedings of the National Seminar on Environmental Biology, April 03-05, 1998
Edited By : A.K. Aditya & P. Haldar
DAYA PUBLISHING HOUSE, DELHI

Acaciasides and Root-knot Nematode Extract Suppress *Meloidogyne incognita* Infection in Lady's Finger Plants

S.C. Datta, R. Datta (Nag), S.P. Sinha Babu & N.C. Sukul
Department of Zoology
Visva-Bharati University
Santiniketan - 731 235, West Bengal, India

Abstract

Two triterpenoid saponins, acaciasides A and B, isolated from the funicles of *Acacia auriculiformis*, applied by foliar spray on lady's finger plants at 20 mg/plant reduce *Meloidogyne incognita* infection of the plants and enhance their growth. An ethanolic extract of the females of *M. incognita*, applied by foliar spray on lady's finger plants at 0.14 mg/plant also suppresses nematode infection and enhances plant growth. Treatment with the nematode extract followed by the application of acaciasides gives better control of root-knot disease and higher growth of plants. In both cases protein content of fruits has also increased. Using double the dose of acaciasides alone by foliar spray and soil drench gave no better results with respect to the control of nematode and growth of test plants.

Key words : *Acacia auriculiformis*, triterpenoid saponins, nematode extract, *Meloidogyne incognita*, control.

Introduction

Aqueous extract of *Meloidogyne incognita* females reduces root-knot disease and increases growth of tomato, cowpea and carrot (Roy *et al.*, 1995; Datta *et al.*, In press). The nematode extract (NE) is thought to induce defense responses in the host plants through the expression of pathogenesis-related genes. The extract from the funicles of the plant *Acacia auriculiformis* A. Cunn. or its active principle acaciasides reduce *M. incognita* infection of plants and enhance their growth (Paul *et al.*, 1995a, b; Sinha Babu *et al.*, 1992). The botanical compound has been found to be highly toxic to the nematode (Sinha Babu *et al.*, 1992). The purpose of the present study is to see whether acaciasides and NE together could reduce root-knot nematode infection of lady's finger more effectively than either of the two agents alone.

Materials and Methods

Preparation of nematode extract (NE)

M. incognita females were collected from the roots of tomato plants grown in an experimental garden. They were washed with sterile distilled water, homogenized in a tissue homogenizer and extracted with 90% ethanol at room temperature (28±2°C) for 5 days. The extract was allowed to evaporate at room temperature and the residue was kept over anhydrous calcium, chloride for dehydration and stored at 4°C. The residue was mixed with sterile distilled water just before application on plants.

Isolation of acaciasides

Air-dried and powdered funicles of *A. auriculiformis* were successively extracted with petrol (60 80°C) and 90% ethanol. The ethanol extract, on removal of the solvent under reduced pressure, yielded a viscous dark brown mass. The extract was chromatographed on silica gel with petrol, petrol-chloroform (1:1), chloroform, chloroform-methanol (9:1, 7:3, 3:2, 1:1 and 2:3) as successive eluents. The chloroform-methanol (7:3 and 3:2) eluates were then combined. The combined fraction was found to be composed mainly of two compounds which were separated by repeated preparative HPLC employing Spherisorb S-10-ODS reversed phase column with the solvent system methanol-water (7:3) at a flow rate of 4 ml/min and refractive index detector as white amorphous solids. These two solid compounds, designated as acaciaside A and acaciaside B (Fig. 1) according to their increasing order of polarity, were found to be triterpenoid saponins by Liebermann-Burchard, molisch and froth tests (Mahato *et al.*, 1992; Ghosh *et al.*, 1993).

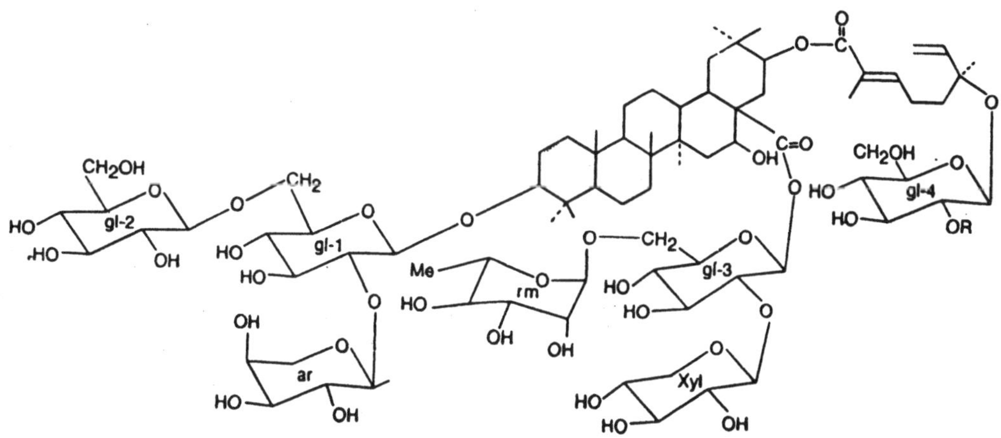

Acaciaside A R = H
Acaciaside B R = Xyl (p)

Fig. 1 : Molecular structure of two triterpenoid saponins (Acaciaside A and Acaciaside B) from *A. auriculiformis*

Inoculation and treatment

Aseptically germinated seeds of *Abelmoschus esculentus* (L) Moench (Pusa Ruby) were sown one seed / pot (32 cm. diameter) containing a mixture of clay soil and composted manure (2:1 v/v). The soil-filled pots were treated with boiling water 4 times. The pots were divided into the following groups, each numbering 10 : uninoculated untreated, inoculated untreated, uninoculated NE-treated, NE-treated before inoculation and acaciasides-treated after inoculation, NE-treated after inoculation, acaciasides-treated after inoculation and acaciasides treated after inoculation by foliar spray and soil drench. Except the last group all other treatments were by foliar spray.

Plants were inoculated at 4-leaf stage with the 2nd-stage larvae of *M. incognita* at 2328±85 larvae/plant. Dehydrated NE was mixed with sterile distilled water at the rate of 0.06 mg/ml and sprayed on leaves of test plants at 0.14 mg/plant. Acaciasides were dissolved in sterile distilled water at 2 mg/ml and sprayed on plants at 20 mg/plant. In the last treatment group each plant received 40 mg of acaciasides, 20 mg from foliar spray and 20 mg from soil drench. Acaciasides solution was sprayed on leaves in 5 installments spanning 2.5 hr. Pre-and Post-inoculation treatments were given 3 days before and 3 days after inoculation with nematodes.

Harvesting

All the plants were uprooted 76 days after sowing of germinated seeds and the following parameters in all the 8-groups of plants were recorded : biomass of shoot and root, no. of leaves/ plant, no. of root galls/plant, nematode population/2g root and 200g soil and percentage of root and fruit protein. Nematodes were extracted from soil and roots by the modified Baermann method. Roots were finely torn into pieces by a blender before nematode extraction. Three samples of fruit and root pieces were taken at random from each treatment and the total protein fraction in each sample was estimated by Folin-Phenol method (Chatterjee and Sukul, 1981; Lowry *et al.*, 1951). The experiment was repeated with similar results. Data from the second experiment are presented.

Results

Data were analyzed by the analysis of variance and the results were given in Table-1. Treatment with NE before inoculation and acaciasides reduced nematode infestation in terms of root-gall number and nematode population in root to a minimum among all the treatment groups (Table-1). Nematode population in rhizospheric soil was maximum with the group treated with NE + acaciasides and minimum with the untreated group (Table-1). Plant growth in terms of biomass of shoot and number of leaves was maximum with un-inoculated and NE-treated group followed by NE+acaciasides-treated and un-inoculated and untreated groups (Table-1). Protein content in fruits was maximum with NE+ acaciasides-treated and un-inoculated and NE-treated groups (Table 1). In general all the treatment reduced nematode infection of the plants significantly (Table 1).

Discussion

The present study once again confirms our earlier observation that nematode extract could significantly reduce *M. incognita* infestation of plants and that pre-inoculation treatment was more effective than post-inoculation ones. It is possible that NE could activate natural defence mechanism in the test plants. Systemic resistance can be induced in different crop plants by non-pathogenic and pathogenic micro-organisms or their culture filtrates. Root invasion by the cyst nematode produced systemic accumulation of new proteins in the leaves of potato plants. Lectins accumulated in gall regions of roots of lady's finger plants infected with *M. incognita.* Root-knot nematodes alter patterns of plant gene expression within cell destined to become the feeding site for them. Plants are known to possess receptors for salicylic acid, the endogenous signal required for the systemic acquired resistance response of plants. They have also receptors for other hormones like ethylene and auxins. Systemic defence response is induced locally by pathogen or pest attack and is mediated by salicylic acid which induces expression of pathogenesis-related

Table 1 : Effect of the ethanolic extract of *Meloidogyne incognita* females and acaciasides from *Acacia auriculiformis* on the biomass, no. of leaves, no. of root galls, nematode population in soil and roots and root and fruit protein content (%) of lady's finger plants inoculated with *M. incognita* J2.

Treatment groups	Biomass (g)		Total leaves	Number of galls/ plant	Nematode population		Protein Percentage	
	Shoot	Root			Soil (200g)	Root (2g)	Root	Fruit
Uninoculated	128.3b	15.10d	11.10ab	Nil	Nil	Nil	1.14h	1.86c
Untreated	±5.86	±0.32	±0.41				±0.07	±0.24
Inoculated	120.00b	27.80b	9.70b	349.30a	29.00e	706.00a	2.88a	1.10f
Untreated	±4.76	±0.18	±0.31	±16.90	±1.12•	±36.04	±0.12	±0.04
Uninoculated	159.50a	17.80cd	12.80a				1.42d	2.30a
NE-treated	±5.56	±0.73	±0.23	Nil	Nil	Nil	±0.05	±0.06
(foliar)								
NE-Pretreated	104.50c	13.10d	11.30ab	49.80d	968.00b	23.50e	1.17g	2.11b
Inoculated	±3.35	±0.17	±0.37	±1.04	±39.70	±0.76	±0.03	±0.03
NE-Pretreated,	108.0c	50.10a	12.70a	24.20e	1322.50a	19.00f	1.83c	2.26a
Acaciaside	±2.21	±1.76	±0.37	±0.84	±68.94	±0.50	±0.07	±0.03
Posttreated								
Inoculated	108.0c	18.20cd	11.80a	50.90c	476.50c	41.50d	1.35e	1.42d
NE-Posttreated	±3.10	±0.44	±0.50	±1.15	±24.95	±0.47	±0.05	±0.05
Acaciaside,	109.20c	20.90c	9.60b	58.40b	938.50b	204.00b	1.94b	1.21c
Posttreated	±2.62	±0.67	±0.28	±2.10	±45.80	±11.01	±0.07	±0.02
Acaciasides by	105.50c	15.50cd	10.60ab	24.90e	159.00d	165.00c	1.2lf	2.18b
Foliar spray	±1.23	±0.57	±0.40	±0.89	±7.28	±7.74	±0.04	±0.05
And Soil drench								

a,b,c... : different small letters in a column indicate significant difference ($P<0.05$) by ANOVA. Except the last group all treatments are by foliar spray.

genes. Besides reducing root-knot disease, acaciasides also suppress foliar diseases caused by virus, fungus and insects in mulberry plants. Treatment with *Acacia*-extract leave no toxic residues in plants. Thus acaciasides would serve as a safe environment friendly pesticide. Use of this botanical compound with NE would go a long way in tackling various pests of crops in a safe way.

Acknowledgement

The work described here has been supported by a grant from the University Grants Commission, New Delhi and Central Silk Board, Govt. of India.

References

Chattejee A, Sukul NC. (1981). Total protein content of galled roots as an index of root-knot nematode infestation of lady's finger plants. *Phytopathology* 71 : 372-374.

Das S, Sukul NC, Mitra D, Sarkar H. (1989). Distribution of lectin in nematode infested and uninfested roots of *Hibiscus esculentus*. *Nematol. Medit.* 17 : 123-125.

Datta SC, Sinha Babu SP, Sukl NC. (1997). Improved growth of silkworms from effective treatment of mulberry diseases by *Acacia auriculiformis* extract. *Sericologia* 37(4) : 707-712.

Datta SC, Sinha Babu SP, Banerjee N, Ghosh K, Sukul NC. In press. *Meloidogyne incognita* extract reduces *M. incognita* infestation of tomato. *Indian J. Nematol.*

Descalzo RC, Rahe JE, Mauza B. (1990). Comparative efficacy of induced resistance for selected disease of greenhouse cucumber. *Can. J. Plant Pathol.* 12 : 16-24.

Ghosh M, Sinha Babu SP, Sukul NC. (1993). Antifilarial effect of two triterpenoid saponins. Indian J. Exp. Biol. 31 : 604-606.

Hammond-Kosack KE, Atkinson HJ, Bowles DJ. (1989). Systemic accumulation of novel proteins in the apoplast of the leaves of potato plant following root invasion by the cyst nematode *Globodera rostochiensis. Plant Pathol.* 35 : 495-596.

Jones AM. (1994). Surprising signals in plant cells. *Science* 263 : 183-184.

Kuc' J, Strobel NE. (1992). Induced resistance using pathogens and non-pathogens. In : Biological control of plant Diseases. (Eds. E.C. Tjamos) *Plenum Press,* New York pp. 295-300.

Lowry OH, Rosebrough JJ, Farr AL, Randall RJ. (1951). Protein measurement with the Folin-phenol reagent. *J. Biol. Chem.* 193 : 265-275.

Mahato SB, Pal BC, Nandi AK. (1992). Structure elucidation of two acylated triterpenoid bioglycosides from *Acacia auriculiformis* Cunn. *Tetrahedron* 48 : 6717-6728.

Merra MS, Shivanna MB, Kageyma K, Hyakumachi M. (1994). Plant growth promoting fungi from zoysiagrass rhizosphere as potential inducers of systemic resistance in cucumbers. *Phytopathology* 84 : 1399-1406.

Opperman CH, Taylor CG, Conkling MA. (1994). Root-knot nematode-directed expression of a plant root specific gene. *Science* 163 : 183-184.

Paul A, Sinha Babu SP, Sukul NC. (1995a). Effect of nematode infected mulberry plants on the growth and silk production of *Bombyx mori L. Indian J. Seric.* 34(1) : 18-21.

Paul A, Sinha Babu SP, Sukul NC. (1995b). Effect of *Meloidogyne incognita* infected mulberry plants and their treatment with nematicides on feeding silk-worms, *Bombyx mori L. Sericologia* 35(3) : 577-581.

Roy D, Sinha Babu SP, Sukul NC. (1995). Root-knot nematode extract increases growth of plants are reduces nematode infection. *Environ. Ecol.* 13 : 775-779.

Sinha Babu SP, Roy D, Sukul NC, Mahato SB. (1992). Nematicidal principle from *Acacia auriculiformis.* Proceeding of the First National Symposium *'Allelopathy in Agroecosystem'* eds. P. Tauro and S.S. Narwal, pp. 181-183, Haryana Agric. Univ. Hisar, India : Indian Society of Allelopathy.

Proceedings of the National Seminar on Environmental Biology, April 03-05, 1998
***Edited By* : A.K. Aditya & P. Haldar**
DAYA PUBLISHING HOUSE, DELHI

Studies on the Bio-diversity in Acridids (Orthoptera : Acridoidea) at Santiniketan, West Bengal, India

P. Haldar, Amlan Das & R.K. Gupta
Entomology Laboratory
Department of Zoology
Visva-Bharati, Santiniketan - 731 235

Abstract

Faunal richness, seasonal occurrence and seasonal fluctuation in the population of short-horned grasshopper from three distinct vegetational sites viz., forest ecosystem, open grassland ecosystem, and agro ecosystem at Santiniketan and its vicinity, West Bengal, India were studied. A total of eighteen species were recorded from the study area. Habitat preference, relative density and relative frequency of all the species are also discussed. Population estimation for sex-ratio revealed that the number of female and male in acridids are almost equal in nature. Study on the seasonal occurrence and relative abundance of acridids revealed that *Spathosternum prasiniferum prasiniferum* (Walker) being widely distributed and occupied top position. Findings of this study also reflected that in the grassland ecosystem, the species namely, *Phlaeoba panteli* (Bolivar), *Gesonula punetifrons* (Stal.), *Trilophidia annulata* (Thunberg) and in the forest ecosystem the species namely, *Catantops innotabilis* (Walker), *Acridella nasuta* (Linn.) and *Trilophidia annulata* (Thunberg), where as, in agro ecosystem only *Aiolopus thalassinus tamulus* (Fabricius) are the opportunistic species.

Key words : Acridids, relative frequency, relative density, Seasonal occurrence.

Introduction

When a single animal is needed to represent the class insecta, the invariable choice is the acridids, commonly known as grasshoppers. Its selection for a detailed study is favoured because of several reasons as it is abundant in all parts of the world and easily secured. These grasshoppers & locusts are commonly noticed as agricultural pests. The ecological studies on Indian acridids are mainly limited to the works of Coleman (1911), Coleman & Kannan (1911), Pruthi and Nigam (1939), Roonwal (1945), Agarwal (1955), Katiyar (1955, 56, 61) Banu & Kushwaha (1974), Ananthakrishnan et al (1985), Tandon et al (1969), Hazra et al (1983), and Bhowmik & Haldar (1983, 1985). The present study on the biodiversity of acridids was initiated at Santiniketan and its vicinity, W.B., in order to determine the faunal richness, incidence, seasonal occurrence, relative frequency, relative density and distribution in relation to different ecosystems, viz. grassland, forest and agro-ecosystems.

Santiniketan is located at approximately 23°29′N latitude and 87°42′E longitude in the district of Birbhum, W.B. with an average altitude of 58.9 m above the sea level. Soil of this place is laterite, reddish in colour and coarse sandy in texture. The temperature ranged between 6.7°C (during coldest weather) to 44.1°C (during hottest weather). The average annual rainfall was 1402.3 mm, having its distribution pattern 1198.33 mm. during June to October, 65.39 mm during November to March and 138.58 mm. during April to May. The southwest monsoon lasts from June to September. October and the first half of November constitute the post monsoon season. Data recorded at the observatory, Sriniketan, within 2 kms. of the study plots.

Materials and Methods

Three habitats viz. open grassland ecosystem, forest ecosystem and agro ecosystem were selected for the present studies. Regular collection of short horned grasshoppers were made from each habitat at an interval of 7 days during different seasons such as summer, monsoon & winter, in the year 1989, using insect net of conventional design of 30 cm in diameter. Efforts were made to standardize the sampling technique by sweeping each site for half an hour on every occasion. Catch and Count method thus applied for the present studies.

Description of Habitat

a. *Open Grassland Ecosystem*

The area was 36m^2, near Santiniketan. It was covered with green common grasses throughout the study period. The *Cynodon dactylon, Cyperus* sp. *Chrysopogon aciculatus* are abundant species in the study area.

b. *Forest Ecosystem*

This area was also 36 m^2 at Ballavpur Reserve Forest at the vicinity of Santiniketan. These are some dry sandy, hard uneven fields with laterite soil and pebbles. The tropical trees such as *Butea monosperma* (Palas), *Shorea robusta* (Shal), *Cassia fistula* (Sodal), *Mangifera indica* (Mango), *Acacia auriculiformis* (Sonajhuri) and various types of shurbs and herbs were abundant there.

c. *Agro Ecosystem*

This area was also 36 m^2, in the agricultural farm at Sriniketan, 2 kms. away from Santiniketan. The paddy (*Oryza sativa*) was the main crop in this agricultural fields. Besides paddy, wheat *Triticum aestivum*, mustard (*Brassica nigra*). Sorghum (*Sorghum vulgare*) and other crops are also cultivated here. Collections were made from the fallow land between cultivated plots.

Results

Throughout the study period, a total number of 1128 specimens belonging to eighteen (18) species were collected from the study areas. Out of these 18 species viz. *Spathosternum prasiniferum prasiniferum* (Walker), *Oxya fascovittata* (Marschall), *Acrida exaltata* (Walker), *Acrotylus humbertianus* (Saussure), *Oedaleus abruptus* (Thunb.) *Gastrimargus africanus africanus* (San.) *Aulaeobothrus luteipes* (Walker), *Pyrgomorpha conica* (Oliv.) *Catantops innotabilis* (Walker), *Acridella nasuta* (Linn.), *Trilophidia annulata* (Thunberg), *Phlaeoba panteli* (Bol), *Aiolopus thalassinus tamulus* (Fab.), *Mesopsis cylindricus* (Uvarov), *Gesonula punctifrons* (Stal.), *Cyrtacanthaeris tatarica* (Linn.), *Chrotogonus trachypterus trachypterus* (Blanchard), *Hieroglyphus banian* (Fab.) were recorded. Only 5 species viz. *Spathosternum prasiniferum prasiniferum, Oxya fuscovittata, Acrida exaltata, Oedaleus abruptus,* & *Pyrgomorpha conica* are common in these three ecosystems. *Acrotylus humbertianus, Aulacobothrus luteipes* & *Trilophidia annulata* are common both in grassland and forest ecosystem, and *Phlaeoba panteli* is common in Forest & Agro-ecosystem. *Aiolopus thalassinus tamulus, Gesonula punctifrons,* and *Hieroglyphus banian* are common in grassland and agro ecosystem. *Gastrimargus africanus africanus, Catantops innotabilis, Acridella nasuta, Mesopsis cylindricus, Cyrtacanthacris tatarica and Chrotogonus trachypterus trachypterus* are exclusively found in forest ecosystem. In total, 15 species are found in forest, 11 species are found in grassland and 9 species are found in agro ecosystem. (Table-1).

Table 1 : Occurrence of Accridid Species in Three Eco-systems

	Species Name	Forest	Grass Land	Agro
1.	*Spathosternum prasiniferum prasiniferum* (Walker)	+	+	+
2.	*Oxya fuscovittata* (Marschall)	+	+	+
3.	*Acrida exaltata* (Walker)	+	+	+
4.	*Acrotylus humbertianus* (Saussure)	+	+	–
5.	*Oedaleus abruptus* (Thunb.)	+	+	+
6.	*Gastrimargus africanus africanus* (Sauss.)	+	–	–
7.	*Aulacobothrus luteipes* (Walker)	+	+	–
8.	*Pyrgomorpha conica* (Oliv.)	+	+	+
9.	*Catantops innotabilis* (Walker)	+	–	–
10.	*Acridella nasuta* (Linn.)	+	–	–
11.	*Trilophidia annulata* (Thunberg)	+	+	–
12.	*Phlaeoba panteli* (Bol.)	+	–	+
13.	*Aiolopus thalassinus tamulus* (Fab.)	–	+	+
14.	*Mesopsis cylindricus* (Uvarov)	+	–	–
15.	*Gesonula punctifrons* (Stal.)	–	+	+
16.	*Cyrtacanthacris tatarica* (Linn.)	+	–	–
17.	*Chrotogonus trachypterus trachypterus* (Blanchard)	+	–	–
18.	*Hieroglyphus banian* (Fabricius)	–	+	+
	Total	**15**	**11**	**9**

Relative Density and Relative Frequency of Grasshoppers in Three Ecosystems

Throughout the study period it was observed that, the relative density of *S. prasiniferum prasiniferum* is highest in all the ecosystems among the 18 species recorded. Its average relative density (RD) was 44.41 having RD 44.60 in grassland, 48.37 in forest and 41.19 in Agro ecosystem. (Table 2).

It was also observed that the relative frequency (RF) was highest in case of *S. prasiniferum prasiniferum* among all, having RF 16.66 in grassland, 21.05 in forest and 19.04 in Agro ecosystem. (Table-2). *C. tatarica* and *C. trachypterus trachypterus* both showed the least RD (0.97 each) in forest ecosystem, *G. punctifrons* & *A. thalassinus tamulus* showed the least RD (0.49) in agro ecosystem and *A. thalassinus tamulus*`also showed the least RD (0.47) in grassland ecosystem. The least RF showed by other species in all the three ecosystems (Table 2). Maximum and minimum relative density was observed in *S. pr. prasiniferum* and *P. panteli* respectively in all the ecosystems. (Fig. 1).

Fig. 1 : Relative Density (RD) of Acridids

Table 2 : Relative Density (RD) & Relative Frequency (RF) of Acridids

Sl. No.	Species Name	Grassland		Forest		Agro	
		RD	RF	RD	RF	RD	RF
1.	*S. pr. prasiniferum*	44.60	16.66	48.37	21.05	41.19	19.04
2.	*O. abruptus*	19.42	12.50	13.96	15.78	4.46	9.52
3.	*A. exaltata*	6.71	8.33	7.79	10.52	16.37	14.28
4.	*H. banian*	4.07	8.33	–	–	18.61	19.04
5.	*O. fuscovittata*	8.39	4.16	6.49	5.26	7.69	9.52
6.	*P. conica*	7.19	8.33	1.94	5.26	7.44	9.52
7.	*P. panteli*	0.95	4.16	1.62	5.26	3.22	9.52
8.	*A. humbertianus*	1.67	4.16	4.87	10.52	–	–
9.	*M. cylindricus*	0.95	8.33	4.22	5.26	–	–
10.	*G. africanus*	1.43	4.16	2.69	5.26	–	–
11.	*A. luteipes*	0.71	4.16	2.92	5.26	–	–
12.	*C. innotabilis*	1.43	4.16	0.97	5.26	–	–
13.	*T. annulata*	1.19	4.16	0.97	5.26	–	–
14.	*G. punctifrons*	0.71	4.16	–	–	0.49	5.26
15.	*A. nasuta*	0.47	4.16	1.29	5.26	–	–
16.	*A. thalassinus tamulus*	0.47	4.16	–	–	0.49	5.26
17.	*C. tatarica*	–	–	0.97	5.26	–	–
18.	*C. tr. trachypterus*	–	–	0.97	5.26	–	–
	Total	**36.96**		**27.30**		**35.72**	

RD = Relative Density
RF = Relative Frequency

Discussion

Studies on the habitats of acridids in Santiniketan, W.B. revealed that, *S. prasiniferum prasiniferum*, *O. fuscovittata*, *A. exaltata*, *O. abruptus* and *P. conica* were found in all the ecosystems where as, *G. africanus africanus*, *C. innotabilis*, *A. nasuta*, *C. tatarica* & *C. trachypterus trachypterus* were restricted to forest ecosystem. The occurrence of *O. fuscovittata* in agro and grassland ecosystems during summar, monsoon and winter seasons was also observed by Hazra et. al. 1983. Out of eighteen species found in the study area, maximum number (15 species) were found in forest ecosystem, where as, only eleven and nine species were found in grassland and agro-ecosystem, respectively (Table 1).

When the sex ratio of acridids was taken into consideration it was observed that the sex-ratio in grassland, forest and agro-ecosystem were 5:4, 9:10 and 25:24 respectively (Table-3). The

Table 3 : Sex Ratio in Acridids

	Grassland	Forest	Agro	Total
Sex ratio (Male : Female)	5:4	9:10	25:24	13:12.66
X^2 value	**2.8****	**0.46***	**0.04***	**0.58***

** Significant (at 0.05% level)

* Nonsignificant

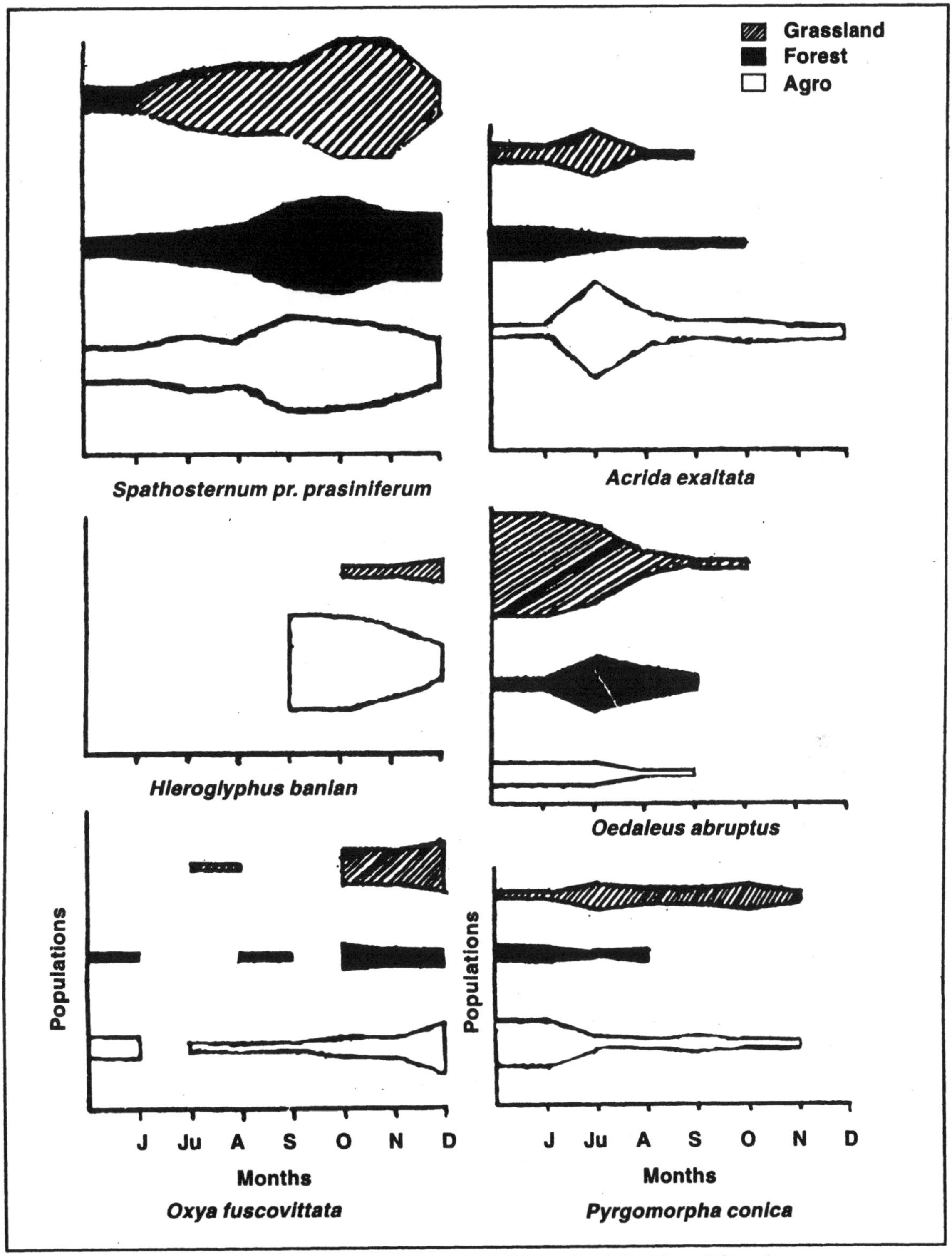

Fig. 2 : Seasonal Population Fluctuation of Six Dominant Species of Grasshoppers

female was dominant in number only in the forest ecosystem, where as, in other two ecosystems male dominancy was observed. This female dominancy in forest was probably due to its liking the forest ecosystem as suitable breeding ground of most of the species. However when the sex-ratio of all the ecosystems considered together it revealed that the number of female and male in acridids were almost equal in nature.

Both the relative density and the relative frequency were highest in *S. pr. prasiniferum* in all the ecosystems among the six domiant species viz. *O. abruptus, A. exaltata, H. banian, O. fuscovittata* and *P. conica* (Fig. 1). In the study area *G. punctifrons, A. nasuta, A. thalassinus tamulus, C. tatarica* and *C. tr. trachypterus* were rare species. (Table-2).

Among the six dominant species, population of *S. pr. prasiniferum* increased maximum during the post monsoon period (September to November) in all the ecosystems. Population richness of *A. exaltata* was observed in the month of July in agro-ecosystem, where as, in other ecosystems the number of population was moderate though out the study period (Fig. 2). The *H. banian* was found abundantly in agro ecosystem during the month of September to December. Similar observation on the abundance of population of *H. nigrorepletus* was recorded by Roonwal, 1945. Its absence in forest ecosystem and low population density in grassland was probably due to its migration from grassland to the crop field as it is recognised as a major paddy pest. The occurrence of *O. abruptus* was observed in all the ecosystems during the study period though their population richness was observed during June to August in grassland ecosystem and in other areas their number were moderate. Similar observation on population richness in *A. punctatus* during late summer was noticed by Katiyar, 1955. The presence of *O. fuscovittata* in all the ecosystems almost throughout the study period indicates its wide range of distribution. Inspite of their wide range of habitat they preferred the cropland probably due to their food habit. *P. conica* maintained almost an uniform population in all the habitats throughout the period of observation (Fig. 2).

Acknowledgements

Authors are thankful to the Department of Sciences & Technology (DST), New Delhi, India, for financial assistance. Laboratory facilities provided by Professor A.K. Aditya, Head of the Department of Zoology, Visva-Bharati University is also gratefully acknowledged.

References

Agarwal, N.S. (1955). Bionomics of *A. crenulata* Fab. (Orthoptera : Acridoidea). Indian J. Ent. 17(2) : 230-240.

Ananthakrishnan, T.N. et al (1985). Behavioral response in terms of feeding & reproduction in some grasshoppers (Orthoptera : Insecta) Proc. Indian Acad. Sci. (Anim. Sci), 94(5) : 443-461.

Banu, F. & Kushwaha, K.S. (1974). Incidence of the grasshopper, *T. annulata* Thunberg seriously infesting pastures in Rajasthan and some important features to indentify its various developing stages in the field. J. Zool. Soc. India. 26(1-2) : 47-59.

Bhowmik, H.K. and Haldar, P. (1983). Preliminary distributional records with remarks on little known species of Acrididae (Orth. : Insecta) from the Western Himalaya (Himachal Pradesh). Rec. Zool. Surv. India, 81 : 167-191.

Coleman, L.C. (1911). The Jola or Deccan grasshopper (*C. Sphenarioides* Bol.), Bull. Dept. Agric. Mysore (Ent. Ser.) 2 : 1-43.

Coleman, L.C. & Kannan, K.K. (1911). The rice grasshopper (*H. banian* Fab.) Bull. Dept. Agric. Mysore (Ent. Ser.) 1: 1-52.

Hazra, A.K. et al (1983). Population ecology of some grasshoppers in cultivated and natural grassland ecosystems of W.B. Environment & Ecology, 2(4) : 319-322.

Katiyar, K.N. (1955). Ecology and Life history of *A. punctatus* Drury, a northern spotted grasshopper (Orthoptera: Acrididae) Agra Univ. J. Res. (Sci.), 4(2) : 397-414.

Katiyar, K.N. (1956). The life history and ecology of the short horned grasshopper *P. bilineatus* Bol. (Orthoptera: Acrididae). Agra Univ. J. Res. (Sci.), 5(1) : 179-192.

Katiyar, K.N. (1961). The life history and ecology of the shorthorned grasshoppers, *E. rosea* Uvarow (Orthoptera: Acrididae). Z. Angew. Ent., 48(4) : 395-409.

Pruthi, H.S. & Nigam, L.N. (1939). The bionomics, Life history and control of the A K. Grasshopper, *P. pictus* Fab., a new pest of cultivated crops in North India. Indian J. Agric. Sci. 9(4) : 629-641.

Roonwal, M.L. (1945). Notes on the bionomics of *H. nigrorepletus* Bol. at Banaras, United provinces, India. Bull.Ent. Res. 36(3) : 339-341.

Tandon S.K. et al., (1969). On a collection of Acridiodea (Orthoptera) from the Nagarjuna Sagar Dam area. Oriental insects, 3(3) : 265-267.

Proceedings of the National Seminar on Environmental Biology, April 03-05, 1998
***Edited By :* A.K. Aditya & P. Haldar**
DAYA PUBLISHING HOUSE, DELHI

Polyphagy in Plant-parasitic Nematodes : A Favourable Force for Speciation and Bio-diversity

Amalendu Chatterjee & Debabrata Sen
Zoological Survey of India
M-Block, New Alipore, Calcutta - 700 053

Abstract

Availability of plenty of species represents a healthy bio-diversity of a particular geographical area. In the present work, thirty five genera of plant-parasitic nematodes are randomly selected. The already recorded number, of host-plant species, of these genera and the number of species belong to these genera, have been taken into account. A statistical positive correlation has been established between the number of host-species and the number of parasitic species, i.e., the number of parasitic-species is directly proportional to the number of hosts species. This is concluded that the increase in number of host spp. favours the isolation machanism; a prerequisite for speciation. The isolation mechanism is being brought in three different ways : (i) geographical isolation, (ii) biochemical isolation and (iii) sympatric isolation due to seasonal variation of different host plants, particularly of agricultural crops. Thus, the polyphagous nature of any species is acting as a drifting force and accelarates to give rise different phyletic lines, and ultimately different species.

Introduction

Biodiversity in any geographical area obviously depends upon the availability of plenty of species. Plenty number of species is the product of active speciation. Perfect isolation is the prerequisite for speciation; that may be sympatric or allopatric. This is encountered by almost all workers, that any plant-parasitic nematode genus, where a good number of host-plant species are recorded; usually that genus holds a good number of species also. Karall (1985) had reviewed the family Hoplolaimidae with all its genera and species. Siddiqi (1986) had reviewed and recorded world genera and species of plant-parasitic nematodes (order Tylenchida). Ahmad (1996) also recorded the Indian species of the same (sub-order Tuylenchina). The range of host-plant species were recorded by Goodey, Franklin and Hopper (1965). Firoza and Maqbool (1994) had provided a diagnostic compendium of the genus *Helicotylenchus* and the host range study of *Helicotylenchus* was provided by Taylor (1960, 1961).

To ascertain, whether any correlation exists between the number of host-plant species and the number of parasitic species, thirty five genera of the order Tylenchda were randomly selected and the number of host-plant species, already recorded for each of these genera were taken into consideration. A list of these genera together with their number of species and the number of host-plant species are shown in the Table 1. The number of parasitic species under each genera and the number of host-species are also plotted in a scattered diagram (X and Y axis respectively) as two independent variants. The regression lines and the correlation coefficinet (=r) have been estimated. The value of r=0.955; that is significant at 0.001 (%) level. This indicates a very strong positive correlation exists between these two variable factors i.e., between the number of host-species and the number of parasitic-species. In other word, the number of species of any genera are directly proportional to the number of host-plant speices.

Table 1 : Host Range and Parasitic Species Range of Nametode Genera

Sl. No.	Name of the Genus	No. of The Parasitic Spp.	No. of The Host-Plant Spp.
1.	*Heterodera*	56	240
2.	*Aglenchus*	04	11
3.	*Pratylenchus*	79	355
4.	*Helicotylenchus*	190	452
5.	*Rotylenchulus*	10	36
6.	*Rotylenchus*	37	136
7.	*Hexadorus*	04	08
8.	*Merlinius*	24	63
9.	*Criconemodies*	12	31
10.	*Dolichodorus*	13	24
11.	*Radopholus*	24	42
12.	*Paratylenchus*	88	252
13.	*Anguina*	18	37
14.	*Trophurus*	11	20
15.	*Filenchus*	45	104
16.	*Scutellonema*	38	78
17.	*Tylenchulus*	02	16
18.	*Polenchus*	01	01
19.	*Hirschmanniella*	22	50
20.	*Meloidodera*	03	07
21.	*Criconema*	58	127
22.	*Boleodorus*	22	66
23.	*Neobakernema*	02	05
24.	*Aorolaimus*	07	19
25.	*Tylenchorhynchus*	71	184
26.	*Globodera*	13	36
27.	*Discotylenchus*	02	02
28.	*Ecphyadophora*	06	11
29.	*Tylenchus*	20	39
30.	*Ditylenchus*	52	136
31.	*Irantylenchus*	02	02
32.	*Hoplolaimus*	18	55
33.	*Sakia*	06	16
34.	*Hemicycliophora*	103	252
35.	*Macroposthonia*	83	220

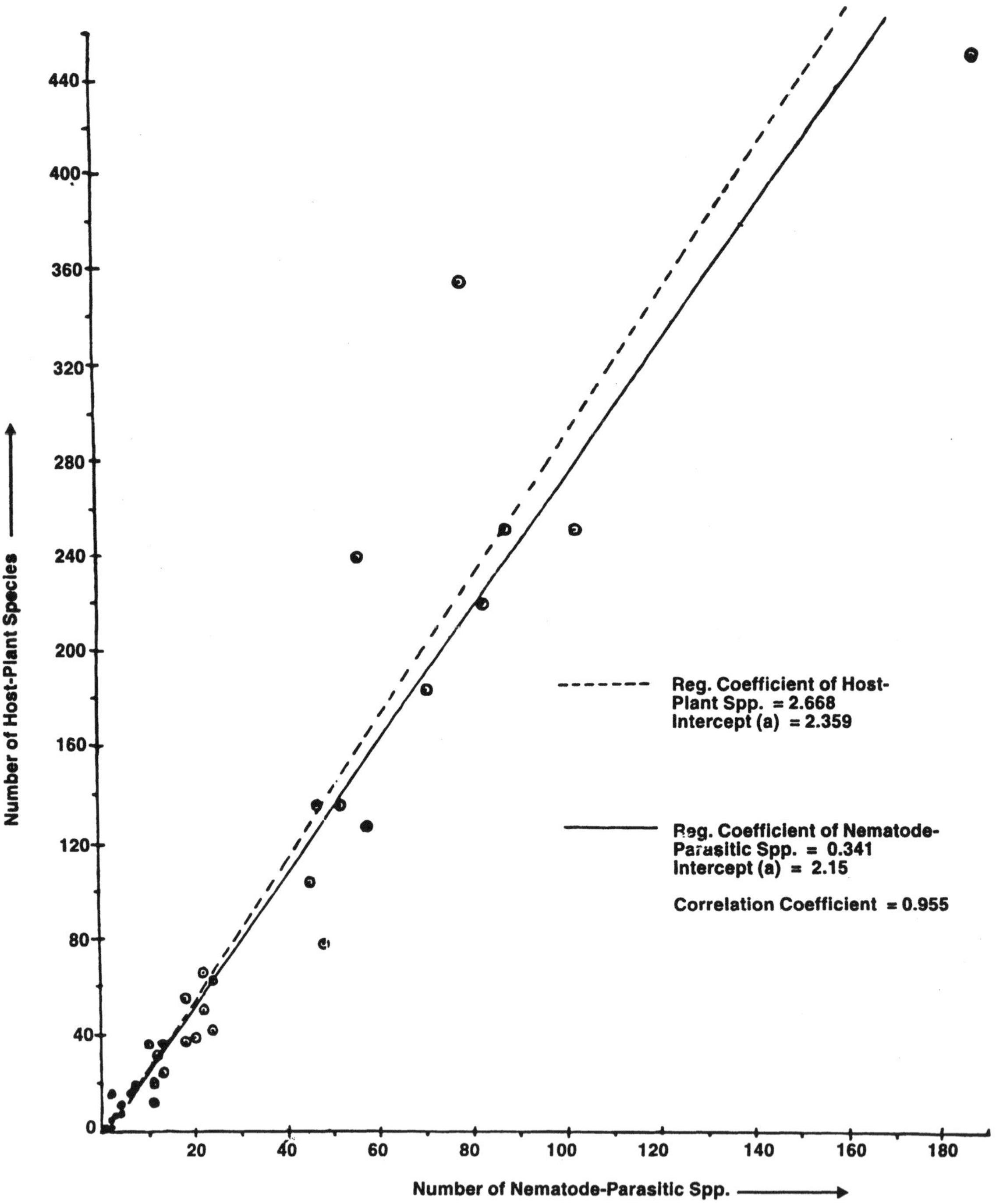

Fig. 1 : Scattered Diagram of Distribution of Host-Plant Species Against Parasitic Species

Discussion

This becomes evident that, in case of polyphagous genera, i.e., where the range of host-species are relatively higher, those particular genera also hold higher number of species. Now the species are taken as the product of speciation, and all the species under a common genus, are of monophytic origin. The speciation i.e., the branching of different phyletic lines from a common ancestral group can not occur; unless some kind of isolation mechanism does not act. In other words, a significant genetic drift is necessary between the isolated phyletic lines. Here in case of plant-parasitic nematode genera, where there are a good range of host-plant species, the isolation may be brought in three different ways : (i) Geographical isolation, (ii) Biochemcial isolation and (iii) Isolation due to seasonal variation of host-plants, i.e., some kind of temporal isolation. Usually a particular geographical territory presents a particular pattern of floral combination; different from other terriotry, with different ecological parameters. Thus any nematode species having many plant species as its host, usually are distributed in geographically and ecologically far different areas and an isolation mechanism acts between them. Second type of isolation is of bio-chemical nature; because in case of polyphagous species, any two different host-species must have biochemical constituent, significantly different from each other. Such differences in micro-habitats have definite but different impacts on the physiology, of the parasitic-species and gradually follows to bring about physiological as well as morphological differences between them. In fact, such type of isolation, are far more effective, than that done by even the perfect geographical isolation. Even the texture and physical nature of a particular plant species exert adaptive force and initiates morphological changes on the parasites. The third type of isolation, is common among nematode parasites of agricultural crops or sometimes those of general plants also. For a particular polyphagous nematode species, the host-plants are not grown in the same season, but at different times, and thus a time barrier acts between them, preventing them to take share in the same genetic pool. Such type of isolation may be termed as the temporal isolation. So in totality, this is observed that polyphagy itself acts as a drifting force and favours adaptive radiation.

Acknowledgement

The authors are very much grateful to the Director, Zoological Survey of India, Calcutta for allowing the authors to publish this paper.

References

Ahmad W. (1996). Plant parasitic nematodes of India. Dept. of Zoology, A.M.U., Aligarh, U.P. pp. 347.

Firoza K. and Maqbool M.A. (1994). A diagnostic compendium of the genus *Helicotylanchus* Steiner, 1945 (Nematoda : Hoplolaimidae). Pak. J. Nematol 12(1) : 11-50.

Goodey J.B., Mary T. Franklin and David J. Hopper (1965). The nematode parasites of plants catalogued under their hosts. C.A.B., Farnham Royal, Bucks, England. pp. 214.

Krall E.L. (1985). Root parasitic nematodes, Family Hoplolaimidae. Oxonian Press Pvt. Ltd., New Delhi, pp. 580.

Siddiqu M.R. (1986). Tylenchida, parasities of plants and insects. Commonwealth Agricultral Bureaux, Slogh, U.K. pp. 645.

Taylor D.P. (1960). Host range study of the spiral namatode, *Helicotylenchus microlobus*. Pl. Dis. Reptr., **44** : 747-750.

Taylor D.P. (1961). Biology and host-parasitic relationship of the spiral nematode *Helicotylenchus microlobus*. Proc. Helminthol. Soc. Wash., 28 : 60-66

Proceedings of the National Seminar on Environmental Biology, April 03-05, 1998
Edited By : A.K. Aditya & P. Haldar
DAYA PUBLISHING HOUSE, DELHI

A New Method of Kitchen Waste Composting by *Perionyx excavatus*

P.S. Chaudhuri & Gautam Bhattacharjee
Department of Zoology, M.B.B. College
Agartala - 799 004, Tripura, India

Abstract

Utilizing the surface feeding habit of *Perionyx excavatus*, a cicrular bamboo made sieve (5-6 mm mesh size) having 50 cm diameter and 5 cm border was used as a container for chopped air dired kitchen waste in an earthen bowl (60 cm diameter at the top) having a vermibed (10 cm thick) made up of sand, garden soil and dry cowdung (5:4:1 ratio). Vermibed was maintained at 50-60 % moisture by periodical sprinkling of water over the kitchen waste. It was observed that 200 *Perionyx excavatus* (weight 100 g) reduced the C:N ratio of 2.5 kg kitchen waste from 14.85 to 9.00 within 30 days under laboratory conditions (27-30°C, 75 -80 % RH). Use of the bamboo made sieve allowed easy harvesting of pure vermicompost without disturbing the earthworms.

Introduction

Vermicomposting is a process of production of vermicompost through stabilization of organic wastes by earthworm activity. Vermicompost is less able to undergo rapid biological decomposition (Mitchell, 1997) and because of its fertilizer value it provides a good plant growth media. Vermicomposting technology, according to Loehr et al (1988), is environmentally sound, needs, no energy, capital or expensive equipment and extensive management.

Epigeic earthworms, on account of their surface activity, their ability of colonize organic wastes quickly to convert it into useful compost and to minimize malodur formation, are considered the best candidates for the above process. Of many species of earthworms tested for mass cultivation all over the world, *Eisenia fetida* (European worm), *Eudrilus eugeniae* (African worm) and *Perionyx excavatus* (Asian worm) come in order of preference for degradation of wastes (Kale 1993). *Perionyx sansibaricus* is another good compositing worm that can survive on slaughter house waste, sericulture and other agricultural wastes (Kale 1996). *Polypheretima elongata*, a tropical earthworm has also been claimed to be used in "vermistabilization" of municipal solid wastes (Edwards 1998).

Household wastes viz. kitchen waste is one of the major source of municipal wastes. Our experiment is designed for compositing of kitchen wastes, utilizing the surface feeding behaviour of indigenous species of earhtworm, *Perionyx excavatus*.

Materials and Methods

Perparation of vermibed : Three large earthen bowls 60 cm diameter at the top were used as containers for vermibed. Four to five minute holes were made at the bottom of the bowls to drain out water. The floor of the bowls was covered with a layer of sand upto a height of 5 cm followed by sieved garden soil layer (4 cm thickness) and dry cowdung layer (1 cm depth) to make the vermibed of 10 cm thicknes. Moisture level of the vermibed was maintained at 50 to 60 per cent by periodical sprinkling of water.

Inoculation of earthworms : Locally collected 200 live *Perionyx excavatus* weighting about 100g were inoculated in the vermibed of each earthen bowl 8-10 days prior to application of wastes.

Organic wastes : Household kitchen wastes were collected from local sources. After separation of polythene and other nonbiodegradable material, the wastes were shredded by a chopper to reduce the size and volume by about 50 to 70%. The wastes were then air dried for 10 days. The following refuses were the chief ingredients of kitchen wastes:

(i) Vegetables skins (potato, onion and green vegetables)
(ii) Fruit skin (banana, lemon etc.)
(iii) Egg shell, fish scales etc.

Vermicomposting : Three circular bamboo made sieves (mesh size 5-6 mm) having 50 cm diameter and surrounding border of 5 cm height were used as a container for the kitchen wastes. Ten days old well mixed 2500g dired kitchen wastes were spread uniformly over each container. Each waste loaded container was set over the moisturised vermibed inside the earthen bowl separately, taking care that the container was in direct contact with the vermibed, thereby allowing free movement of earthworms from vermibed to the organic wastes through the pores of the bamboo sieve (Fig 1). Three experimental sets came in operation from the same day i.e. 1st June, 1997. Composting was continued for 30 days under laboratory conditions (27 -30°C temperature, 75 -80% RH). Water was sprinkled gently to maintain 25-30% moisture content of the wastes. The upper organic waste layer was turned over periodically for better aeration. After 27 days, spraying of water was stopped. The upper manure layer was brought to the centre of the sieve in the form of a small heap and left for remaining experimental period. Vermicompost was harvested after 30 days and sieved through the same sieve that was used in the experiment.

Fig. 1 : Process of kitchen waste compositing by *P. excavatus*.

For chemical analysis, kitchen waste and vermicompost (20g each) were taken from the container on the day of loading (0 day) and harvesting day (30 days). Total nitrogen was estimated by Kjeldahl apparatus. Organic carbon was determined by Walkley and Black's (1934) rapid titration method.

Result and Discussions

Earthworms appeared in the kitchen wastes within the day of its introduction. However, their activity increased during 2nd and 3rd week of decomposition and this was indicated both by the presence of increased number of earthworms in the wastes, and appearance of a good quantity of dark brown loose granular casts over the wastes. Fungal mycelia, a large number of Collembola and Oribatid mites also appeared during decomposition of kitchen wastes. The malodour that was present at the beginning of the experimentation disappeared quickly corresponding with the increase in earthworm activity within the organic wastes. The transformation of the refuse into a soft, spongy, sweet smelling, dark brown compost was noticeable on the 27th day. Discontinuation of the watering and making a pile of compost at the centre compelled the worms to migrate from the feeding zone to the deeper layer of resting zone i.e. the vermibed due to decrease in moisture content. Use of bamboo sieve as a separate container for vermicomposting facilitated easy harvesting of compost (on the 30th day) without disturbing the earthworms. (Fig. 1).

The changes in pH, organic carbon, nitrogen and C/N ratio during vermicomposting of kitchen wastes are give in Table 1. Data reveals that there is nearly 60% reduction in the C: N ratio by the activity of *Perionyx excavatus* during composting of kitchen waste in a period of 30 days.

Table 1 : The changes in chemical characteristics during conversion of kitchen waste into vermicompost

Days	pH	C(g%)	N(g%)	C/N
0 (Kitchen waste)	8.83	31.19	2.10	14.85
30 (Vermicompost)	7.34	13.95	1.55	9.00

The fact that the earthworms have a great role in lowering the C:N ratio is now established (Lee 1985, Dash and Senapati 1986, Julka and Mukherjee 1986, Singh 1997, Kale 1998). The lowering of C:N ratio is mainly achieved by the combustion of carbon substances during respiration. This lowering of C:N ratio is very important as it enhances the utilisation of nitrogen by plants. A C : N ratio of 20 : 1 or less is very favourable for plants to utilise nitrogen in different froms (Dash 1994).

Precautions to be Taken During Kitchen Waste Composting

1. Moisture level in the vermibed should be maintained around 50-60%. Excess water chases the worm away due to anaerobic condition and deviation of pH of the medium.
2. Fresh kitchen wastes should not be used druing vermicomposting because its alkalinity (pH>8) kills the earthworm, *P. excavatus*.

3. Vermicomposting system should be kept free from ants, termites, rats etc.
4. Pungent, spicy and spoiled food should be avoided (Kale and Banu 1995).

Tips to Increase the Efficiency of Vermicomposting System

1. Addition of 1-5% cowdung slurry and straw enhance the decomposition process and quality of compost.
2. Use of vermibed increases the efficiency of the vermicomposting system, although its presence is not essential for waste feeding epigeic earthworms.
3. To maintain aerobicity in the waste combined with optimal moisture and temperature condition (Edwards and Bater 1992).
4. Vermicomposting system operates at the maximum efficiency if precomposting is kept to a minimum time consistent with effective sanitisation of the waste (Frederickson et. al. 1997).

Acknowledgement

Authors express their sincere thanks of Prof. J.P. Roychaudhuri, Head Deptt. of Zoology M. B.B. College for providing laboratory facilities and constant encouragement. Our thanks are due to Mr. Tapan K. Paul, Tripura Rubber Borad, Agartala for chemcial analysis. Financial assistance from I.C.A.R., New Delhi is greately acknowledged.

References

Dash MC (1994). *Earthworm ecology to vermitechnology. Changing Villages.* 13(2) : 18-26.

Dash MC. and Senapati BK. (1986). Vermitechnology, an option for organic waste management in India. pp. 157-172, In : *Proc. Nat. Sem. Org. Waste. Utiliz. Vermicomp. Part B- Vermas and Vermicomposting,* Das, Senapati and Mishra, eds., Sambalpur University, Orissa.

Edwards CA and Bater JE. (1992). The use of earthworms in environmental management. *Soil Biol. Biochem.* 24(12) : 1683-1689.

Edwards CA. (1998). The use of Earthworms in the Breakdown and Management of Organic Wastes. pp. 327-354, In : *Earthworm Ecology.* Edwards CA., CRC Press, Florida.

Frederickson J, Butt, KB, Morris RM, and Daniel C. (1997). Combining Vermiculture with traditional green waste composting systems. *Soil Biol. Biochem.* 29(3/4): 725-730.

Julka JM, and Mukherjee RD. (1986). Preliminary obervations on the effect of *Amynthas diffringens. (Oligochaeta : Megascolecidae)* on the C/N ratio of the soil. pp. 66-68 In : *Proc. Nat. Sem. Org. Waste Utiliz. Vermicomp, Part-B Verms and Vermicomposting.* Das Senapati and Mishra, eds., Sambalpur University, Orissa.

Kale RD. (1993). Vermiculture : Scope for new Biotechnology. pp. 105-108, In: *Earthworm Resources and Vermiculture.* Zoological Survey of India, Calcutta.

Kale RD. (1996). Earthworms - The signifcant contributors to organic farming and sustainable agriculture. pp. 52-57, In : *Organic Farming and Sustainable Agriculture.* Veeresh, Shivashankar and Singlachar, eds., APOF. Bangalore.

Kale RD. (1998). Earthworm : *Cinderella of organic farming.* pp. 1-88, Prism Books Pvt. Ltd., India.

Kale RD. and Banu K. (1995). Earthworm culturing techniques and cultivation pp 44-49, In : *Alternatives to fertilizers in sustainable agriculture,* Shivashankar, eds., UAS. Bangalore.

Lee KE. (1985). *Earthworms, the ecology and relationship with soils and land use.* Academic Press, Sydney.

Loehr RC., Martin JH., Neuhauser EF., Malecki MR. (1988). Stabilization of liquid muncipal sludge using earthworms. pp. 1345-1350, In : *Earthworms in Waste and Environmental Management,* Edwards and Neuhauser, eds., SPB. The Hague.

Mitchell A. (1997). Production of *Eisenia fetida* and Vermicompost from feed-lot cattle manure. *Soil Biol. Biochem.* 29(3/4) : 763-766.

Singh J. (1997). Habitat preferences of selected Indian earthworm species and their efficiency in reduction of organic materials. *Soil Biol. Biochem.* 29(3/4) : 585-588.

Walkely A. Black IA. (1934). Determination of organic carbon in soil. *Soil Sci.* 37 : 29-38.

Proceedings of the National Seminar on Environmental Biology, April 03-05, 1998
Edited By : A.K. Aditya & P. Haldar
DAYA PUBLISHING HOUSE, DELHI

Biodiversity of Aphids in India (Homoptera : Aphididae)

R.C. Basu & L.K. Ghosh*
M-Block, New Alipur, Calcutta-700 053
Zoological Survey of India
* Former Deputy Director

Abstract

An analysis of available data on Aphididae of Indian region reveals that a great diversity of aphid fauna is found in the eastern and northeastern Himalayas. Here, there is admixture of Palaearctic, Chinese and Malayasian fauna. In the western Himalaya 4 phytogeographical zones are observed viz. Tropical, Subtropical, Temperate and Alpine zones. Subtropical zone is represented chiefly by rich aphid faunas.

In Indian Peninsula as well as Gangetic, India plains and Desert region represents poor aphid fauna. Some aphids exhibit peculiar diversity of life cycle also. Out of 653 species found in India about 417 species (63.8%) and 208 genera (20%) are endemic. Aphid diversity depends both on vegetational deiversity, as well as, on biotic and abiotic factors.

The concentration of specific type of plant community varies, which determines the availability of abundance of aphid in different altitudinal ranges.

This paper is an attempt to provide an account of the important aspects as above with regard to the biodiversity of Indian aphids.

Introduction

Aphids or plant lice are a small homopteran plant sap sucking insects infesting both aerial and subaerial plant parts including the economically important ones. These insects are considered as obligatory parasites of plants. Various damage symptoms are noticed in plants as a result of infestation of aphids. They are also remarkable for their ability to transmit many a plant viral diseases.

The aphids enjoy more or less cosmopolitan distribution and are found in plants occurring in temperate climate. They have attracted the interest of scientific community on global basis.

Of the 4702 aphid species recorded world over (G. Remaudiere and M. Remaudiere, 1997), 1015 thrive well in oriental region (Agarwala and Ghosh, 1984) of which India accounts for 653 species belonging to 200 genera. Of these 150 spp reported in India are definite pests of cereals, vegetable crops, fruit trees, fibre crops, pulses, oil seeds, sugarcane, ornamental plants and forest trees as against 250 species which are economically important pests of agricultural and horticultural crops in the world.

Results and Discussion

It is known that total area of India is only 2% of world's landmass. However, India harbours nearly 5% of known species of animals and plants.

The aphid fauna of India constitutes about 13.8% of the World aphids (about 4702 species), including a large number of rare and endemic species. Out of total Indian species, nearly 63.8% species and 20% genera are endemic (Table-1). The richness and diversity of aphid fauna is largely due to diverse flora in different kinds of ecosystem prevailing in India. The data reveal that the food plants of Indian aphids cover 1250 species belonging to 137 plant families, 86 Orders and Suborders (Agarwala and Ghosh 1985). Divesity of aphid fauna (about 400 spp.) is found in Eastern and North Eastern Himalayas (Raychaudhuri, D.N. ed. 1980) where the aphid fauna is an admixture of Palaearctic, Chinese and Malayan fauna. Similar is the case with the Western Himalayas (about 250 species) and North West Himalaya (about 280 species). Peninsular, Gangetic plains and Indus plains represent about 150 species, whereas due to lack of vegetabtion Desert region is poorly represented by only 19 species (Table 2).

Table 1 : Subfamily Wise Break up of Indian Aphids Showing Endemism

Subfamilies	World species	INDIA species (No)	Genera (No)	Endemic Species	Endemism(%)
Anoeciinae	44	16	2	12	75
Aphidinae	3144	328	104	198	60.3
Drepanosiphinae	489	56	30	36	64.2
Chaitophorinae	55	20	4	14	70
Greenideinae	141	84	10	67	79.7
Hormaphidinae	178	48	24	27	56.2
Lachninae	376	36	14	26	72.2
Pemphiginae	275	65	20	27	56.7
Total	**4702**	**653**	**208 (20% endemic)**	**417**	**63.8**

Table 2 : Biodiversity in Indian Aphids

Zone	No. of spp./Sub spp.
Eastern Himalayas	400
Western Himalayas	250
North West Himalayas (Simla Himalaya)	280
Peninsula, Gangetic and Indus Plains	150
Desert	19

Among 4 phytogeographical zones (Tropical, Subtropical, Temperate and Alpine), Subtropical zone is represented chiefly by rich aphid fauna. The Subfamily Aphidinae among 6 other Subfamilies viz., Anoeciinae, Drepanosiphinae, Greenideinae, Hormaphidinae, lachninae and Pemphiginae has higher frequency of occurrence than any other subfamilies. The Indian Peninsula having Eastern and Western Ghat and Bindhya range as well as Gengetic and Indus plains has poor diversity of aphid fauna. Aphid diversity depends both on vegetational diversity, as well as, biotic and abiotic factors. It is observed that the morphology, karyomorphology, host plant relationship, pattern of life cycle, predator, parasites, ant association etc. help in the identity of aphids more precisely.

It is interesting to note that out of 653 species of aphids recorded from India, sexuales have been recorded for only 120 species, the rest are known by pathenogenetic forms. A majority (115 spp.) of species known by sexual forms are distributed in the temperate regions and only 5 spp. are so far known to produce sexuals in tropical regions of India.

The biogeographical divesity in India is reflected in the distribution and abundance of flora and fauna. Aphid-host plant association, as for instance, is a typical example where about 653 spp. of aphids infest about 1250 plant species belonging to 700 genera and 75 families against a total plant speices of 45,000 found in India (Table-3).

Table 3 : Aphid Fauna and Flora

Indian Hemiptera	6500	Species
Indian Aphididae	653	Species
World Aphididae	4702	Species
Total Plants in India	45000	Species
Aphid Infested Plants in India	1250	Species

Inspite of vast differences in floral distribution and aphid fauna composition in different biogeographical regions in India, there is no such attempt to analyse the biogeographical pecularities in respect of abundance endemism in differnet regions, host plant association as polyphagism of Aphids. Melicott and Blandford (1879) were one of the earliest workers to remark on the phenemenon of discontinuous distribution of fauna and flora. Discontinuity is observed between southern range including Sri Lanka and Eastern Himalayas, Assam, Burma and Malaya. Many workers have observed the phenoemenon in serveral groups including aphids.

Depending upon the climatic condition in different zoogeographical regions, the concentration of specific type of plant community varies which determines the availability of aphid abundance in different altitudinal ranges. It is observed that with the availability of particular types of plant species only 30% aphid species are available in tropical climate, 64% available in Subtropical region, maximum concentration of 83 per cent in the temperate and minimum concentration of 23% in alpine region (Table 4).

Table 4 : Common Host Plants In Stratigraphic Zones

Tropical *C* 300m-900m	Subtropical *C* 900m-2000m	Temperate *C* 2000m-3600m	Alpine *C* 3600m-5500m
Ageratum	*Alnus*	*Hydrangea*	*Betula*
Artemisia	*Artemisia*	*Impatiens*	*Rhododendron*
Browallia	*Berberis*	*Lonicera*	*Sorberi*
Costrum	*Cestrum*	*Salix*	*Viburnum*
Chrusanthemum	*Geranium*	*Rhododendron*	
Dolichos	*Prunus*	*Prunus*	
Grasses etc.	*Rosa, Rubus* etc.	*Viburnum* etc.	
Aphids 30% out of 653 spp.	Aphids 64%	Aphids 83%	Aphids 23%

In India, it is known that some aphids exhibit peculiar diversity of life cycle also. The density of these insects depend not only vegetational diversity alone but also on temperature and relative humidity.

Interestingly, out of 653 species found in India about 417 species (63.8%) are endemic (Table 1).

India is known to have representation of every major ecosystem starting from tropical froests, mountain, fresh water, estuary, mangrove, marine, desert to island ecosystem. As such India is regarded as the country of Megadiversity both in forms of species and ecosytems.

Acknowledgement

Authors are grateful to Dr. J.R.B. Alfred, Director, Zoological Survey of India, Ministry of Environment and Forests for his constant encouragement and facilities rendered for this scientific work.

References

Agarwala, BK, Ghosh, AK. (1984). A checklist of Aphididae of India *Rec. Zool. Surv. India,* Occasional paper No. 50: 1-71.

Agarwala, BK, Ghosh, MR. (1985). Bioecological Considerations of Indian Aphididae (Homoptera). *Insecta matsumurana* N.S., 31 : 81-96.

Medlicot, HB, Blaford, WT. (1879). Faunal Geology of India, IXX: 374-375.

Raychaudhuri, DN (ed.) (1980). Aphids of Northeast India and Bhutan. The Zoological Society, Calcutta : 1-521.

Remaudiere, G. Remaudiere, M. (1997). Catalogue of the World's Aphididae, Paria : 1-426.

Proceedings of the National Seminar on Environmental Biology, April 03-05, 1998
***Edited By* : A.K. Aditya & P. Haldar**
DAYA PUBLISHING HOUSE, DELHI

Studies on Cocoons of Some Tropical Earthworms

Gautam Bhattacharjee & P.S. Chaudhuri
Department of Zoology, M.B.B. College
Agartala - 799 004, Tripura, India

Abstract

Cocoon morphology, incubation, hatching pattern and juveniles in indigenous species of earthworms viz. *Perionyx excavatus, Lampito mauritii, Polypheretima elongata, Pontoscolex corethrurus, Eutyphoeus gammiei, Dichogaster modigalianii and Darwida nepalensis* have been studied. Studies on the rate of cocoon production, incubation period, hatching success etc. in the worms understudy indicate the possible utilization of *P. excavatus, D. modiglianii, P. elongata, P. corethrurus* and *L. maruitii* in vermiculture based biotechnology (i.e. vermicomposting and vermiconservation).

Introduction

The population dynamics, productivity and energy flow in earthworm cannot be fully understood unless the life cycle of earthworms are known adequately. Besides these, studies on the life cycle of earthworms are also necessary for the academic interest, as well as, growth and development of vermiculture based biotechnology.

Studies on the reproductive biology of tropical earthworm is very scanty and information that is available on Indian earthworm is absolutely due to the pioneer work of Dash and Senapati (1980), followed by Kale and Banu (1985), Senapati and his associates (Sahu and Senapati, 1986, 1988, 1991; Sahu et. al., 1988, Senapati and Sahu, 1993).

The aim of our paper is to describe the cocoon morphology, incubation and hatching pattern etc. in some of the indigenous species of earthworms under laboratory conditions.

Materials and Methods

(a) Species

1. *Perionyx excavatus* (Megascolecidae; Size : 100-180 mm x 5-6 mm).
2. *Lampito maruritii* (Megascolecidae; Size : 140-160 mm x 5-6 mm).
3. *Polypheretima elongata* (Megascolecidae; Size : 200-250 mm x 5-7 mm).
4. *Pontoscolex corethrurus* (Glossoscolecidae; Size : 70-90 mm x 4-5.2 mm).
5. *Drawida nepalensis* (Moniligastridae; Size : 40-50 mm x 2-3 mm).
6. *Eutyphoeus gemmiei* (Octochaetidae; Size : 200-400 mm x 7-10 mm).
7. *Dichogaster modiglianii* (Octochaetidae; Size : 25-40 mm x 2-3 mm).

The above worms were collected from Sadar Sub-division of Tripura a north eastern state located between latitude 22°51′ - 24°32 ′N and longitude 90°21′ E during the month of June-July, 1997 and studies on cocoon was started from August, '97 to July '98.

(b) *Preparation of containers and food media for culture maintenance*

Earthen pots (diameter 30 cm, height 28-30 cm in case of *E gammiei;* diameter 14-15 cm, height 7-10 cm in case of other worms) free from crevices were used for rearing of earthworms.

Perionyx excavatus inhabits loose compost heaps with high organic matter. It has greater preference to animal wastes. Cowdung dust prepared by sieving (1-2 mm mesh size) of dried and grinded cowdung constitutes suitable feed for the species.

Dichogaster modiglianii is found both in animal waste and in soil. Therefore, a mixture (v/v) of cowdung dust and garden soil (in 1:1 ratio) was used as feed for this species.

The rest of the worms are geophagous. Therefore, only garden soil grinded properly and sieved through 1-2 mm mesh was used as feed.

The moisture level was kept>50% by sprinkling of water inside the container. In each container 1 pair of healthy, clitellate species of earthworms were introduced, except in case of small sized *Dichogaster modiglianii* where 6 adults were kept together in each container in order to increase the mating chance.

The containers were being covered with jute cloth to check the entry of light inside the container and also the escape of worms (especially night traveller *P. excavatus*) from the same. Three replications were maintained for each experimental set.

(c) *Cocoon collection*

The culture media were checked every alternate day (to know more or less the exact date of cocoon production) after passing a jet of water through 0.5-1 mm mesh sieve in order to find out cocoons. Just after collection, size and weight of the cocoons were measured.

(d) *Incubation of Cocoon*

After isolation, the freshly laid cocoons were kept on moist filter paper spread over water soaked cotton inside a petridish. Precautions were taken not to allow excess water inside the petridish. Cocoons were observed daily to determine whether any hatchlings emerged. Incubation period is the time lapse (in days) from cocoon formation until the first hatching appears. Number of hatching per cocoon and size of hatching were also recorded. Average room temperature and relative humidity (RH) were noted at the date of hatching.

Results and Discussion

Earthworms are hermaphrodite and each worm produces both ova and spermatozoa. Generally cross fertilization between two adult worms results into the formation of cocoon. Earthworm cocoons are in general ovoidal or spheroidal capsules (Fig 1) containing fertilized ova and nutritent albuminous fluid. The wall of the cocoon is produced by the clitellar secretion and contains protein and a chitinoid material probably chitin (Needham, 1969). Freshly laid cocoons are soft, translucent and lightly coloured (in some species whitish eg. *Pontoscolex*). Later, the wall of cocoon becomes harder (probably due to deposition of chitinoid substance) with exposure to air. This is probably related to protection and prevention of evaporative water loss from the cocoon. With the progress of incubation time the colour of the cocoon changes. Ultimately, the cocoon becomes opaque red due to increase in vascularisation in the developing embroyo prior to hatching. Rhythmic circulation of blood is clearly visible from outside during late phase of development in the cocoons of *Pontoscolex* and *Lampito.* The hatchlings of all species of earthworms come from one of the pointed ends of the cocoon, following the appearence of a hole at the point of emergence. The cocoon morphology, incubation period and hatching pattern of seven different species of earthworms are given in Table 1.

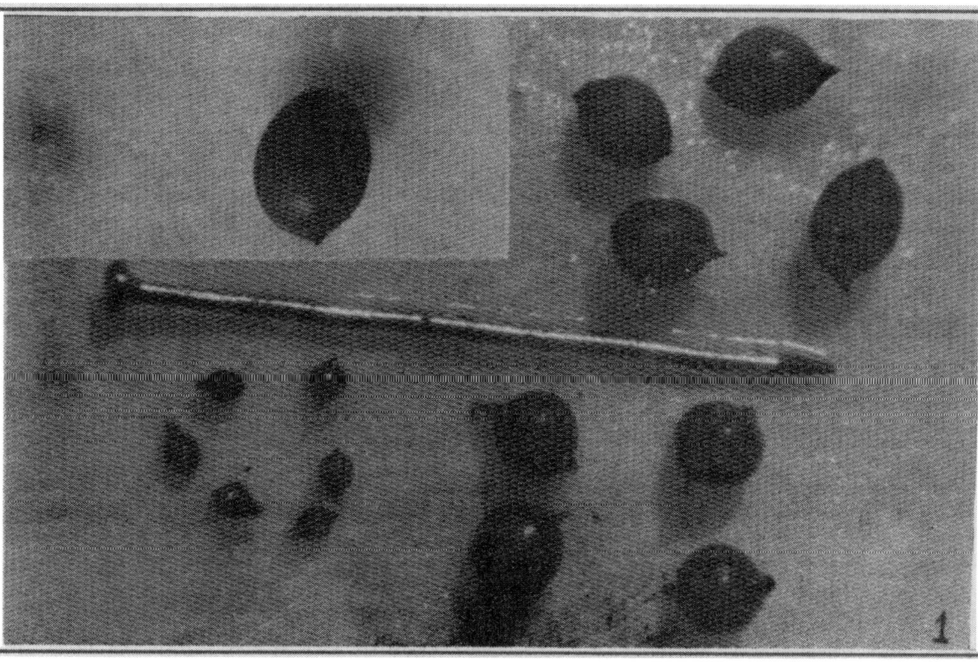

Fig. 1 : (i) *Eutyphoeus gammiei* (ii) *Polypheretima elongata* (iii) *Drawida nepalensis* (iv) *Dichogaster modiglianii*

From the given data (Table 1) it reveals that the highest and lowest size and weight of cocoon are found in gaint worm, *E. gammiei* and small earthworm, *D. Modiglianii* respectively. Interestingly, Senapati and Sahu (1993) reported a significant positive correlation between worm diameter to cocoon diameter, worm biovolume to cocoon biovolume and worm dry weight to cocoon dryweight. Earlier, Lee (1985) also advocated that cocoon size is closely related to adult size. Tuft of bristles at the pointed ends of cocoon in case of *P. excavatus* and *d. modiglianii* are the adapitve features of these epigeic worms that enable them to adhere to the litter in their surrounding. Viljoen and Reinecke (1989) also reported fibrous tips at both ends of cocoon in African epegeic worm, *Eudrilus eugeniae.* Incubation period of cocoon varies with species. Short incubation period and high rate of cocoon production in epigeic worms like *P. excavatus* (13 days) and *D. modiglianii* (12 days) are probably the adaptive strategies of 'r'-selected worms to cope with drastic environmental changes, especially heat, drought, predation etc. Lee (1985) correlated the higher risk of mortality in early life with the higher rate of cocoon production. Low rate of cocoon production coupled with long incubation period (i.e. prolonged emergence pattern) in deep burrowing large endogeic worms, *Polypheretima elongata* and *Eutyphoeus gammiei* are

Table 1 : Cocoon morphology, incubation, emergence pattern etc. in some tropical earthworms

Parameter	*P. excavatus*	*L. mauritii*	*P. elongata*	*P. corethrurus*	*E. gammiei*	*D. modigtianii*	*D. nepalensis*
1. Morphology (i) Shape	Spindle shaped	Oval	Spheroidal	Spherical	Spheroidal	Pear shaped	Spherical
(ii) Size	4 × 1.5 (mm)	5.5 × 3.00 (mm)	4.5 × 2.6 (mm)	5 × 3.5 (mm)	8 × 6 (mm)	2 × 1.5 (mm)	5 × 3 (mm)
(iii) Colour	Dark straw	Straw colour	Yellowish	White	Light straw	Off white	Reddish
(iv) Texture	Rough	Smooth	Smooth	Smooth	Rough	Smooth	Smooth
(v) Ornamentation	Bristles at the pointed end.	Absent	Present in form of short curved pointed ends.	Absent	Absent	One end with single pointed end and other bears a circlet of bristles	Well developed pointed end on either side
2. Average weight (g)	0.006 (n = 49)	0.02 (n = 20)	0.03 (n = 13)	0.021 (n = 6)	0.1 (n = 5)	0.0015 (n = 12)	0.0145 (n = 5)
3. Time of cocoon production	Continuous	Discontinuous	Discontinous	Continous	Discontinous	Continuous	Discontinuous
4. Rate of coccon production (per pair per month)	40.00	13.00	10.00	6.00	1.5	23.00	1.45
5. Average Incubation period	13.00 days (n = 13)	16.25 days (n = 16)	49.53 days (n = 15)	28.36 days (n = 11)	100 days (n = 3)	12.00 days (n = 12)	23.00 days (n = 8)
6. Number of hatchling per cocoon.	1 (Sometimes 2)	1 (Sometime 2, rarely 3)	1 (Sometimes 2)	1	1	1	2 (rarely 1 or 3)
7. Hatching success	36.1% (n = 36)	46.00% (n = 22)	68.18% (n = 22)	61.11% (n = 18)	50.00% (n = 6)	86.00% (n = 14)	80% (n = 10)
8. Size of hatchling	L : 4 – 8 mm B:0.5 – 1mm	L:20 – 25mm B:1 – 1.5mm	L:25 – 30mm B:1 – 1.5mm	L:6 – 10mm B:2 – 2.5mm	L:45 – 50mm B:2 – 2.5mm	L:5 – 7mm B:0.5 – 1mm	L:10 – 17mm B:0.5 – 1mm
9. Average temperature during hatching	26.69 °C	28.00 °C	30.56 °C	31.37 °C	31.00 °C	31.58 °C	28.5 °C
10. Relative humidity during hatching	80–90%	75–85%	75–85%	75–85%	75–85%	80–90%	75–85%

charecteristics of species, under K - selection where environment is predictable and competition is intense (Wallwork, 1983; Sahu and Senapati 1991). According to Lee (1985) cocoon production is influenced by population density, age structure and external factors especially soil temperature and the energy content of available food. Both moisture and temperature have considerable influence on cocoon incubation and also the emergence pattern of the juveniles (Das and Senapati, 1980). According to Lee (1985) hatching is delayed when moisture content is low.

Gaint worm, *Eutyphoeus gammiei* that shows very long incubation period, discontinuous and very low rate of cocoon production is not a suitable species for vermiculture.

High hatching rate and possibility of more than one hatching emergence per cocoon in case of *Polypheretima elongata* and *Drawida nepalensis,* and more or less continuous as well as, high rate of cocoon production coupled with short incubation period in *Perionyx excavatus, Dichogaster modiglianii, Pontoscolex corethrurus* and *Lampito mauritii* are indicative of their possible utilization in vermiculture based biotechnology.

Acknowledgement

Authors express their sincere thanks to Prof. J.P. Roychaudhuri, Head, Deptt. of Zoology, M.B.B. College for providing laboratory facilities and Prof. D.K. Nanda, Dept. of Zoology, Calcutta University for encouragement to work on developmental biology of earthworms. Financial assistance from I.C.A.R., New Delhi is gratefully acknowledged.

References

Dash M.C. and Senapati B.K., (1980). Cocoon morphology, hatching and emergence pattern in torpical earthworms. *Pedobiologia.* 20 : 317-324.

Kale R.D. and Banu K. (1985). Laboratory propagation of some indigenous species of earthworms. *J. Soil. Biol. Ecol.* 5(1) : 20-25.

Lee K.E. (1985). Earthworm. The ecology and relationship with soils and land use. Academic Press. Sydney. pp. 67-88.

Needham A.E. (1969). Growth and development. pp. 372-441, In : Floricin and Scheer, eds., *Chemical Zoology* iv. Academic Press. London.

Sahu. S.K. and Senapati B.K. (1986). Population density, dynamics, reproductive biology and secondary production of *Dichogaster bolai* (Michaelsen). pp 97-110, In: Dash, Senapati and Mishra, eds., *Proc. Nat.Sem.Org.Waste Utiliz. Vermicomp, Part B- Verms and Vermicomposting.* Five Star Printing Press, Burla.

Sahu S.K. and Senapati B.K. (1988). Alternative proposals for quantification of reproduction in tropical earthworm. *Tropical Ecology.* 29: 6-14.

Sahu S.K. and Senapati B.K. (1991). Reproductive strategy of a 'r' - selected and 'K' - selected earthworm from tropical pastures of Orissa, India. pp 665-669, In: Veeresh and Rajagopal eds., *Advances in Management and conservation of Soil Fauna.* Oxford and IBH Publ. Co. Pvt. Ltd. New Delhi.

Sahu S.K., Mishra S.K. and Senapati B.K. (1988). Population biology and reproductiove strategy of *Dichogaster bolai* (Michaelson) in two tropical agroecosystem, *Proc. Ind. Acad.Sci.(Animal Sci.).* 97(3) : 239-250.

Senapati B.K. and Sahu S.K. (1993). Reproductive biology (cocoon morphology, life cycle pattern and life table analysis) in earthworms. pp. 79-96, In: The Director, Z.S.I. eds., *Easrthworm Resources and Vermiculture.* Zoological Survey of India, Calcutta.

Vil join S.A. and Rainecke A.J. (1989). Life cycle of the African night - crawler, *Eudrilus eugeniae* (Oligocheta). *S. Afr. J. Zool.* 24(1). 27-32.

Wallwork J.A. (1983). *Earthworm Biology.* Arnold - Heinemann. London pp 47-52.

Proceedings of the National Seminar on Environmental Biology, April 03-05, 1998
Edited By : A.K. Aditya & P. Haldar
DAYA PUBLISHING HOUSE, DELHI

Implications of the Spatial Distribution Pattern of the Plant Community on the Insect Diversity in a Scrub Jungle Ecosystem

D. Suresh Chand, K.P. Sanjayan & M.C. Muralirangan
G.S. Gill Research Institute
Guru Nanak College
Chennai - 600 042

Abstract

The plant community complex of the Forest, a natural scrub-jungle ecosystem was studied in terms of the dispersal and spatial distribution patterns of 24 plant species that are relatively common in the habitat. The generalization that tropical species are relatively uniformly dispersed did not hold. All plant species were either clumped or randomly distributed and none showed uniform distribution reflecting on the reproductive biology of the species. Monthly surveys undertaken using well designed insect trapping methodologies resulted in enumerating the insect species of the forest and analysing the changes in the diversity in relation to season. Thirteen orders of insects were encountered in the forest. Among them, Coleopteran insects comprised the maximum number of individuals collected followed by insects belonging to the orders Lepidoptera, Orthoptera, Hemiptera. In terms of the number of species collected, as many as 81 species of Coleopterans, 73 species of Lepidopterans and 52 species of Orthopterans were recorded. Neuropteran insects were the least trapped both in terms of the number of species as well as individuals. As a paradigm in the study of insect-plant interaction, the *Euproctis fraterna* (Moore) – *Lannea coromandelica* (Houtt.) model was analysed.

Introduction

Each forest has a distinct floristic composition which varies in its species richness and the abundance of different species. Environmental variability in terms of climatic factors, soil resource, grazing by herbivores, and human interference are the critical factors which regulate the spatial and temporal patterns of vegetation of an ecosystem (Chapin *et al.*, 1993). Community ecology studies have generalized that there is a latitudinal gradient of increase in biological species richness and diversity from the temperate regions to the tropics (Krebs 1985, Colinavaux 1986). The dispersion of host plants bear an intimate relationship to densities of phytophagous insects (Kareiva, 1983). While the tropical forests are known to be, in general, rich in their species diversity, and have been extensively investigated, there is hardly any information on the tropical scrub-jungle ecosystems. Scrub-jungle are seasonal or monsoon forests growing in areas with a well-defined dry season. Several such natural forests occurring in Tamil Nadu have not received any attention so far. The spatial distribution of plants in a natural forests influence the efficiency of resource discovery by herbivores (Schowalter *et al.*, 1986) for feeding and oviposition (Crawley,

1983). Information on spatial patterns in concept with knowledge on the nutritional ecology of herbivores is therefore necessary for a proper appreciation of the ecological relationship between insects and their hosts. Hence, the present study was made with the objective to analyze the plant community structure and the dispersion patterns of common species in some detail and interpret these findings with regard to their insect fauna in terms of species richness and individual abundance.

Materials and Methods

The natural scrub-jungle forests selected for the study is located in Chengai-MGR district of the state of Tamil Nadu (13° N latitude, 80° E longitude). The forest has a total area of 676 hectares and lies 233 m above sea level in a slightly hilly terrain. The forest has a typical tropical climate with annual temperatures ranging between 19-32° C (minimum) and 32-41° C (maximum), 64-87% relative humidity and a total annual rainfall of about 700-900 mm. Rainfall generally occurs during October-November during the northeast monsoon, but a few showers are also brought by the southwest monsoon during June-July. Table 1 lists the species typical of the scrub-jungle ecosystem.

Table 1 : Plant Species, their Index of Dispersion (ID) and Green's Index (GI) of the Plant Community in the Scrub Jungle Forest (Tamil Nadu)

S.No.	Plant species	Family	Dispersion	ID	χ^2*	GI
1.	*Gymnosporia montana* Benth.	Celestraceae	Random	1.34	26.86	0
2.	*Wilkstromia viridiflora* Meissen	Thymelaeaceae	Clumped	2.66	53.38	0.08
3.	*Phoenix sylvestris* Roxb.	Palmaeceae	Random	1.43	28.76	0.02
4.	*Glycosmis mauritiana* (Lam.) Y.T.	Rutaceae	Clumped	4.59	91.94	0.51
5.	*Ziziphus mauritiana* Lamk.	Rhamnaceae	Random	0.9	1	0.05
6.	*Enretia microphylla* Lamk.	Boraginaceae	Clumped	2.05	41.19	0.26
7.	*Atalantia monophylla* Coorea.	Rutaceae	Clumped	8.63	172.6	0.13
8.	*Flacourtia indica* (Burm.f.) Merr.	Flacourtiaceae	Clumped	5.93	118.68	0.12
9.	*Cassia marginata* Linn.	Caesalpinaceae	Random	0.89	17.92	0.05
10.	*Cassia occidentalis* Linn.	Caesalpinaceae	Clumped	2.24	49.14	0.07
11.	*Pavetta indica* Linn.	Rubiaceae	Random	1	20	0
12.	*Grewia hirsuta* Vahl.	Tiliaceae	Random	1.61	32.3	0.1
13.	*Glycosmis pentaphylla* (Retz.) DC.	Rutaceae	Random	0.9	18	0.05
14.	*Gymnosporia wallichiana* Lawson	Celestraceae	Random	0.84	16.94	0.05
15.	*Vitex negundo* Linn.	Verbanaceae	Clumped	4.69	93.92	0.36
16.	*Flueggea leucopyrus* Willd.	Euphorbiaceae	Random	0.94	18.94	0.05
17.	*Aerva lanata* (L.) Juss.	Amaranthaceae	Random	1.63	32.76	0.05
18.	*Pamburus missionis* (Wight) Swingle	Rutaceae	Clumped	4.33	86.76	0.41
19.	*Sartewa* sp.	Acanthaceae	Random	1.21	24.36	0.05
20.	*Tephrosia purpurea* (L.) Pers.	Fabaceae	Random	1	20	0
21.	*Cissus quadrangularis* L.	Vitaceae	Clumped	2.15	43.08	0.23
22.	*Opuntia coccinellifera* Mill.	Cactaceae	Random	1.6	32.1	0.3
23.	*Lannen coromandelica* (Houtt.) Merr.	Anacardiaceae	Random	1	20	0
24.	*Dodonaea angustifolia* L.F.	Sapindaceae	Random	1	20	0

* Critical values of χ^2 are : 9.59 at $P = 0.975$ and 34.17 at $P = 0.025$.

Analysis of spatial patterns

The forest community being a continuous or non-discrete habitat, a line transect running east to west across the forest was laid in order to avoid the influence of the sampling unit. Along this transect, 21 quadrats of 20'×20' area were randomly selected and the number of individuals of all the species within the quadrats were counted. Vegetation was analyzed using the methods of Phillips (1959) and Krebs (1985). The distribution patterns of the species were analyzed by computing the indices of Dispersion (ID) and the Green's Index (GI, Green 1966) as follows :

$$\text{Index of Dispersion (ID)} = \frac{S^2}{\overline{X}}$$

where $\overline{X}$ = mean and S^2 = variance.

$$\text{Green's Index (GI)} = \frac{(S^2\sqrt{\overline{X}})\text{-}1}{n\text{-}1} = \frac{IC}{n\text{-}1}$$

where 'n' is the total number of individuals in a sample.

χ^2 test statistic was computed using the formula, $\chi^2 = \text{ID (N-1)}$

where N is the total number of sampling units.

The following vegetative data for *L. coromandelica* were calculated :

$$\text{Species abundance} = \frac{\text{Total no. of individuals of the species}}{\text{Total no. of quadrats in which the species has occurred}}$$

$$\text{Species density} = \frac{\text{Total no. of individuals of the species}}{\text{Total no. of quadrats studied}}$$

$$\text{Relative frequency} = \frac{\text{Total no. of quadrats in which the species occurred}}{\text{Total no. of qudrats studied}} \times 100$$

Impotance value index = The relative value of frequency, abundance and density are summed to represent the importance index.

The observed spatial pattern was measured using ID.χ^2 statistic help in interpreting ID values. If the values fall between 0.974 and 0.025 probabilities, than random distribution is accepted. Green's index help is comparing samples since it is independent of 'n'.

Insect-sampling methods

Quantitative sampling of insect species was undertaken using a package of traps set up within an area of 10 × 10 m randomly chosen quadrat. The package included light trap, scented trap, pitfall trap, sticky trap, sweepnet and searchout techniques. Data collected from five such replicates were pooled for further analysis. The insects collected were classified into various orders and within each order they were grouped into a recognizable taxonomic unit (RTU) based on distinct morphological differences. Each such unit is referred in this paper as a spcies.

A fluorescent light (7.5 inch; 6 watts) was used in the light trap which was switched on at dusk and allowed to fade itself out as the batteries drained after 6 hours. A plastic jar of 2.5

litre capacity was used to fabricate a scented trap. The jar was baited with 200ml. saturated jaggery (unrefined cane sugar) solution with two tubes of baker's yeast, 0.05% (final concentration) Methyl parathion and 0.5ml of pineapple essence. The trap was hung at about 2-3 m from the ground on a branch of a tree. The pitfall trap consisted of a 2.5 litre plastic jar with an opening 9 cm in diameter, buried at ground level and protected from rain by a tripod stand carrying a plastic plate of about 30 cm diameter placed at a distance of 15 cm above the ground level. Each jar carried 25 ml of 0.05% Methyl parathion. The traps were set up between 15.00-17.00 hrs and were collected the next morning. The sticky trap consisted of a 1 sq. ft. cardboard sheet, on either side of which vaseline was applied evenly. A thread was tied at one corner of the cardboard and hung from a wooden tag or a branch of a tree at a distance of about 2-3 m above ground. Sweep nets were used to collect insects of the vegetation. The nets were made up of thick nylon cloth with a diameter of 30 cm at the mouth and a bag length of 60 cm. The entire ground level vegetation in the chosen quadrat was covered during the sweeping which was usually carried out between 8.00-12.00 hrs. Searching was carried out by means of hand picking the insects directly from the vegetation after locating them. Searching was always done between 8.00-16.00 hrs.

Feeding preference and food utilization by Euproetis fraterna

Leaves containing egg masses of *E. fraterna* were handpicked from randomly selected trees of *L. coromandelica* from the forest and incubated at room temperature (37° C) in a plastic jar containing a wet cotton wad. Upon hatching the larvae were transferred into growth chambers (boxes) measuring 7.7 × 7.7 × 2.6 cm. Three replicates of 10 larvae were separated and feeding preference studies conducted through no-choice experiments. Leaves of *L. coromandelica* were removed from the side branches (25-30 feet above ground level) of trees selected at random during the early morning hours and separated based on maturity into young (light green and soft texture), mature (dark green and thick texture), and senescent (yellowish and thick texture) leaves. The leaves were then weighed and placed in a box with their petioles stuck through a hole in the box and into a plastic vial filled with water. Scriber (1977) has shown that excised leaves kept in water trays maintain full water content for at least 3 days. After two days of feeding by the larvae, the leaves were weighed again and the quantity of leaf consumed calculated and represented as a measure of feeding preference.

Because *E. fraterna* preferred feeding on mature leaves, experiments tc evaluate damage potential were conducted using mature leaves. Potential damage to the tree was assessed by evaluating the weight of foliage consumed by the insect. Preweighed mature fresh leaves of *L. coromandelica* were provided to developing larvae *ad libitum* every day throughout their larval period. At the end of each 24 hours, the leftover leaf and faeces were collected and weighed. Larvae were checked daily for moulting, maturity, pupation, weight gain and adult emergence. The amount of food consumed, assimilation, excreta and other nutritional indices (AD-approximate digestibility; ECI-efficiency of conversion of ingested food; ECD-efficiency of conversion of digested food; GR-growth rate) were determined by standard gravimetric procedure as described by Waldbauer (1968). Adults were provided with 2% sucrose solution. For nitrogen estimation, leaves and faeces of larvae were dried and estimated by the micro kjeldhal method of Humphries (1956).

Results and Discussion

The spatial pattern of plants is an important characteristic of ecological observations made in viewing any community. It facilitates an understanding of the structure of ecological communities (Williams 1976). It has been generally held that tropical trees species are relatively uniformly dispersed and appear in low densities. If this was true, then there would be profound consequences for the reproductive biology, population structure and evolution of tropical tree species. Several theories have been proposed to explain both the causes and consequences of such a phenomenon (Janzen 1970, Connell 1970). This generalization however does not agree with the present observation on the scrub-jungle forest. The scrub-jungle plant species of the forest are not uniformly dispersed in the forest as shown in Table 1, for the 24 species on which information was computed. Of these species 15 (6.25%) exhibit significant random dispersion. The remaining 9 species (37.5%) have clumping. No species has a significant uniform dispersion. Random patterns in the population imply environmental homogeneity and/or non-selective behavioral patterns, while clumping suggests that individuals are aggregated in more favourable parts of the habitat; this may be due to, among several factors, environmental heterogeneity and reproductive mode (Pemberton and Frey, 1984). Nature is multifactorial and many interacting biotic and abiotic processes may contribute to the existence of patterns (Quinn and Dunham, 1983).

The Dispersion Indices (ID) of various species in the scrub-jungle forest are given in Table 1. Analaysis of ID values with Chi-square test statistic shows that *Atalantia monophylla* is the most highly clumped species followed by *Flacourita indica, Vitex negundo* and *Pamburus missionis.* On the contrary *Gymnosporia wallichiana* and *Sartewa* sp. showed highly random distribution.

ID is not very useful as a comparative measure of the degree of clumping because of its dependence on 'n'. Thus Green's index (GI) can be used to compare samples that vary in the total number of individuals and the number of sampling units. GI varies between 0 (for random) and 1 (for maximum clumping). Accordingly, *P. missionis* is the most clumped species followed by *V. negundo, Gymnosporia montana, Pavetta indica, Tephrosia purpurea, Lannea coromandelica* and *Dodonaea angustifolia* are highly randomly distributed. These analyses indicate with certainty that *P. missionis* and *V. negundo* have distinctly clumped distribution. Clumped dispersion of the plant species may reflect on the breeding system of the plants, their pollination biology and dispersal pattern (Frankie, 1976).

Insect fauna

A summary of the monthly insect catch data in terms of orders, number of species and number of individuals is provided in Fig. 1 & 2. This data was obtained by pooling all the insects collected from the various trapping techniques employed. Some patterns became apparent in this figures. All the orders of insects enlisted are available in the forest throughout the year except Neuroptera which were not caught during September. Over the year, we encountered 12015 individual insects belonging to 379 species and 13 orders. Among the orders encountered, Coleopterans comprised the maximum number of individuals collected (2383) followed by Lepidoptera (2292), Orthoptera (1588) and Hemiptea (1524). In terms of the number of species collected, as many as 81 species of coleopterans, 73 species of lepidopterans and 52 species of orthopterans were recorded. Neuropteran insects were the least trapped both in the number of species as well as individuals.

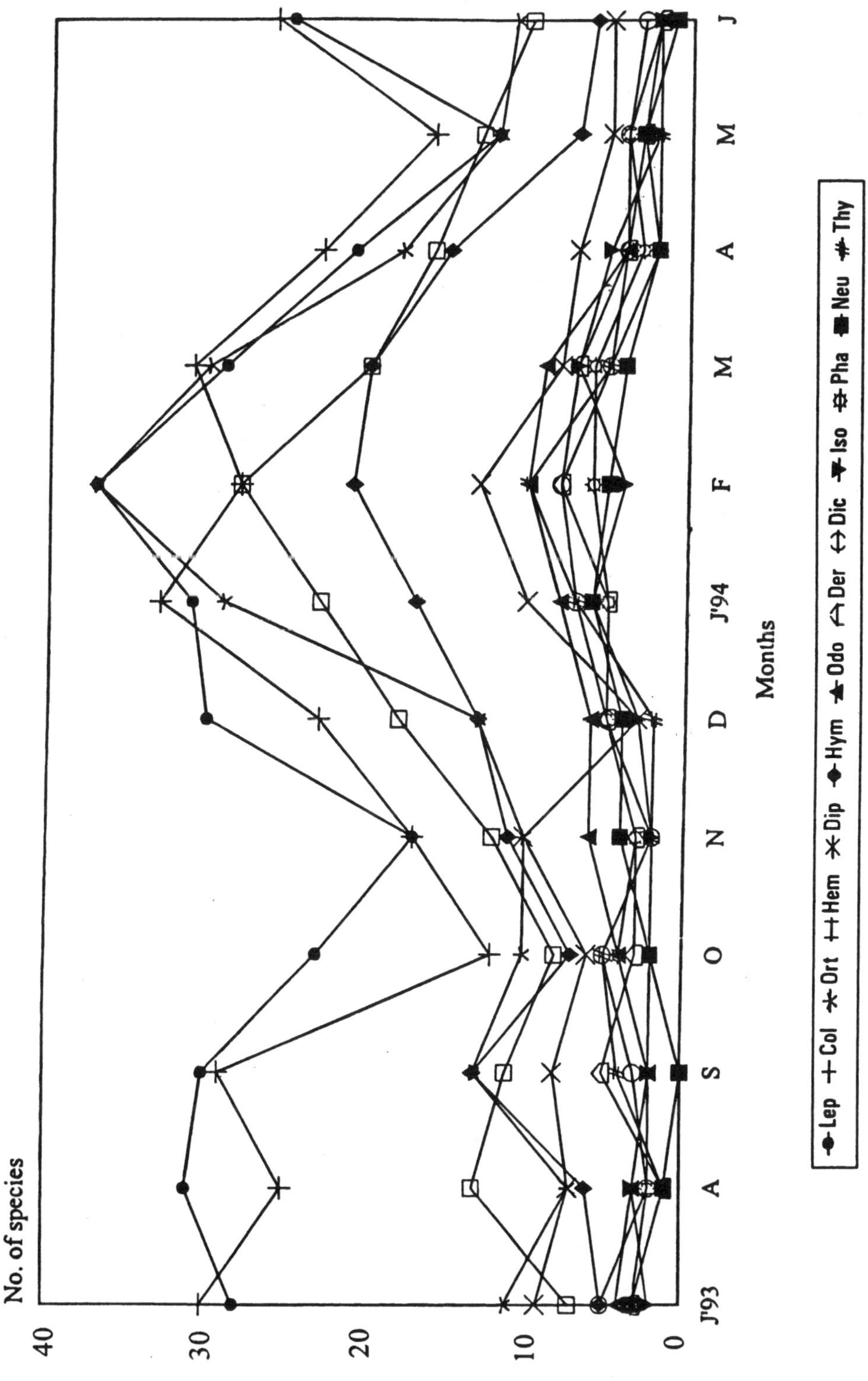

Fig. 1 : Species abundance of the insect community in the scrub jungle ecosystem.

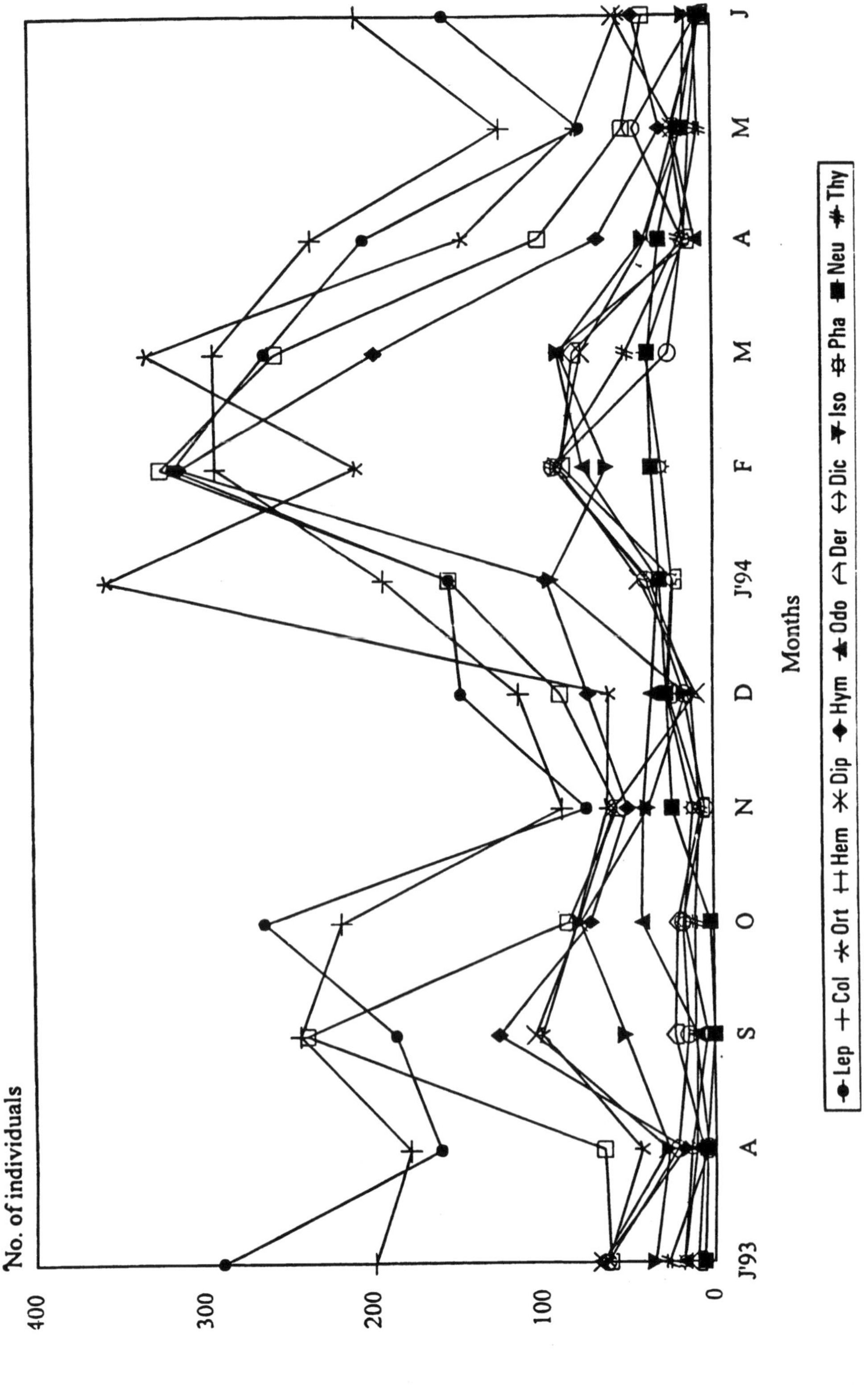

Fig. 2 : Individual abundance of the insect community in the scrub jungle ecosystem.

Collection of insects is largely determined by the type and efficiency of the traps employed. Fig. 3 provides a comparative picture of the efficiency of various traps in terms of the number of orders, species and individuals. It is evident that each method yielded quite a different sample of insect, so much so that when all these methods are employed together within the habitat, the resulting data would represent the insect fauna of the area efficiently.

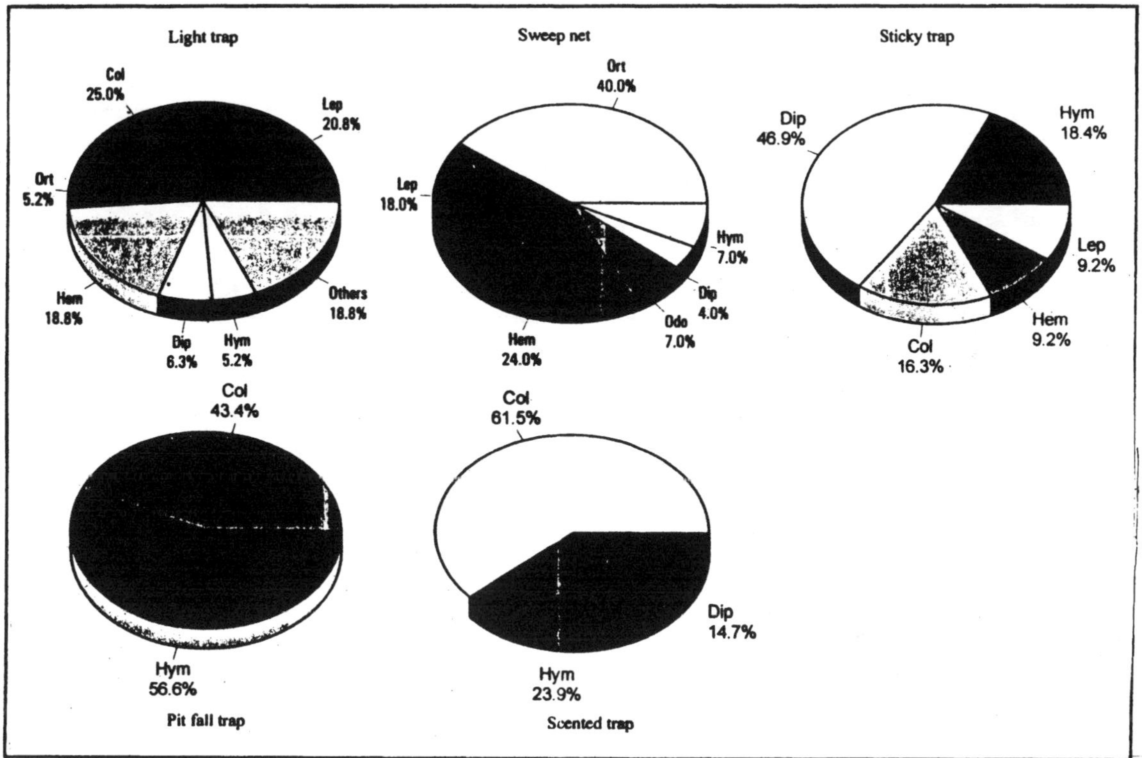

Fig. 3 : Pie chart showing orderwise proportion (%) of various insects trapped by different methods.

Light traps resulted in the capture of Coleoptera (25%), Lepidoptera (20%), Hemiptera (18%) and Diptera (6%) in significant number. Scentred traps attracted Coleoptera (61%), Hymenoptera were trapped in the pitfall, while small sized insects particularly those of Diptera, Hymenoptera and Hemiptera were trapped in the sticky trap. Sweep nets were good for collection of Orthoptera, Hemiptera and Lepidoptera.

Spatial distribution of L. coromandelica

Lannea coromandelica has a randon distribution pattern in the forest with an index of dispersion (ID) equal to 1.00 (Table 2). The oberved spatial distribution pattern was further tested using the Chi-square test; and it was found the χ^2 value falls well within 0.975 and 0.025 indicating the *L. coromandelica* has significant random distribution within the Forest. We hypothesize that the random distribution of the host makes individual trees more prone to attack by *E. fraterna*. Conversely, tree species with a clumped disperison have an increased number of individuals that escape insect attack. This appears especially important for insect species attacking the canopy of trees. Previous studies have shown that isolated palms and palms of irregular heights are particularly liable to *Oryctes rhinoceros* attack (Hill, 1987). In the scrub jungle *L. coromandelica* is not abundant; it is, however, the dominant tree species for colonization by *E. fraterna*.

Table 2 : Indices of Spatial Distribution of *L. coromandelica* in the Scrub Jungle

ID	X^2	0.975*	0.025*	Density	Abundance	Relative frequency	Importance index
1	20	9.591	34.17	0.04	1	0.28	0.26

* Critical value of X^2

Stand density has been reported to influence herbivory in the least three ways: First, the distance between hosts determines the period of time that dispersing insects are exposed to between-tree mortality agents (Schultz, 1983). Lance (1983) observed that host switching by gypsy moth larvae increases with increasing tree density. Second, the degree of canopy exposure to climatic extremes can affect folivore survival and foliage conditions (Whitham, 1981). Third, the numbers of competetions for resources influence tree condition and patterns of resource allocation (Cates *et al.*, 1983). The low values of density, abundance and relative frequency of *L. coromandelica* observaed in the forest requires that the larvae of *E. fraterna* utilise the resources available on the tree on which the eggs were laid, rather than risking mortality during between-tree dispersal.

However, *E. fraterna* is polyphagus, feeding on rose and acacia and on the foliage of numerous shrubs and trees. Therefore, populations would not depend on *L. coromandelica* alone for survival.

Feeding preference, growth and development of E. fraterna

There is substantial heterogeneity in the temporal and spatial distribution of several components of leaf quality including water and nitrogen content, toughness and the concentration of defensive chemicals (Whitham, 1981; Schultz, 1983). If plant defenses vary through time, herbivores have the option of adjusting consumer phases of their life histories to periods when host defenses are minimal. *E. fraterna* infests *L. coromandelica* during September-November when the new leaves of the perennial tree mature. Feeding by *E. fraterna,* like other defoliators, was shown to be selective with respect to the maturity of leaf tissue (Table 3). *E. fraterna* consumed significantly higher proportion of mature leaves than the senescant and young leaves. This

Table 3 : Nitrogen and water contents of different leaf stages of *L. coromandelica* and the relative feeding preference of *E. fraterna*

	Young	Mature	Senescent
% Nitrogen	4.1 ± 0.86 a	3.6 ± 0.62 a	2.2 ± 0.29 b
% water	65.4 ± 6.77 a	64.6 ± 7.44 a	56.0 ± 7.50 a
Amount of leaf consumed in 48 hours mg/ind.*	8.3 ± 2.2 a	17.2 ± 3.6 b	6.2 ± 1.3 a
Water-Nitrogen index	15.9	12.95	25.46

Values followed by similar letters within a row are not significantly different at $P > 0.05$ *n = 8 feeding preference may be partially explained by the chemistry of the leaf, particularly its nitrogen and water content. Phenological studies indicated that both water and nitrogen content of the leaves decreased with the maturity of the leaves. *E. fraterna* thus preferred feeding on the mature leaves of *L. coromandelica* which registered a low water-nitrogen index. Palatability may also vary depending on the change in the presence and/or concentration in leaves of terpenes, tannins and other secondary plant products (Brattsten and Ahmad, 1986) and may account for the periodicity infestation on *L. coromandelica* by *E. fraterna*. Feeny (1970) reported that winter moth larvae feeding on oak leaves in the spring abruptly cease feeding in mid-June and turn to other trees, despite there being no obvious change in the nutritional content of the leaves. He hypothesised that a concomitant sharp increase in leaf tannins complexed the leaf protein *in vivo*, thereby lowering the nutritional value of the leaves. Physiological changes during tree development influence its susceptibility and suitability as a host (Coley, 1980; Niemela *et al.*, 1980). There exists a negative correlation between dietary nitrogen and various fitness parameters (Broadway and Duffey, 1986; Schroeder, 1986). Brewer *et al* (1985) obtained an optimum level of foliar nitrogten beyond which the western spruce budworm, *Choristoneura accidentalis* showed a decline in its development.

The data on the growth and development of *E. fraterna* is presented in Table 4. The larvae passes through five instar stages for both males and females. The total average larval period for the male and female in 30.1 and 35.6 days, respectively. The pupal duration did not differ among the sexes, being 24.5 days for the males and 25.2 days for the females. The adult longevity was 5-6 days. However, the females took longer time to complete their life span (67.3) when compared to the males (60 days).

Table 4 : Growth and development of *E. fraterna* on *L. coromandelica*

Sex	I	II	III	IV	V	Total larval period	Pupal period	Adult longevity	Total life span
Male	4.5	5.4	5.2	8.5	6.5	30.1	24.5	5.4	60
Female	4.5	6.5	5.8	10.3	8.5	35.6	25.2	6.5	67.3

Values indicate mean number in days

Food Utilisation

The total food consumed and the different growth indices are presented in (Table 5; Fig. 4 & 5). The fresh weight consumption index is generally taken as a measure of the behavioral response of the insect towards the food (Waldbaure, 1968). The consumption increased with increase in age. The highest consumption was seen in the V instar stage (3963.84 mg/ind/day) (Table 5). There was sudden increase of weight gain from the II instar stage. The approximate digestibility, taken as the index of digestibility, was found to increase from the first instar to the third instar stage. The efficiency of conversion of digested food the body substance (ECD) increased during the early instar (I&II) stages and decreased subsequently (Fig. 4). Thereafter the growth rate (GR) decreased with increase in age, except in the III instar where there was slight increase (Fig. 5).

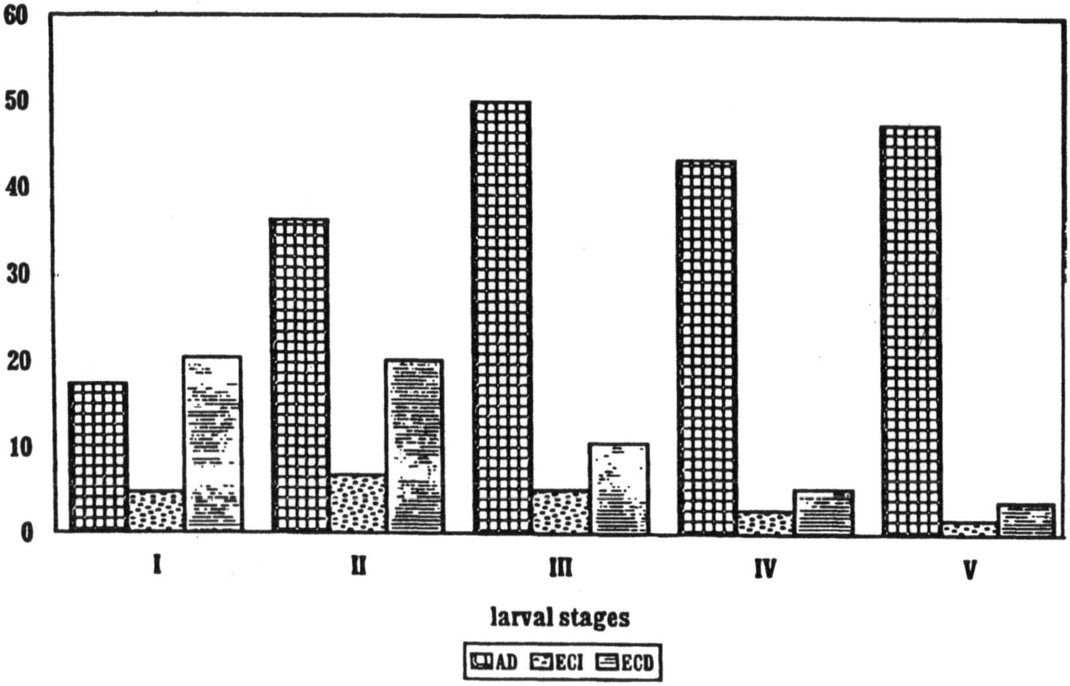

Fig. 4 : AD, ECI and ECD of *E. fraterna* when fed on *L. coromandelica*

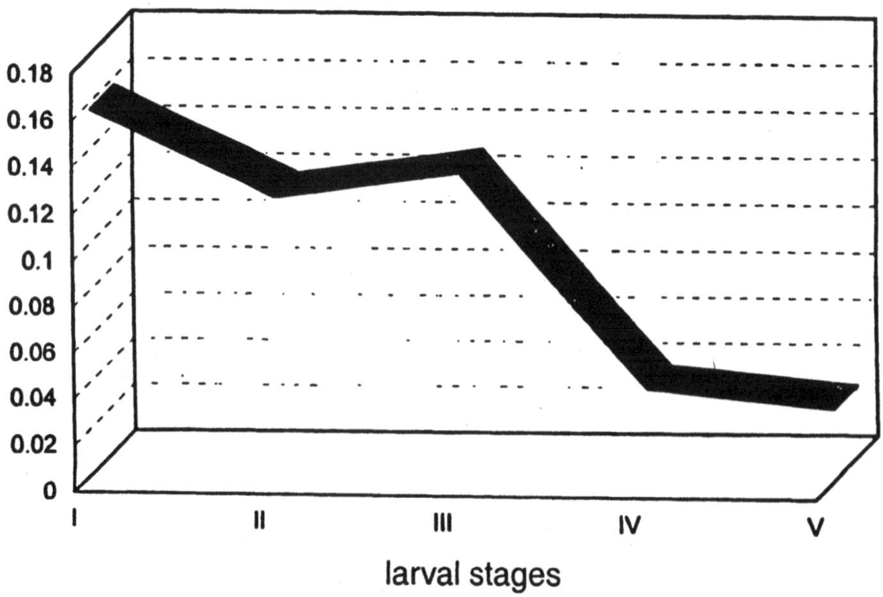

Fig. 5 : Growth rate of *E. fraterna* when fed on *L. coromandelica*

Table 5 : Ecological efficiency of *E. fratern* when fed on *L. coromandelica*

Instar	Wt. of insect	Food consumed	Excreta	Wt. gain
I	3.81	55.63	44.38	2.64
II	16.62	138.4	83.62	12.81
III	88.43	980.51	456.38	71.81
IV	169.54	2050.33	1093.86	81.11
V	253.26	3963.84	1968.36	83.72

Values indicate mg/ind.

Survivality

The ecological success of an insect is determined by the reproductive potential of the adults and survibility of the offspring. We recorded a 8.62% egg mortality with about 60% larval pupating (Table 6). The total mortality was only 45.79% with 12.3% mortality during the first instar. This low mortality, however, relates to observation under laboratory conditions and the value might be much higher under field conditions where several different ecological and biological factors operate.

Table 6 : Percentage mortality of *E. fraterna* at different stages when fed on *L. coromandelica*

Egg	I*	II*	III*	IV*	V*	Total larval period	Pupal period	Adult longevity
8.62	12.31	10.2	7.64	5.7	3.54	39.39	3.6	2.8

* Larval stages

As *Lannea coromandelica* sheds its leaves in June every year, fresh leaves develop immediately after the first (June-July) monsoons. During September to November, when its leaves mature, *E. fraterna* infests the perennial tree. The present study therefore indicates that *L. coromandelica* may serve as a suitable host for *E. fraterna* in the absence of other hosts. Therefore caution needs to be exercised in using this tree in social forestry programmes.

Acknowledgements

This work was carried out during the tenure of research project no. 14/35/91. Under the Man and Biosphere/Research (MAB/RE) project funded by the Department of Environment and Forests, the receipt of which is gratefully acknowledged. Our thanks are also due to the Tamil Nadu Forest Department for permitting us to work in this reserve forest. Thanks are also due to Mr. V. Selvaraj, for helping us in field collection.

References

Brattsten, L.B. and Ahmad, S. (1986). *Molecular aspects of insect-plant associations.* Plenum press: New York and London: 346.

Brewer, J.W., Capinera, L.L., Deshan, R.E. Jr., and Walmsley, M.L., (1985). Influence of foliar nitrogen levels on survival, development and reproduction of western spruce budworm, *Chroistoneura occidentalis* (Lepidoptera : Torticidae). *Can Entomol.*, 117: 23-32.

Broadway, R.M. and Duffey, S.S. (1986). The effect of dietary protein on the growth and digestive physiology of larval *Heliothis zea* and *Spodoptera exigua. J. Insect. physiol.*, 32: 673-680.

Cates, R.G., Redak, R.A. and Henderson, C.B. 1(1983). Natural products defensive chemicals of Douglasfir, western spruce budworm success, and forest management practices. *Z. Angew. Entomal.*, 96: 173-182.

Chapin, F.s. III., Rincon, E. and Haunte, P.(1993). Environmental responses of plants an ecosystems as predictors of the impact of global change. *J. Biosci.*, 18(4): 515-524.

Colinvaux, P. (1986). Ecology. John Wiley, New York: 725.

Connell, J.H. (1970). On the role of natural enemies in preventing competitive exclusion in some marline animals and in rainforest trees: 298-312. In: Den Boer, P.I. and Gradwell, G.R. (Editor) *Dynamics of populations.* Center for Agricultural Publishing and Documentation (PUDOC), Wageningen: Netherlands.

Coley, P.D. (1980). Effects of leaf age and plant life history patterns on herbivory. *Nature.* 284: 545-546.

Crawley, M.J. (1983). *Herbivory: The dynamics of animals plant interactions.* University of California Press: Berkeley and Los Angeles: 437.

Feeny, P.D. (1970). Seasonal changes in oak leaf tannins and nutrients as a cause of spring feeding by winter moth caterpillars. *Ecology.* 51: 565-581.

Frankie, G.W. (1976). *Tropical Trees: Variation, Breeding and conservation.* In: Burley, J. and Styles, B.T. (Editors) Linnean society symposium series: 151-159.

Green, R.H. (1966). Measurement of non-randomness in spatial distributions. *Researches in population ecology.* 8: 1-7.

Hill, D.S. (1987). *Agricultural insect pests of tropics and their control.* Cambridge University. Press : 740.

Humpries, E.C. (1956). *Mineral compunds and ash analysis in modern methods of plant analysis:* 237-248 (Edited by Peach, K. Tracey, M.V.) Springer Verlag: Berlin.

Janzen, D.H., (1970). Herbivores and the number of tree species in tropical forests. *American Naturalist.* 104: 501-528.

Kereiva, P. (1983). Influence of vegetation texture on herbivore population: Resource concentration and herbivore movement: 259-289. In: Denno, R.F. and McClure, M.S. (Editors) *Variable plants and herbivores in natural and managed systems.* Academic Press: New York.

Krebs, C.J. (1985). The experimental analysis of distribution and abundance. Harper and Row, London : 694.

Lance, D.R. (1983). Host -seeking behavour of gypsy moth: The influence of polyphagy and highly apparent host plants: 201-224. In: *Herbivorous Insects: Host seeking behaviour and mechanisms* (Edited by Ahmad, S.). Academic Press: New York.

Niemela, P., Tuomi, J. and Haukioja, E. (1980). Age specific resistance in trees: Defoliation of Tamaracks (*Larix laricina*) by larch bud moth (*Zeiraphera improbana*) (Lepidoptera: Tortricidae). *Re. Kevo. Subarct. Res. Stn.*, 16: 49-57.

Pemberton. S.G. and Frey, R.W. (1984). Quantitative methods in Ichnology: spatial distribution among population. Lethaia 17: 33-49.

Philips, E.A. (1959). Methods of vegetation study. Henry Holt and Co., New York: 107.

Qinn. J.F. and Dunham, A.E. (1983). On hypthesis testing in ecology and evolution. *American Naturalist.*, 122: 603-617. Schowalter, T.D. Hargrone, W.W. Crossley, D.A. Jr. (1986). Herbivory in forested ecosystem. *Ann. Rev. Entomol.*, 31: 177-196.

Schroeder, L.A. (1986). Protein limitations of a tree feeding lepidopteran. *Entomol. Exp. Appl.*, 41: 115-120.

Schults, J.C. (1983). Habitat selection and foraging tactics of caterpillars in heterogenous trees, pp. 61-91. In: *Variable plants and herbivores in natural and managed systems* (Edited by Denno, R.F. McClure, M.S.). Academic Press: New York.

Scriber, J.M. (1977). Limiting effects of low leaf-water content on the nitrogen utilization energy budget and larval growth of *Hylaophora cecropia. Oecologia*. 28: 269-287.

Waldbauer, G.P. (1968). The consumption and utilization of food by insects. *Adv. Insect Physiol.*, 5: 229-288.

Whitham, T.G. (1981). Individual tress as heterogenous environments: Adaptation to herbivory or epigenetic noise: 9-27. In: *Insect life history patterns; Habitat and geographic variations.* (Edited by Denno, R.F. and Dingle, H). Springer-Verlag: Berlin and New York.

Williams, W.T. (Editor). (1976). *Pattern Analysis in agricultural science.* Elseiver: New York: 412.